Toxicogenomics

Toxicogenomics

Principles and Applications

Edited By

Hisham K. Hamadeh, Ph.D. DABT
Cynthia A. Afshari, Ph.D.

A John Wiley & Sons, Inc., Publication

Library of Congress Cataloging-in-Publication Data:
Toxicogenomics : principles and applications / edited by Hisham K. Hamadeh, Cynthia A. Afshari.
    p.   cm.
Includes bibliographical references and index.
    ISBN 0-471-43417-5 (Cloth)
    1. Genetic toxicology. 2. Genomics. 3. Proteomics. 4. Gene expression. I. Hamadeh, Hisham K.   II. Asfshari, Cynthia.
RA1224.3.T6975   2004
616′.042—dc22

                                                                    2003024163

Printed in the United States of America
10 9 8 7 6 5 4 3 2 1

Contents

Contributors

CYNTHIA A. AFSHARI, Department of Toxicology, Amgen Inc., Thousand Oaks, California

RUPESH P. AMIN, Merck Research Laboratories, West Point, Pennsylvania

STEVEN P. ANDERSON, Investigative Toxicology and Pathology Group Safety Assessment, GlaxoSmithKline Research and Development, Research Triangle Park, North Carolina

GARY A. BOORMAN, National Center for Toxicogenomics, National Institute of Environmental Health Sciences, National Institutes of Health, Research Triangle Park, North Carolina

H. ROGER BROWN, Investigative Toxicology and Pathology Group Safety Assessment, GlaxoSmithKline Research and Development, Research Triangle Park, North Carolina

PIERRE R. BUSHEL, National Center for Toxicogenomics, National Institute of Environmental Health Sciences, National Institutes of Health, Research Triangle Park, North Carolina

J. CHRISTOPHER CORTON, Toxicogenomics, Chapel Hill, North Carolina

LYNN M. CROSBY, *No affiliation*

XUDONG DAI, Department of Molecular Profiling, Rosetta Inpharmatics LLC, Kirkland, Washington

THOMAS J. DOWNEY Jr., Partek Inc., St. Charles, Missouri

ROBERT T. DUNN II, Department of Toxicology, Amgen Inc., Thousand Oaks, California

HISHAM K. HAMADEH, Department of Toxicology, Amgen Inc., Thousand Oaks, California

YUDONG HE, Department of Molecular Profiling, Rosetta Inpharmatics LLC, Kirkland, Washington

DAVID R. HOUCK, SCYNEXIS Inc. Research Triangle Park, North Carolina

ZAID JAYYOSI, Aventis Pharmaceuticals, Bridgewater, New Jersey

DANIEL C. LIEBLER, School of Medicine, Vanderbilt University, Nashville, Tennessee

ROBERT E. LONDON, Laboratory of Structural Biology, National Institute of Environmental Health Sciences, National Institutes of Health, Research Triangle Park, North Carolina

PEK YEE LUM, Department of Molecular Profiling, Rosetta Inpharmatics LLC, Kirkland, Washington

JEANELLE M. MARTINEZ, Laboratory of Computational Biology and Risk Analysis, National Institute of Environmental Health Sciences, Research Triangle Park, North Carolina

B. ALEX MERRICK, National Center for Toxicogenomics, National Institute of Environmental Health Sciences, National Institutes of Health, Research Triangle Park, North Carolina

KEVIN T. MORGAN, Aventis Pharmaceuticals, Raleigh, North Carolina

EMILE F. NUWAYSIR, NimbleGen Systems, Inc., Madison, Wisconsin

CHRISTOPHER J. ROBERTS, Department of Molecular Profiling, Rosetta Inpharmatics LLC, Kirkland, Washington

RONALD C. SHANK, Department of Community and Environmental Medicine, University of California, Irvine, Irvine, California

CHARLES J. TUCKER, National Center for Toxicogenomics, National Institute of Environmental Health Sciences, National Institutes of Health, Research Triangle Park, North Carolina

ROGER ULRICH, Department of Molecular Profiling, Rosetta Inpharmatics LLC, Kirkland, Washington

NIGEL J. WALKER, Laboratory of Computational Biology and Risk Analysis, National Institute of Environmental Health Sciences, Research Triangle Park, North Carolina

JEFFREY F. WARING, Department of Cellular and Molecular Toxicology, Abbott Laboratories, Abbott Park, Illinois

Foreword

Toxicology is an old discipline that dates back to the eleventh or twelfth century. Records show that early humans were concerned about the toxic effects of snake venoms, hemlock, and metals such as lead and arsenic. Toxicology is an important science that impacts both environmental health regulatory policy and the practice of medicine. However, this field of medical research is not on par with genetics and other fields of biomedical research in terms of public support and enthusiasm among scientists. In large measure, this deficiency is due to the pervasiveness among toxicologists of a scientific culture that is wedded to old problems and outdated technologies and model systems. The time-honored way of determining which drugs or environmental xenobiotics are toxic to humans is to expose hundreds of animals to the specific compound and observe them, months or years later, for adverse health outcomes (e.g., cancer or developmental anomalies). Such studies take years to complete, cost millions of dollars, require hundreds of animals, and provide little information with respect to mechanism(s). But, like many other branches of biomedical research, toxicology is now experiencing a renaissance fueled by the application of "omic" technologies to gain a better understanding of the biological basis of toxicity of drugs and other environmental factors.

Two major problems in drug discovery and environmental health risk assessment are: (i) the paucity of mechanistic data to explain toxic responses, and (ii) the lack of knowledge of intrinsic toxicity for the majority of the high-production volume chemicals introduced into the environment during the last half of the 20th century. Also, biomarkers of early molecular alterations that lead to toxicity or chronic disease do not exist for most phenotypes. While the most sought after data in human risk assessment are population studies that associate exposure to toxicity or adverse health outcome, such information is rarely available. So, the foundation of many risk assessments rests on experimental studies performed at high doses in rodents, and on decisions that frequently require the use of default assumptions because of these and other limitations in scientific knowledge.

This quagmire in assessing risk of toxicity for drugs and environmental xenobiotics exists because of the use of simplistic models and reductionist approaches to understand the development of a complex phenotypes. For example, toxicity is likely the result of intricate networks involving hundreds of genes and proteins operating in concert, as individual proteins and pathways are assembled into complex biochemical networks. Because of this complexity, our understanding of toxicity remains grossly descriptive and the molecular mechanisms elusive. Descriptive studies in genetically inbred animals do not portray genetic or biological differences in the human population that surely influence individual response to drugs and environmental xenobiotics. But, thanks to the recent advances in genomics, proteomics, and metabolomics, the interactions between multiple genes, proteins, and pathways can now be investigated with more rigor and specificity. By combining these new "omic" approaches with conventional toxicology and pathology databases, one can develop experimental models and strategies to evaluate: (i) the diverse structure and properties of various chemicals; (ii) the relationship between the time of exposure, dose, and health outcomes; (iii) the influence of genetics and behavioral factors; and (iv) interactions between multiple components of biological systems in development of the toxic response. This systems biology approach will allow for monitoring of multiple molecular events, pathways, and interactive networks simultaneously—a requirement for elucidating toxic mechanisms.

To promote the further development and application of the "omic" technologies to toxicology and environmental health risk assessment, the National Institute of Environmental Health Sciences of the National Institutes of Health developed the National Center for Toxicogenomics in the year 2000, after approximately three years of development by intramural scientists. The editors and several authors of various chapters in this textbook are current or former members of this Center. Center investigators are using a combination of genomics, proteomics, metabolomics, toxicology, pathology, and informatics approaches to survey the entire human genome and the various organ systems for toxic responses to specific drugs or environmental xenobiotics. The objective is to determine whether gene, protein or metabolite expression profiles or "signatures" can serve as markers to predict toxicity. Current efforts are underway to establish "best practices" and perform proof-of-principle experiments to phenotypically anchor altered patterns of expression to conventional parameters of toxicity. These studies are also attempting to define dose and time relationships critical for expression of "signatures" that forecast the development of overt toxicity. Such a database of chemical effects on biological systems can be used to predict toxicity and elucidate common mechanisms of toxicity and drug action.

The ultimate promise of toxicogenomics lies in its potential ability: (i) to identify sources of interindividual variability in response to drugs and environmental xenobiotics, both in terms of efficacy and toxicity; (ii) to provide a database for the development of high-throughput and low-cost platforms for screening substances for toxicity, and (iii) to improve the process of discovering new targets for drug action. However, the field of toxicogenomics is currently in its infancy, and the promise in informing risk assessment is yet to be realized. However, its development and application are certain to be expedited by bringing researchers from multiple disciplines together to develop robust databases and best practices. However, developing molecular approaches to predict toxicity and elucidate the relationship between genetic variation and toxic response is only the first step. Determining how to use this information in risk assessment represents yet another daunting challenge.

Editors Hamadeh and Afshari should be congratulated on preparing an excellent resource for anyone in toxicology seeking to address these challenges. This publication is the most concise and authoritative source of information on concepts and technologies in toxicology that have been impacted by investments in genomic sciences over the past 25 years.

KENNETH OLDEN, PH.D., SC.D., L.H.D.
Director, National Institute of
Environmental Health
Sciences and the
National Toxicology Program
National Institutes of Health

This material was written by Dr. Olden in his private capacity. No official support or endorsement by the National Institutes of Health or the United States Department of Health and Human Services is intended or should be inferred.

Preface

The field of toxicology has been formally practiced and taught for many decades. Toxicologists are charged with assessing risk and safety of diverse agents in complex biological systems as well as the environment. Traditionally biochemical and cell-based observations were used to assess the status of a system postexposure to an agent. For the past few decades we have seen the advancement of molecular methods to further probe the complexities of cellular responses to compound exposure.

With the advance of the recent completion of the sequencing of the human, mouse, and other genomes, we have now officially entered into the postgenome era where application of genomic knowledge and techniques will penetrate into all areas of science, including the classical field of toxicology. We are now poised to begin full-fledged indoctrination of a new science known as toxicogenomics. This term describes the application of new genomics information and methods to toxicology studies, with the goal of advancing our understanding of the mechanism of action of compounds, the response of organisms to these exposures, and the ultimate application of contributing to the development of more accurate risk assessment models.

In this new discipline of toxicogenomics we see molecular biologists, toxicologists, statisticians, engineers, pathologists, chemists, and mathematicians coming together to build the framework for the indoctrination of this new field. The requirement for cooperation of these diverse groups requires some cross-disciplinary training of scientists from these arenas in order to facilitate communication. We have put this book together in this spirit.

This book is organized to provide a solid background in the principles and examples of the types of work covered under the umbrella of toxicogenomics. The reader will find informative introductory chapters on the concepts and principles of toxicology followed by detailed descriptions of such technologies as microarray/high-density gene expression profiling, quantitative RNA measurement, proteomics, and analysis of metabolites (metabonomics). This book should provide an up-to-date guide on the state of the art for these technologies and their current applications to the field of toxicology. We hope that new toxicology graduate students and even seasoned toxicologists will find it useful to provide a roadmap to critically evaluating new research papers and conducting new studies. In addition, it is our intention that this book will be a useful resource to basic molecular and cellular biologists, informatics and computational scientists, and engineers who are interested in applying their expertise to the area of toxicology and risk assessment.

We thank you for taking the time to take a look at this first edition and hope that you find it to be a useful guide. We wholeheartedly thank all of our coauthors in generously providing their time to contribute. We look forward to the continuing advancement of new postgenome technologies and the emergence of powerful examples of application in the field of toxicology solidifying the integration practice and principles of toxicogenomics into mainstream toxicology.

HISHAM K. HAMADEH
CYNTHIA A. AFSHARI

Thousand Oaks, California
April 2004

1

General Toxicology

INTRODUCTION

Toxicology is a multidisciplinary field. This overview of toxicology presumes a strong background in anatomy, histology, physiology, and biochemistry. There are several detailed reference books in toxicology available (e.g., Klaassen, 2001).

Toxicology is the qualitative and quantitative study of the adverse effects of chemicals on living organisms. A toxicant is a chemical that causes harmful effects in living organisms. Toxicity is the adverse biological response to a chemical reaction. The biological activity of a compound is dependent on its chemical reactivity. It is helpful to think in terms of chemical reactivity producing the biological activity that equals toxicity.

Following the law of mass action (chemical equilibrium), the rate of a chemical reaction depends on the concentrations of the two reactants (the active toxicant and the biological receptor or target)—that is, the more active toxicant present (the greater the dose), the more toxicity (the greater the response). This is a statement of the dose-response relationship, and toxicologists rely on this relationship to assure that the toxicity observed in an experiment or in the environment is caused by the chemical being considered and not by some artifact.

Exposure to a single high concentration is likely to elicit an immediate (acute) response that is qualitatively different from repeated exposure (chronic) to the same chemical at much lower concentrations—for example, a

Toxicogenomics: Principles and Applications. Edited by Hamadeh and Afshari
ISBN 0-471-43417-5 Copyright © 2004 John Wiley & Sons, Inc.

single exposure to 1 milligram of aflatoxin B_1 to a rat will result in the killing of large numbers of liver cells and death due to liver failure within 5–6 days; exposure to a few nanograms of aflatoxin B_1 for several months does not kill liver cells but does result in liver cancer.

THE DOSE–RESPONSE RELATIONSHIP

Dose

Toxicants react chemically with cells, and the biological activity of a toxicant is due to its chemical reactivity. The amount of the toxicant that reacts with the cells in the body is the dose. A dose is expressed as a unit of weight or volume per unit body weight or surface area. Dose refers to the amount of a chemical at the target organ in the chemically reactive form to induce toxicity. This is contrasted to exposure, which refers to the amount of chemical in the immediate vicinity, not necessarily in its chemically reactive form; it is the concentration of the chemical in the air that is breathed, in the water that is drunk, in the food that is eaten, and in the material applied to the skin.

A dose is often expressed as a concentration, especially for toxicants in the air and drinking water; more accurately, this is an exposure. The term *parts per million* (ppm) is often used; for toxicants in water, ppm is equal to milligrams per liter, but for toxicants in air the conversion from ppm to milligrams per cubic meter of air (mg/m^3) is slightly more complex. The following formulae (Phalen, 1984) can be used to convert between ppm and mg/m^3 for particulates in air:

$$ppm = \frac{v \times d(22.4/MW)(T/273)(760/P)}{V}$$

$$mg/meter^3 = \frac{(ppm)(MW)}{24.04}$$

where

v = volume of material used (ml)
d = density (μg/ml)
MW = molecular weight of the substance (μg/μM)
T = absolute temperature (°C + 273)
P = pressure (mm Hg)
V = volume in liters

and at 0°C and 760 mm Hg, 1 gram-mole of a gas occupies 22.4 L.

Response

The response is the biological reaction of the cell, tissue, organ, and/or body to the chemical reaction by the toxicant. The intensity of the response is usually a function of the magnitude of the dose. This is the basis for the dose-response relationship. The toxic response may be an adverse change in morphology, physiology, growth, development, biochemistry, or life span of an organism that results in the impairment to function normally. Responses may be immediate or delayed, reversible or irreversible, local or systemic, direct (resulting from the chemical agent itself without involvement of an intermediate agent) or indirect (resulting from a derivative—e.g., metabolite—of the chemical agent), and graded or quantal. Toxic responses depend on the route (inhalation, ingestion, topical, etc.) and frequency of exposure.

The toxic response results from the chemical reaction between the toxicant and a molecule in or on the cell; the cellular molecule is a target or receptor. The toxicant and the target form a chemical bond between them; that bond can be a covalent bond, an ionic bond, or van der Waals forces. Therefore, the basis of the dose-response relationship is the law of mass action in chemistry that states the rate at which the chemical reaction takes place is a function of the concentration of the toxicant and the concentration of the cellular molecule (target). Since the concentration of the target molecule is a given, the higher the concentration of the toxicant, the faster will be the rate of interaction between the toxicant and the target and the greater will be the toxic response.

A typical dose-response relationship is represented in Figure 1-1. For low doses, represented in the figure as zone A, the amount of the toxicant in the blood/body is insufficient to elicit a detectable response. In zone B the range of doses is sufficient to elicit the response, and the dose-response relationship follows the law of mass action. At high-dose ranges, zone C, the reaction kinetics and ability of the biological system are no longer responsive to increases in dose. The interface between zones A and B represents a threshold in the dose-response relationship; this threshold is dependent in part on the size of

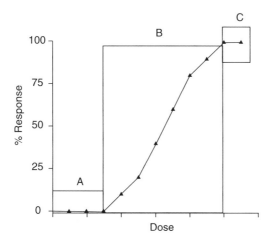

Figure 1-1 *Typical dose–response relationship.*

the population exposed. The relationship comes closer to linear when the log dose is plotted and is most useful when the log dose is plotted versus the probability (probits) that a response will occur, as this makes the relationship less dependent on the size of the population used to determine it. This also leads to the use of risk assessment to predict responses at doses well below those used experimentally. Figure 1-2 illustrates a log dose probability of response relationship.

The concept of threshold is important, as it means there can be doses of toxicants that are safe—that is, doses that do not elicit a toxic response. It also means that all chemicals can elicit a toxic response if the dose is high enough. There are doses so low that there are not enough chemical reactions to have a toxic effect on a cell or human being. A small number of chemical reactions take place, but the net effect on the cell is nil. Consider the example of poisoning of a cell by hydrogen cyanide, a gas that blocks the use of oxygen to make energy in a cell. In an unpoisoned cell, hydrogen atoms react with molecular oxygen by a series of controlled chemical reactions to eventually produce energy and water; hydrogen cyanide blocks this reaction. Small concentrations of hydrogen cyanide in the cell are insufficient to block this pathway completely, and the cell can continue to generate enough energy to survive until the cyanide is excreted (Fig. 1-3). As the concentration of hydrogen cyanide

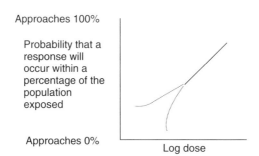

Figure 1-2   *Log dose-response (probits) relationship. Solid line represents experimental data and dotted lines represent extrapolations to responses at lower doses predicted by various models.*

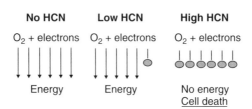

Figure 1-3   *Scheme illustrating importance of dose of hydrogen cyanide (shaded oval) on energy production and viability in mammalian cells.*

TABLE 1-1 Estimated toxicity of hydrogen cyanide in humans

HCN conc. (mg/m^3 air)	Duration of Exposure	Observable Effect
Less than 20	Several hours	None
24–48	Several hours	Feel tired
50–60	1 hour	Feel ill; consciousness
112–150	30–60 min.	Unconsciousness
240–360	5–10 min	Death
420	5 min	Death
1000	Less than 1 min	Death

in the cell increases due to increased exposure, the capability of the cell to continue producing energy is compromised, and the cell perishes.

There is no observable effect in humans exposed to hydrogen cyanide at levels of less than 20 mg/m^3. Table 1-1 summarizes the relation between exposure (hydrogen cyanide concentrations in inhaled air), duration of exposure, and toxic response that can be observed. Humans exposed to 24–48 mg HCN/m^3 for several hours feel tired. Since these data are based on a great deal of clinical observations, it is reasonable to conclude that a threshold for hydrogen cyanide exposure exists and that it is approximately 20 mg/m^3 of air for an exposure lasting several hours.

Time is an important factor in the dose-response relationship. Exposure to a given amount of toxic agent at one time, such as ingestion of a bolus, can result in a toxic response, whereas the same amount of toxicant divided into several daily doses may not elicit a toxic response at all. This is due in part to the rate of absorption and the blood concentration of the toxicant, which are time-dependent functions.

ABSORPTION, DISTRIBUTION, AND EXCRETION

When an exposure occurs, the toxicant must first cross from the environment and enter the body; transfer across cells composing the body surfaces (skin, lungs, gastrointestinal tract) is called absorption—more specifically, absorption from the environment into the blood or lymphatic circulatory systems. From these circulatory systems, the toxicant passes to several or all of the tissues in the body; this process is called distribution. The transfer of the toxicant from the circulatory system into the tissues is also called absorption; it is similar to the process for the transfer of a chemical from the body surface into the circulation. Therefore, two aspects of absorption must always be considered: (1) transport from the body surface into the blood (or lymph) and (2) transport from the blood into the tissues. The removal of toxicant from the body is excretion, and this process is usually accomplished by specific action of the kidney (formation of urine), liver (formation of bile), and lung (exhalation of volatile compounds). The entire process of absorption of a chemical into the circulatory systems and tissues, distribution within the body, biotransformation of the chemical, and excretion of the chemical and/or its metabolites from the body is called disposition.

Cell Membrane

To understand the process of absorption of chemicals from the body surfaces into the blood and from the blood into the tissues, it is important to consider the structure and chemical nature of the cell membrane, because in most cases the toxicant must pass through this membrane to reach the target site and elicit the biological (toxic) response. There are special features of this membrane that make the processes of absorption and excretion much easier to understand. The membrane behaves as a film of lipid in an aqueous environment. Globular proteins in the lipid film are free to move along the plane of the membrane. Some of these proteins completely traverse the membrane, providing aqueous channels through it. Small water-soluble molecules and ions can diffuse through these channels, while lipid-soluble molecules diffuse freely through the phospholipid component of the cell membrane. Large water-soluble molecules cannot readily cross these membranes except by special transport mechanisms. Proteins can cross, both in absorption and secretion, by pinocytosis. Specialized transport mechanisms for absorption (active and facilitated transport, etc.) are not considered in this overview.

Because the majority of the surface area of the cell membrane is phospholipid, lipid-soluble compounds cross the cell membrane at greater rates than do water-soluble compounds, which are restricted to crossing the membrane only where protein channels occur. Thus, on the basis of the structure of the cell membrane, a generalization regarding absorption can be made: lipid-soluble compounds are absorbed from the body's surfaces at faster rates (usually much faster) than water-soluble compounds, unless the water-soluble compound crosses the cell membrane by a specialized transport mechanism.

Some xenobiotics act directly on the exterior surface of the plasma membrane and bind to a specialized protein (receptor) in the membrane. Reaction with that membrane receptor can cause an endogenous compound to move from the plasma membrane to other organelles in the cell, such as the nucleus, to effect a biological response.

Rate of Absorption

Recalling the importance of the dose-response relationship, the toxic effect depends heavily on the concentration of the toxicant in the blood; this concentration drives the distribution of the toxicant into the tissues where the chemical can act on targets and receptors to initiate the toxic response. Here, the rate of absorption is important. As the rate of absorption increases, the concentration of the toxicant in the blood and tissues increases.

The rate of absorption of a chemical through a cell membrane and through a tissue depends on the size of the molecule and the lipid:water partition coefficient for that molecule. The lipid:water partition coefficient is determined by dissolving the chemical in a mixture of lipid and water and measuring the concentration of the chemical in each solvent. The lipid:water partition coefficient is the ratio of the concentration of the chemical in octanol to the concentration in water: coeff. = [conc. in octanol] ÷ [conc. in water]. The relation between the rate of absorption for a chemical and the logarithm of the coefficient, log K_{ow}, for that chemical is given in Figure 1-4.

In most cases in environmental toxicology where the concentration of the xenobiotic in the air, water, food, etc., is low, absorption takes place by passive diffusion, and therefore at a rate proportional to the concentration gradient between the site of absorption and the blood. In these cases, the rate of absorption is described by exponential or first-order kinetics.

$$\ln C_t = \ln C_o - k_a t$$

where

C_o = initial concentration of xenobiotic at the absorption site
C_t = concentration of xenobiotic at time t
k_a = rate constant for absorption, equal to $0.693 \div t_{1/2}$
$t_{1/2}$ = time taken for 50% of C_o to be absorbed; time when $C_t/C_o = \frac{1}{2}$

The rate constant, k_a, is the natural log of the proportion of the xenobiotic that has been absorbed in one unit of time. If half of the xenobiotic is absorbed in 1 hour, the concentration at the site of absorption will have decreased by a factor of 2; in this case, k_a is ln 2 ÷ 1 hr, or 0.693/hr. Therefore, $k_a = (0.693) \div t_{1/2}$. For example, the concentration of an ingested toxicant in the stomach determines, for the most part, the rate at which the toxicant is absorbed from the stomach into the blood; as the concentration of the toxicant in the stomach decreases due to absorption into the blood, the rate at which more toxicant is absorbed also decreases.

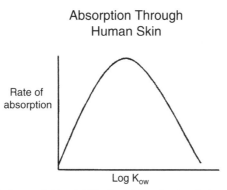

Absorption Through
Human Skin

Rate of
absorption

Log K_{ow}

Figure 1-4 *Relation between lipid solubility (x-axis) and rate of absorption (y-axis) through human skin into the circulation.*

For most toxicants at high dose, the concentration of the xenobiotic at the site of absorption may be so high that the amount being absorbed per unit time has little effect on the concentration at time t, C_t. The rate of absorption will be independent of C_t until the concentration decreases to a much lower value. Under these conditions, the rate of absorption is constant and follows zero-order kinetics.

Absorption of Toxicants Through the Skin

For a toxicant to be absorbed through the skin into the circulation, it must pass through several layers of cells (Fig. 1-5). The epidermis, which has no blood or lymph circulation, is the rate-limiting barrier to absorption through the skin. Polar toxicants appear to diffuse through the outer surface of the keratin filaments of the hydrated stratum corneum. Nonpolar toxicants dissolve in and diffuse through the nonaqueous lipid matrix between protein filaments; the rate of diffusion is related to lipid solubility and inversely to molecular weight. Most toxicants, if absorbed, pass through the epidermal cells; sweat glands and hair follicles account for less than 1% of the total skin area, and little toxicant overall is absorbed at these points. If absorption takes place at hair follicles, such as in the scalp, the rate of absorption at these sites is greater than through typical epidermis, due to the lack of the stratum corneum in the follicles.

The rate of clearance of toxicants from the dermis into the systemic circulation depends on skin thickness, effective blood flow, interstitial fluid movement, lymphatics, and other factors. The faster the absorption, the higher the

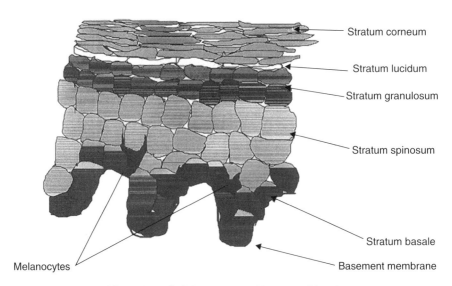

Stratum corneum

Stratum lucidum

Stratum granulosum

Stratum spinosum

Stratum basale

Basement membrane

Melanocytes

Figure 1-5   *Cellular structure of human epidermis.*

blood concentration, the greater the distribution and diffusion pressure to drive the toxicant into the cells of the body.

Absorption of Toxicants by the Lungs

The human lung has more than 50 square meters of alveolar surface area at the terminus of the airway; the distance between the alveolar epithelium and blood capillary wall is approximately 10 microns. Water-soluble gases dissolve in the mucus lining the respiratory tract and can accumulate there, causing local damage; lipid-soluble gases diffuse across the alveolar membrane at a rate dependent on the lipid:water partition coefficient and solubility of the gas in blood. Aerosols are absorbed according to the size of the particles and their aerodynamic properties, lipid:water partition coefficients, and several other factors.

Particles more than 10 μm in diameter usually impact in the upper airways, especially the nasal turbinates and trachea. Particles between 1 and 5 μm in diameter impact in the lower bronchioles and alveolar ducts. Particles less than 1 μm in diameter usually reach alveolar sacs. Particles that lodge in the airways above the terminal bronchioles are usually removed by the ciliary epithelium and are then expectorated or swallowed; ciliary transport rates are rapid and are measured in mm/min or cm/min depending on the location in the respiratory tract and on the nature of the particle. Often more than 90% of the particles deposited on the mucosa can be cleared out of the lung within 1 hour. Particles that lodge in the lower bronchioles can be carried to the ciliary area by macrophages and fluid from alveoli; this circulation is dependent on lymphatic flow, capillary action, the respiratory motion of alveolar walls, the cohesive nature of respiratory tract fluid, and the propelling power of ciliated bronchioles. Approximately half of the total deposit can be cleared in 1 day; the remainder is cleared in days or years, depending on the compound. The least soluble particles in ductal fluid are cleared the slowest. Soluble particles that lodge in the alveolus diffuse directly into the pulmonary blood circulation; insoluble particles may slowly penetrate into the interstitial spaces and reach the blood via the lymphatic system.

Absorption of Toxicants by the Gastrointestinal Tract

Absorption can take place from the mouth to the rectum; generally, compounds are absorbed in the portion of the gut in which the compound exists at its highest concentration and in its most lipid-soluble form. Toxicants similar in structure to nutrients and electrolytes may be actively transported into blood (e.g., 5-fluorouracil by pyrimidine transport; lead by calcium transport). Particles of several hundred angstroms or Å ($1\,Å = 10^{-8}\,cm = 0.1$ nanometer) in diameter enter the intestinal epithelium, are carried through the cytoplasm within intact vesicles, are discharged into the interstices of the lamina propria, then gain entrance into the lymphatics of the mucosa, much like fat absorption (e.g., azo dye particles, several hundred Å in diameter; latex particles, up to 2,200 Å in diameter, botulinum toxin).

Important biochemical transformations can take place in the GI tract that can alter absorption or toxicity. These transformations occur in tract bacteria (flora) and/or in epithelial cells lining the tract.

Many toxicants are weak acids or weak bases and exist in solution as a mixture of ionized (protonated) and nonionized forms. The less polar, nonionized forms are usually more lipid soluble and will diffuse rapidly across a lipoid membrane. The proportion of a toxicant that exists in the nonionized form depends on the dissociation constant of the compound and on the pH of the solution in which it is dissolved. The relationship is given by the Henderson-Hasselbach equation:

$$\text{for a weak acid:} \quad pK_a - pH = \log\frac{[unionized]}{[ionized]}$$

$$\text{for a weak base:} \quad pK_a - pH = \log\frac{[ionized]}{[unionized]}$$

When a weak acid is half ionized, the concentration of the ionized form, [*ionized*], is equal to the concentration of the nonionized, [*nonionized*], so the log expression above becomes the log of 1, which is zero; therefore, at half ionization the pK_a is equal to the pH. It is the nonionized form of the compound that is less polar and therefore more rapidly absorbed. The nonionized form predominates at low pH for weak acids and at high pH for weak bases. Therefore, weak acids are usually absorbed faster from the acidic gastric juice and slower in the lower, less acidic intestine. The reverse is true for weak bases.

Distribution and Excretion of Toxicants in the Body

The distribution and excretion of toxicants depends on

- Water compartments in the tissues of the body
- Lipid compartments in the tissues of the body
- Macromolecular binding
- Passage through the placenta
- Passage in the brain and cerebrospinal fluid
- Rate of pulmonary excretion
- Rate of renal excretion
- Rate of biliary excretion
- Rate of metabolism
- Lactation, perspiration, salivation, lachrymation; reproductive tract secretions

Protein binding of toxicants is analogous to enzyme-substrate binding and drug-receptor binding, except there is no decomposition of the substrate or any biological response to the process. There are no covalent bonds involved; the bonds are ionic and therefore the process is readily reversible. (Covalent binding is usually referred to as an alkylation of arylation and is not considered here.)

Binding takes place on both plasma and tissue proteins; not all proteins bind each toxicant to the same extent, and the degree of binding depends on the type and number of binding sites on the protein and the pH of the solution (controls ionization). Albumin is most important (constitutes 50% of the plasma protein); it has approximately 100 positive and 100 negative potential binding sites. At pH 7.4, albumin has more negative charges than positive charges.

The biological response caused by the toxicant is dependent on and parallels the concentration of the unbound toxicant in the plasma. Toxicants that form stable bonds with blood proteins will accumulate in the body and can be potentially dangerous. The plasma-bound compounds can be released suddenly by exposure to new compounds that can compete for the same binding sites—for example, in some newborns the liver may lack the enzyme glucuronyl transferase, which conjugates bilirubin, a water-insoluble hemoglobin breakdown product; as a result, large amounts of bilirubin accumulate on the plasma albumin (hyperbilirubinemia). If such infants are given sulfonamides or vitamin K, which displace the bilirubin from albumin, enough bilirubin can be freed in a short period of time and enter the brain to cause kernicterus (widespread destruction of nerve cells in the brain).

The process of excretion is basically the same as that of absorption: the transfer of chemicals across biological membranes according to chemical concentration gradients; the chemicals move from compartments of high concentration to compartments of low concentration. Water-soluble compounds can be rapidly excreted by the kidney via the urine (compounds with molecular weights usually less than 400) and by the liver via the bile (compounds with molecular weights usually greater than 300). Both urine and bile are aqueous (water) systems. As in absorption, there are specialized processes that can act against concentration gradients and can move water-soluble compounds rapidly across lipid membranes; the secretion of bile (an aqueous fluid) by the liver is an example of a tissue removing water-soluble chemical compounds from cells.

Lipid-soluble compounds can be excreted only slowly, if at all, into the body's aqueous waste routes: urine and bile; therefore, lipid-soluble compounds are retained in the body for long times or until they are metabolized to water-soluble derivatives. Lipid-soluble compounds that are filtered from the blood by the kidney are rapidly reabsorbed back into the blood by the kidney before the urine leaves that organ. The kidney can eliminate only water-soluble compounds and in special cases compounds that are characterized chemically as organic anions or cations, for which there are active transport systems to move these compounds from the plasma into the urine.

Volatile compounds can be excreted by the lung in the expired air. Some lipid-soluble compounds can leave the body as components of lactation fluid, seminal fluid, dead skin cells, and hair.

Metabolism of Xenobiotics

Xenobiotics that are highly soluble in water are often excreted from the body so fast as to provide little time for their metabolism. Xenobiotics that are highly soluble in lipid cannot be excreted by the body (unless highly volatile) and therefore stay in the body until they are metabolized to water-soluble derivatives.

Xenobiotics can be made less toxic or more toxic by metabolism. Increases in toxicity are often the result of oxidative phase I metabolism of lipid-soluble xenobiotics to more reactive intermediates in preparation for addition of water-soluble groups as part of phase II metabolism. Xenobiotics can be metabolized along more than one pathway. The faster pathways are more important in removing the compound from the body, especially at low doses, but all pathways must be considered for toxicological importance. Metabolism of xenobiotics occurs mainly in the liver, kidney, skin, lungs, and to a much more limited extent in other tissue. Chemical transformation of the xenobiotic usually makes the compound more polar, thus facilitating excretion of the compound in the urine or bile.

Xenobiotic metabolism is divided into two steps or phases. Phase I reactions include oxidation, reduction, and hydrolysis; they usually make the lipid-soluble toxicant more reactive chemically so that phase II reactions can covalently add water-soluble groups to facilitate excretion by the kidney and liver. Phase II reactions are conjugations.

Phase I Metabolism

Oxidations. Most oxidations of lipid-soluble xenobiotics are carried out by enzymes that use one atom of oxygen from molecular oxygen and are therefore called monooxygenases; other names for this enzyme system include mixed-function oxidases, microsomal hydroxylases, aryl hydrocarbon hydroxylase, polyaromatic hydrocarbon hydroxylase, and cytochrome P450. The appropriate name is cytochrome P450 monooxygenase; it is a large class of oxidative proteins containing many isozymes. Other oxidations are carried out by flavin monooxygenases, oxidases, peroxidases, and dehydrogenases, as discussed later.

The cytochrome P450 monooxygenases are heme-containing enzymes embedded in the smooth endoplasmic reticulum or are strongly attached to this membrane. The xenobiotic binds to the enzyme when the iron atom in the heme moiety is in the oxidized state; the iron atom is oxidized by way of a two-electron reduction of an oxygen atom. A molecule of oxygen binds to the enzyme when the iron atom in the heme moiety is in the reduced state; the iron atom is reduced by the coenzyme nicotinamide adenine dinucleotide phosphate in the reduced form (NADPH) via a flavoprotein intermediate. In most cases the oxidation of a lipid-soluble xenobiotic by a cytochrome P450 monooxygenase is a hydroxylation—that is, the end result is the addition of a hydroxyl group to the toxicant. Addition of an oxygen atom across a carbon-carbon double bond results in an epoxide rather than a hydroxyl group.

There are several families of cytochrome P450 monooxygenases. Each family is identified by the nature of the protein and is ascribed a Roman numeral; the genes are designated cyp and given Arabic numerals (e.g., cyp IIA1).

Another group of enzymes uses flavin adenine dinucleotide (FAD), rather than heme, as a prosthetic group. These enzymes are the microsomal flavin-containing monooxygenases. They use molecular oxygen and require electrons from NADPH and NADH. NADPH binds to the enzyme and reduces the flavin moiety, then an oxygen molecule associates with the complex. An internal rearrangement results in the formation of a peroxy complex to which the xenobiotic will bind; the xenobiotic is oxygenated and released from the enzyme complex and the enzyme is regenerated. Substrates (xenobiotics) include secondary and tertiary amines, sulfides, thioethers, thiols, thiocarbamates, phosphines, and phosphorates.

Oxidation of many water-soluble compounds can be catalyzed by oxidases, peroxidases, and NADH or NADPH dehydrogenases; these enzymes are not monooxygenases. Simple alcohols can be oxidized to their respective aldehydes by alcohol dehydrogenase and the aldehydes to their acids by aldehyde dehydrogenase. Monoamine oxidase (MAO) and diamine oxidase (DAO) are oxidative deaminases that produce aryl or alkyl aldehydes, which are further oxidized to the corresponding carboxylic acids. MAO is a mitochondrial enzyme in liver, kidney, intestine, and nervous tissue; DAO is a cytoplasmic enzyme in liver, intestine, and placenta. At least two forms of MAO are known: MAO-A oxidizes secondary amines, such as epinephrine, and MAO-B oxidizes primary amines, such as benzylamine.

Reductions. Some cytochrome P450 monooxygenases, some flavin-containing monooxygenases, and some reductases are able to reduce azo and nitro compounds to amines. Azo reductases and nitro reductases contain flavin and require NADPH, but only nitro reductases can also use NADH. Molecular oxygen inhibits nitro reductases but not azo reductases.

Hydrolyses. Hydrolysis of xenobiotics includes the breaking of ester, amide, hydrazide, carbamate, and epoxide bonds by the addition of water to the ester bond, and so on. Substrate specificities are highly dependent on species and tissue where the enzyme is located; within species there can be extensive polymorphism (different forms of the enzyme with different kinetic properties).

Hydrolysis of procaine is rapid in human plasma but slow in the plasma of many other species. Atropine is not hydrolyzed in human or mouse plasma but is hydrolyzed rapidly in rabbit plasma. Genetics-based polymorphism in phase I hydrolyses is common in humans.

Phase II Metabolism. Phase II reactions are conjugations in which a water-soluble endogenous compound is added covalently to the xenobiotic. Phase I reactions apply to lipid-soluble compounds and are regarded as binding a reactive group to a molecule with which a phase II conjugation can take place. The reactive moieties from phase I are usually hydroxyl groups to which the conjugating enzymes add a sugar or an acid, making the final product readily soluble in water and excretable by the liver and kidney. Conjugation reactions result in the enzymatic formation of glucuronides, ethereal sulfates, mercapturic acids through conjugation with glutathione, amino acids conjugates, acetylated amines, and methylated compounds. The overall relationship between phase I and phase II metabolism, solubility, and toxicity is summarized in Figure 1-6.

Sulfate Conjugation. This is a common pathway for conversion of phenols and alcohols to sulfate esters and aromatic amines to sulfamates; this conjugation takes place in three steps and occurs rapidly. Inorganic sulfate ion is activated by ATP to adenosine-5'-phosphosulfate (APS), then to 3'-phosphoadenosine-5'-phosphosulfate (PAPS); sulfotransferases use active sulfate, PAPS, in the sulfation of the xenobiotic. If the dose of the toxicant is greater than can be sulfated by this pathway, the amount to PAPS is soon exhausted (rate of synthesis of PAPS is slow), and other conjugation pathways must be evoked to make the toxic substrate less reactive and more soluble in water to facilitate its excretion.

Glucuronide Formation. Glucuronide formation is an important pathway for the metabolism of phenols, carboxylic acids, long-chain alcohols, primary

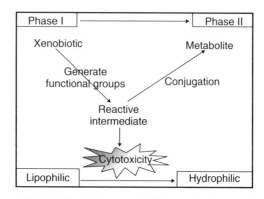

Figure 1-6   *Relation between solubility, metabolism, and toxicity.*

amines, hydroxylamines, and certain thio compounds; the liver is the main site of formation, but the kidney, gastrointestinal tract, and skin are also active. First, glucuronic acid must be converted to an active form, uridine diphosphate-α-d-glucuronic acid (UDPGA). Second, UDPGA conjugates with an oxidized (hydroxylated) substrate. UDP is a metabolic product in addition to the metabolite. This enzyme family includes O-transferases, N-transferases, and S-transferases; there are several forms of UDP-glucuronyl transferases.

Acylation Reactions. These are mitochondrial reactions and form amides from carboxylic acids and amines. The acid must be activated by forming a coenzyme A derivative. The acid may be the xenobiotic (e.g., benzoic acid) or an endogenous acid (e.g., acetic acid). These reactions take place in the liver and kidney. There are many differences between animal species in determining which amino acid takes part in the conjugation.

Glutathione Conjugation. Glutathione (GSH) is a tripeptide: glycine–cysteine–glutamic acid; it is a nucleophile that will react with the highly reactive (and therefore toxic) electrophiles generated by cytochrome P450 oxidations (such as epoxide formation). The liver and most other tissues maintain a cellular pool of glutathione but do not restore diminished supplies rapidly; hence, if the demand for glutathione is great (large dose of xenobiotic with rapid generation of metabolic electrophiles), the glutathione pool becomes depleted and the electrophile is now available to react with other nucleophiles in the cell, such as water, nucleic acids, protein, and so forth. The liver attempts to conserve amino acids and will remove the glycine and glutamic acid from the glutathione conjugate; the resulting cysteinyl derivative is rapidly N-acetylated to form the N-acetylcysteine derivative, which is more often called the mercapturic acid derivative.

Glutathione conjugates are not always less toxic than the original xenobiotic; in some cases GSH conjugation results in a chemically unstable intermediate that breaks down to a reactive derivative more toxic than the parent compound. In other cases the conjugate diffuses from the liver into the blood; some tissues contain a plasma membrane-bound enzyme (γ-glutamyltranspeptidase) that can facilitate the transport of the conjugated metabolite into the cells to be metabolized to toxic intermediates; this is especially true of the kidney and testes.

N-, O-, and S-Methylation Reactions. These reactions are carried out by a methyltransferase that uses S-adenosylmethionine as the methyl group donor for amino, hydroxyl, and thiol groups.

Integration of Metabolic Pathways. Few if any lipid-soluble xenobiotics have single metabolic pathways; the metabolism of such xenobiotics is more complex than a simple phase I reaction followed by a simple phase II conjugation. In most cases there are several phase I and phase II reactions that take place; which pathways are followed depends on (1) the concentration of reactants (dose), including rates of synthesis of conjugating intermediates; (2) the rates of the individual reactions; these can be greatly influenced by the chemical environment, including diet and drugs; (3) the cell type (different cells have different enzymes); and (4) the animal species. The favored pathways are those for which the reactant concentrations are high and the reaction rates are fast. An example of the complexity of a complete metabolic scheme for a xenobiotic is given in Figure 1-7. Iproniazid, an antidepressant, can undergo two different phase I reactions: O-dealkylation to isoniazid and acetone, or hydrolysis to isonicotinic acid and isopropyl hydrazine. Isoniazid undergoes N-acetylation (the rate for this reaction has a strong genetic dependency). Acetone, isonicotinic acid, and isopropyl hydrazine are water soluble and are excreted in the urine. Isopropyl hydrazine can be N-dealkylated to form acetone and hydrazine. Acetylisoniazid can be hydrolyzed to isonicotinic acid and acetylhydrazine, both of which are water soluble and are excreted in the urine. The toxicity of iproniazid has been ascribed to acetylhydrazine, isopropylhydrazine, and hydrazine itself; of the three, hydrazine is the most toxic.

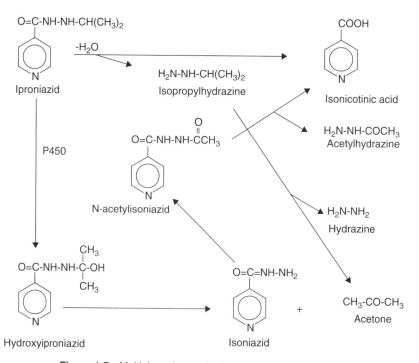

Figure 1-7 *Multiple pathways in the metabolism of iproniazid.*

ACUTE AND CHRONIC TOXICITY

The terms *acute*, *subacute*, *subchronic*, and *chronic toxicity* primarily describe the duration of exposure in toxicity testing protocols. The term *subacute* is seldom used in modern toxicology and has been replaced by *subchronic*, since *sub* refers to "less than" an acute exposure.

Acute toxicity occurs when the exposure to a chemical agent is sufficient to produce an adverse response in minutes, hours, or several days. The exposure is usually a single event, one dermal application, one gavage, or a single short period of inhalation exposure, often of 4 to 6 hours duration. In an investigative situation, the experimenter can determine the target organ(s), the approximate dose range for toxicity, and perhaps the reversibility of the response. For neurotoxins there may be little or no cellular necrosis observed, but for other tissues cell death is often seen within a few hours for direct-acting agents and 6 to 12 hours for agents requiring metabolic activation.

At lower doses repeat administration may be required to elicit the toxic response; when this delayed response takes more than 6 months to become manifest, the toxicity is referred to as chronic. In between acute and chronic is subchronic toxicity, which may take weeks or even months (but less than 6 months) to develop. The response to the toxicant in chronic toxicity is usually different from that in acute toxicity; however, the target organ is usually but not always the same.

In acute toxicity the experimental animal is usually a valid predictor of the response seen in humans. In chronic toxicity a single species may not be a valid predictor of the human response, especially in the case of carcinogens, and testing in more than one species is required. An important limitation to animal testing of chemicals in safety evaluation is extrapolation from high doses given to small populations of test animals to low exposures for large human populations. Most toxicologists believe thresholds for chemical carcinogens exist but at this time cannot be determined by scientific means. Regulators use risk assessment models to estimate low-risk exposure to carcinogens.

Cell Death Due to Chemical Injury

Consider the classical example of acute carbon tetrachloride poisoning and toxicity to the hepatocyte (liver parenchymal cell) by an oral administration of a cytotoxic but nonlethal dose to an experimental animal. The toxicant, a metabolite of carbon tetrachloride, can (1) produce free radicals, (2) react with protein or nucleic acid to inhibit vital enzyme activity, and/or (3) react with structural units to disrupt the cytoskeleton (filaments, tubules, etc.).

Free Radical Injury. A large number of chemical agents cause irreversible damage to cell membranes by producing free radicals, particularly reactive oxygen species: superoxide anion ($\bullet O_2^-$), hydrogen peroxide (H_2O_2,), and hydroxyl radical ($\bullet OH$). Reactive oxygen species (ROS) production appears to be a final common pathway of cell injury not only by chemicals but also by ionizing radiation, tissue hypoxia, aging, microbial killing by phagocytic cells, inflammatory damage, tumor destruction by macrophages, and other cell-damaging processes. In the normal, unchallenged mammalian cell, molecular oxygen undergoes a four-electron reduction to water in the mitochondrion. Partially reduced oxygen species are produced in the cell but have short half-lives; such partial reduction can occur much more frequently in injured cells. For example, the addition of one electron to an oxygen molecule produces the superoxide anion free radical ($\bullet O_2^-$), an oxygen molecule with one oxygen atom containing one unshared electron. Superoxide can be formed by a variety of oxidases (xanthine oxidase, cytochrome P450 monooxygenase, etc).

In the formation of reactive oxygen species, O_2 is converted to superoxide by oxidative enzymes in the endoplasmic reticulum, mitochondria, plasma membrane, peroxisomes, and cytosol. $\bullet O_2^-$ is converted to hydrogen peroxide (H_2O_2) by dismutation (a reaction between two identical molecules in which one gains what the other loses), then to a hydroxyl free radical ($\bullet OH$) by a Cu^{++}/Fe^{++} catalyzed Fenton reaction. Resultant free radical damage to lipids (peroxidation), proteins, and DNA leads to various forms of cell injury.

The major antioxidant enzymes are superoxide dismutase (SOD), catalase, and glutathione peroxidase. The most common pathway for the removal of hydrogen peroxide from a cell is by the action of catalase, which decomposes the hydrogen peroxide to oxygen and water. The hydroxyl radical is highly reactive and reacts with a wide variety of compounds; for toxic reactions, proteins, nucleic acids, and lipids in membranes are particularly important. In the peroxidation of phospholipids in cell membranes, the radical attacks unsaturated fatty acids (e.g., linoleic acid) to create an organic radical at the site of a double bond. These radicals lead to extensive damage to the membrane, especially truncation and cross-linking of fatty acid chains. Free radicals can be removed from cells by several mechanisms. The most important are (1) scavenging by antioxidants such as α-tocopherol and sulfhydryl-containing compounds such as cysteine and glutathione, and (2) enzymatic reduction by glutathione peroxidase, which catalyzes the transfer of the hydrogen atom from the sulfhydryl group in GSH to the radical.

Although much attention has been given here to free radical–induced injury, this is by no means the only mechanism for chemical injury; adduct formation, enzyme inhibition, macromolecular cross-linking, and so forth, are just as important, if not more so.

Necrosis and Apoptosis. Once a cell dies, the lysosomes within release phosphatases, nucleases, proteases, and so forth, digesting the cell by the process of autolytic necrosis; if the extent of the necrosis in a tissue is sufficient, inflammation may result (see below). Quite distinct from the morphological changes that follow cell death due to necrosis is a different pattern of cell death, apoptosis, in which the cell fragments; the apoptotic fragments are taken up by phagocytic cells and degraded, and no inflammation results.

Apoptosis is activated by several stimuli including death ligands (tumor necrosis factor; Fas ligand), withdrawal of hormones and growth factors, and toxic agents to activate proteolytic caspases directly or through changes in mitochondria. The caspases activate latent cytoplasmic endonuclease and proteases to initiate an intracellular degradation and fragmentation of chromatin and the cytoskeleton. This cellular self-destruction is gene directed and serves a biologically meaningful, homeostatic function.

In apoptosis, as the chromatin compacts, DNA strands are cleaved at the linker regions between nucleosomes to form fragments that are multiples of 180–200 base pairs; these fragments give a typical "ladder" in agarose gel electrophoresis. DNA ladders can also be observed in necrosis and are no longer regarded as indicative of apoptosis. In fact, true apoptosis is difficult to demonstrate rigorously.

Apoptosis is not limited to toxic reactions; it is a routine process in multicellular organisms in embryogenesis and maintenance of adult tissues. The complete molecular mechanism for apoptosis is not known, but the process is gene regulated. Many viruses code proteins (e.g., E1b from adenovirus) that suppress apoptosis, thus allowing the infected cells to live and provide a medium for the virus. Humans have a gene, bcl-2, which codes for a protein (Bcl-2) that is a component of the mitochondrial, endoplasmic reticulum, and nuclear membranes; the protein binds to the p53 protein (*p53* is a tumor suppressor gene) and inhibits apoptosis. *p53* gene expression increases in response to DNA damage, which either brings the cell cycle to a halt to allow time for DNA repair, or the cell enters apoptosis. The p53 protein has a role in regulating apoptosis.

Inflammation. Inflammation is a reaction of tissue to local injury involving the vasculature. The reactions include phagocytosis of the injurious agent, entrapment of the irritant by specialized cells (hemocytes) that ingest the irritant, and neutralization of noxious stimuli by hypertrophy of the cell or one of its organelles. Blood vessels respond and lead to accumulation of fluid and blood cells. Inflammation destroys, dilutes, and wards off the injurious agent.

Inflammation is caused by microbial infections, physical agents (burns, radiation, trauma), chemicals, necrotic tissue, and all types of immunologic reactions. In acute inflammation (duration of a few minutes to 1 or 2 days) there is exudation of fluid and plasma proteins (edema) and emigration of leukocytes (mostly neutrophils). In chronic inflammation lymphocytes and macrophages are present, and there is a proliferation of blood vessels and connective tissue. Chemicals can cause the inflammatory response directly, but necrotic tissue itself, the result of chemical injury, can cause this response and not involve the chemical toxin other than its induction of necrosis.

Chemical Carcinogenesis

Genotoxic Mechanisms. Genotoxic carcinogens are chemicals that damage DNA and cause cancer in the same tissue (same cell type) in which the DNA damage occurred. This damage occurs in somatic cells and leads to the activation of growth-promoting oncogenes and/or the inhibition of tumor suppressor genes. The target cells transform to altered genotypes, undergo clonal expansion, further mutations, and eventually a malignant neoplasm.

Types of Chemical Carcinogens. Direct-acting carcinogens are capable of covalent binding to DNA without the need for metabolic activation of the carcinogen; they can produce cancer at the site of application. These chemicals are also called primary, proximate, active, or ultimate carcinogens. Procarcinogens do not necessarily act at the site of application; they are metabolized to direct-acting carcinogens and produce cancer in the tissue where metabolic activation occurs. These compounds are also called secondary, parent, or inactive carcinogens or precarcinogens. Promoters cannot cause cancer by themselves; they act only on initiated or transformed cells to facilitate the multistage process of carcinogenesis. Cocarcinogens cannot cause cancer by themselves at doses at which they serve as cocarcinogens, nor do they promote primary carcinogens; rather they increase the potency of other carcinogens.

Stages in Genotoxic Chemical Carcinogenesis. The development of cancer in a tissue is a multistep process that is still not yet completely understood. The major stages in the carcinogenic process include metabolic conversion of the inactive procarcinogen to a DNA-damaging intermediate, chemical alteration of the genome, promotion of mutagenic processes, and a progressive selection of neoplastic cells to form a frank neoplasm.

Most environmental carcinogens, if they are to persist in the environment for any appreciable time, are relatively stable and therefore not highly reactive chemically (procarcinogens). In order to begin the carcinogenic process, a procarcinogen is first converted to a highly reactive intermediate, a metabolite in the pathway of conversion of the lipid-soluble xenobiotic to a water-soluble end product that can be excreted in the bile or urine. In many, if not most, cases the cells that have the enzymes necessary to carry out this metabolic activation become the target cells for cancer; the metabolism can then determine which tissue will develop the cancer.

Some carcinogens decompose spontaneously under physiological conditions, and these compounds are often local carcinogens—that is, they produce cancer at the site of entry to the body. If, however, the rate of decomposition is slow, tumors may be induced at different sites. The highly reactive intermediate produced in the metabolic activation or decomposition of the xenobiotic reacts with DNA to form a chemical adduct. The reactive intermediate is termed the ultimate carcinogen. If a chemical adduct is formed between the carcinogen and DNA in the target cell, the DNA is said to be damaged. A mutation takes place when the adduct leads to erroneous replication of the genetic material, so the daughter cell(s) does (do) not contain the exact base sequence as in the preexposed parental cell. This may be a point mutation (base-pair transition or transversion) or a frameshift mutation. Once one or more mutations have occurred, the cell is transformed to a cancerous cell. An early consequence of this (these) mutation(s) is a destabilization of the genome, leading to further mutations.

Initiation of the carcinogenic process begins with the formation of the chemical adduct between the ultimate carcinogen and the target cell DNA and covers all intermediate processes until the cell has transformed. In order to form a tumor, the growth of the transformed cell must be stimulated and facilitated so that the tumor has a growth advantage over the neighboring cell population; this process is called promotion. Without promotion the neoplastic cell seems to remain dormant, its growth being held in check by homeostatic factors from nearby normal cells. As neoplastic cells begin to grow and divide, more mutations occur, and selection processes favor the genotype with phenotypic properties that will allow the cell to dominate its environment. This is the progression stage of carcinogenesis. In most cases the cancer arises from a single transformed cell (the tumor is a clonal population with abnormal properties in gene expression).

There are several nucleophilic sites in information molecules with which the electrophilic carcinogens react; the major sites in DNA include guanine (N-1, N^2, N-3, O^6, N-7), adenine (N-1, N-3, N^6, N-7), cytosine (O^2, N-3, N^4), thymidine (O^2, N-3, O^4), and deoxyribophosphate (O). Much remains to be understood about the specificity of these reaction sites, but stereoavailability is known to be important. Adduct formation does not occur at random; there are hot spots in DNA at which adducts are more likely to form than anywhere else, and the chemical nature of neighboring bases can have a profound influence on whether an adduct will form at a particular base in the DNA sequence. Mutations arise from miscoding by the base at which the adduct appears. For example, formation of O^6-methylguanine in DNA alters the base-pairing properties of the purine; instead of pairing with cytosine, the normal complement to guanine, O^6-methylguanine pairs with thymine, the normal complement to adenine. This base-pair transition can lead to a mutation in daughter cells. The mutation can result in a change in the amino acid sequence in the protein that is the product of the gene in which the mutation occurs.

The DNA damage—that is, the DNA adduct—will not cause the mutation to occur until the nucleic acid replicates; this replication must take place before the DNA is repaired. There are several enzymatic repair systems that remove the carcinogen-DNA adduct and restore the genetic material to the normal state.

Protooncogenes, Oncogenes, and Tumor Suppressor Genes. Protooncogenes are DNA sequences that code for normal cellular growth factors. Some of the gene products are first messengers in cell growth that act on the external aspect of the plasma membrane, leading to signal transduction through the membrane; other products are second messengers that transmit the transduced signal from the inner aspect of the plasma membrane across to the nucleus. Protooncogenes can be mutated into oncogenes by chemical carcinogens. The products of oncogenes are proteins, which are different from protooncogene proteins by only one or a few amino acids; the biological activity of the oncogene protein, however, is different, and usually the rate of gene expression increases as the cells transform to cancer cells, giving the cancer cells significant growth advantage over neighboring normal cells. It appears that the activation of one protooncogene to its oncogene form is not sufficient to cause cancer; more than one protooncogene must be activated.

Tumor suppressor genes are DNA sequences that code for proteins that inhibit (suppress) cell proliferation. Some carcinogens can damage DNA and result in the inhibition of the expression of tumor suppressor genes; the consequence is the release from inhibition of cell proliferation genes.

Nongenotoxic Mechanisms. Nongenotoxic carcinogens do not react directly with DNA but indirectly result in an altered genome. Alteration of gene expression and production of chromosomal aberrations are two mechanisms for nongenotoxic carcinogenesis. Chemicals that alter the hormonal balance (some organochlorine pesticides) and result in accelerated cell turnover can induce carcinogenesis; similarly, agents that are cytotoxic in tissues capable of regeneration cause increased cell replication and increased chances of erroneous DNA replication. Asbestos fibers do not react chemically in the cell but do interact with chromosomal material during cell division to result in abnormal distribution of genes to daughter cells.

Altered DNA Methylation. Part of the control of gene expression in mammalian cells involves the normal methylation of the 5-position of cytosine in DNA; the presence of 5-methyldeoxycytidine in a 5-GC-3′ pair inhibits the expression of the gene downstream from this site. This methylation process takes place immediately after the replication fork produces the daughter strand of DNA and uses S-adenosylmethionine as the source of the methyl group. Interference with the methylation process by depleting S-adenosylmethionine, inhibiting the methyltransferase or altering the template used to identify which cytosine bases require methylation, can lead to altered expression of the gene downstream from the methylation site.

Induction of Reactive Oxygen Species. A common response to toxic injury in a cell is the production of reactive oxygen species. Chemicals that generate these reactive intermediates can alter DNA indirectly through oxidative damage. Oxidative deamination converts 5-methyldeoxycytidine to deoxythymidine; thus, a C to T transition occurs, leading to a mutation, even though the carcinogen itself does not react with DNA. The same is true for the oxidative deamination of deoxycytidine to deoxyuridine, deoxyguanosine to deoxyxanthine, and deoxyadenosine to deoxyhypoxanthine.

Liver as a Target Organ

There are two ways to approach the study of toxicology. One is to focus on the chemical and follow its mechanism of toxic action. The second is to focus on specific tissues and see how they respond to toxic insult. This latter approach is called target organ toxicology. The reader is referred to anatomy, histology, and physiology texts for detailed information on the structure and function of the liver and all the other organs covered in this overview. The material presented here focuses primarily on how individual organs respond to toxic insult, rather than providing a list of chemicals that damage various organs.

Most (80%) of the hepatic blood supply enters the liver via the portal vein; the remaining 20% arrives in the hepatic artery. The portal vein and the hepatic artery enter the liver together and divide repeatedly within the liver into small branches that finally enter an anastomosing system of sinusoids. The sinusoids carry mixed venous and arterial blood past the hepatic parenchymal cells (hepatocytes) and drain into central veins. The central veins empty into the hepatic vein that delivers blood to the inferior vena cava.

The major portion of the liver mass is made up of hepatocytes and sinu-soidal capillaries with little connective tissue. As the portal vein and hepatic artery enter the liver and divide, they are joined by a third vessel, the hepatic bile duct, with its flow in the opposite direction. The three vessels together constitute the portal tract that lies at the periphery of hepatic lobules. In a cross-sectional view of the lobule, the portal tract appears as a cross-section of the three vessels; these are portal triads located at three alternating vertices of the six vertices of the lobule. One triad supplies two faces of one lobule and the opposing faces of the two adjacent lobules. At the center of each lobule is a central vein. The blood vessels of the tract form branches, and the smaller vessels pass between adjacent lobules and enter the sinusoids. The sinusoids are surrounded by branching and anastomosing sheets of hepatocytes. Each primary afferent branch of the portal vein and hepatic artery supplies a mass of tissue (sinusoids and hepatocytes) called an acinus. Each small portal tract gives off three branches that leave the connective tissue of the tract and, running between the adjacent lobules, divide into terminal branches that enter the sinusoids.

The acinus is an irregular structure with a broad equator where the blood vessels enter and two narrow extremities that end on two central veins. The blood enters the center of the acinus and diverges there, passing along the sinusoids towards each central vein. There are approximately 20 hepatocytes between the beginning of the acinus and the terminal end (central vein). The acinus is divided into three zones. Figure 1-8 compares the classical lobular view (left) of the liver and its zones with the newer acinar view (right) of the liver.

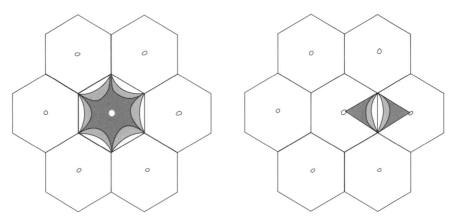

Figure 1-8   *Comparison of lobular (left diagram) and acinar (right diagram) models for the liver. In the modern view of the structural unit of the liver (right diagram), blood entering between two lobules traverses both lobules, moving toward the central vein of each lobule; the area imme- diately adjacent to the interface between the two lobules (light gray) is termed zone 1. In the classical view (left diagram) the blood entering one lobule traverses from the periportal area toward the central vein of the lobule; this area is termed the periportal area. Zone 1 (light gray) receives fresh blood and the highest concentration of solutes coming into the liver. Zone 3 (dark gray) receives blood that has percolated along the sinusoids and lost oxygen and other sub- stances to the adjacent epithelial cells of zones 1 and 2 (medium gray). In the classical model, zone 3 is termed the centrilobular area because of its proximity to the central vein. Cells in zone 1 are rich in respiratory and cytogenic enzymes; zone 3 cells are rich in cytochrome P450 enzymes and are more active in the oxidation of foreign compounds (xenobiotics). Zone 2 (medium gray) is an intermediate or transitional area between zones 1 and 3 or between the periportal and centrilobular areas.*

Toxic Response by the Liver

Lipid Accumulation (Fatty Liver). Accumulation of lipids in the hepato- cytes is one of the most common responses to toxic injury to the liver and is detected by light microscopy; in the vast majority of cases, the accumulated lipid is triglyceride. The lipid accumulates because (1) the rate of lipid syn- thesis is normal, but lipid utilization is blocked; (2) the rate of lipid synthesis is increased and lipid utilization is normal; (3) the rate of lipid synthesis is increased, but lipid utilization is blocked; (4) lipid synthesis takes place in a cellular compartment other than the endoplasmic reticulum where it normally occurs.

Lipid Synthesis and Transport. Free fatty acids are absorbed from the plasma into the hepatocytes and diffuse with glycerophosphate to specialized areas of the endoplasmic reticulum (ER) (liposomes); dietary triglycerides in the plasma are absorbed into the liver, are hydrolyzed to free fatty acids, then follow the same pathway in lipid synthesis. The fatty acids are covalently bonded to the glycerophosphate to form the triglyceride (TG) in the ER;

the TG is weakly bonded to very low-density lipoproteins that serve as the vehicle in the transport of the TG to extrahepatic tissues. Many toxicants inhibit protein synthesis in the liver and therefore inhibit the synthesis of this very low-density lipoprotein. Without this carrier molecule, the newly synthesized TG cannot be transported out of the hepatocyte; thus, fat accumulates.

Carbon tetrachloride, ethionine, phosphorus, puromycin, and so on, seem to cause fatty liver by inhibiting protein synthesis; the carrier lipoprotein is not formed and the TG is not carried out of the hepatocyte in animals treated with these toxicants. Some compounds (orotic acid) cause fatty liver but do not inhibit protein synthesis; phospholipids and cholesterol (esters) are also needed to transport TGs, and toxicants can inhibit the synthesis of these cofactors. In ethanol intoxication there is an increase in the rate of TG synthesis in the liver; not only do TGs build up in the liver (because the rate of secretion is exceeded by the rate of synthesis) but the plasma TG level also increases (hyperlipidemia).

Mobilization of Lipid Stores. Some toxicants (e.g., methyl xanthines such as caffeine) act by mobilizing lipid from compartments in the body, increasing the delivery of TG to the liver. Some toxicants do this indirectly by blocking the use of TG in the liver for synthesis of blood sugar, thereby causing hypoglycemia. In response to hypoglycemia, the body mobilizes fat stores in an attempt to increase blood sugar.

Triglyceride Synthesis in Other Cellular Compartments. The enzymes for TG synthesis are located in liposomes in the ER; however, it is conceivable that a toxicant could cause a redistribution of these enzymes if it disrupts the ER. The end result could be synthesis of TG in the cytoplasm itself without the transportation system to export the TG.

Fibrosis and Cirrhosis. Often in response to repeated necrogenic injury to the liver, connective tissue will begin to proliferate. If the attacks are relatively few and mild, the change in tissue structure can lead to fibrosis; however, if the attacks are frequent and severe, the change can progress to an irreversible cirrhosis. Fibrosis seems to arise by one of two mechanisms: (1) Toxicants can kill hepatocytes, leaving reticulum strands that collapse into bands, forming trabeculae. These "scars" are infiltrated with lymphocytes and macrophages from the blood and bile duct epithelial cells and endothelial cells from the sinusoids. These are the cells that regenerate and fill the regeneration nodules. (2) The other mechanism involves toxicant attacks on connective tissue itself, stimulating myofibroblasts in blood vessels and sinusoidal walls to synthesize new collagen (ethanol does this). Both forms of fibrosis can progress to cirrhosis. Cirrhosis is a chronic, diffuse inflammation of the liver accompanied by proliferation of connective tissue and by degenerative and regenerative changes, resulting in distortion of the liver architecture.

Portal Hypertension and Ascites. In response to toxic injury and resulting cirrhosis, the disruption of normal blood flow in the liver can give rise to greatly increased pressure in the portal vein (portal hypertension). Because the venous and arterial blood supplies mix at the entrance to the sinusoids, occlusion of normal sinusoidal flow causes direct communication of arterial blood pressure to the venous system. Since the portal vein drains the large capillary bed in the peritoneal cavity, the pressure in the thin-walled capillaries also increases, and fluid from the plasma is forced through the capillary wall into the peritoneum, forming the ascites fluid. This is compounded by the liver's failure to maintain albumin in the blood, resulting in lowered osmotic pressure; the liver is the source of blood albumin.

Cholestasis and Jaundice. The average life span of a red blood cell is 120 days; as part of the normal erythrocyte destruction process, hemoglobin from dead red blood cells is degraded to bilirubin in the reticuloendothelial system. One of the functions of the liver is to clear plasma of bilirubin, make the pigment water soluble by conjugating it with glucuronic acid, and secrete the conjugate in the bile. In toxic damage to the liver, this metabolism of bilirubin can become deranged, and the pigment itself (or a bacterial metabolite of bilirubin) may circulate and be stored in the lipid compartments of the body (such as cell membranes). When the plasma level of bilirubin reaches 2 mg/100 ml, the skin, mucous membranes, and sclera of the eyes become colored with the pigment and the condition reflects jaundice. Jaundice often indicates liver injury but not always. Agents that increase the rate of red blood cell destruction, and therefore bilirubin production, can also cause jaundice; and biliary tract obstructions can result in reentry of conjugated bilirubin into the blood, causing jaundice.

Liver Cancer. Cancer of the liver is almost always fatal. A cell in the liver mutates and loses its ability to remain the same kind of cell as its origin; the mutant cell either loses its ability to be regulated by other cells or gains new abilities that allow it to grow with limited or no restraint. The result is the clonal expansion of an abnormal cell in the liver that destroys much of the normal liver tissue. The tumor cells often metastasize to other tissues, especially the lung. In most cases there are three types of liver cancer: (1) hepatocellular carcinomas arise from hepatocytes; (2) cholangiocellular carcinomas from bile duct epithelial cells; and (3) angiosarcomas from sinusoidal endothelial cells. There is a long list of chemicals that induce liver cancer in experimental animals. Aflatoxin B_1 and tetrachlorodibenzo-p-dioxin (TCDD) are the most potent; N-nitrosodimethylamine is among the fastest. Vinyl chloride has been shown to induce angiosarcomas of the liver in workers in the plastics industry.

Kidney as a Target Organ

Each of the two kidneys in the human adult weighs about 120–170 grams. Basically, the kidney is a series of coiled capillaries invaginated into hollow tubes, producing an ultrafiltrate of the plasma and selectively reabsorbing needed solutes and water. This is the principal organ for excreting water-soluble waste products from the blood. The renal cortex is a pale, finely granular outer region, and the renal medulla consists mainly of renal pyramids, which are striated columns of darker tissue continuous with the cortex. The apices of the pyramids project as renal papillae toward the renal pelvis. The ureter receives the urine from the funnel-shaped pelvis; the wall of the pelvis consists of muscle and connective tissue. The pelvis divides into several recesses or calyces that cover the tips of the papillae and are invaginated by them. At the tips of the papillae are the minute openings of the collecting tubules through which urine enters the recesses of the pelvis before passing to the ureter. The flow of the urine is aided by rhythmical contractions of the smooth muscle in the walls of the calyces and pelvis. The kidneys derive all their blood supply from the renal arteries directly from the abdominal aorta; the two renal veins drain directly into the inferior vena cava. The kidneys directly receive 20–25% of the cardiac output.

The basic anatomical and functional unit of the kidney is the nephron; there are approximately 1.3 million nephrons per human kidney. The nephron is composed of a glomerulus, a proximal and a distal tubule, and a collecting duct. The nephron is formed as the afferent arteriole divides directly into a thin-walled vascular bed (glomerulus), filling the capillary with blood under arterial pressure (Fig. 1-9). The blood is thus filtered through the capillary walls, and the filtrate is collected in Bowman's capsule, leading into the proximal convoluted tubule. Filtered blood leaves the glomerulus via the efferent arteriole, then forms a capillary bed in intimate contact with the tubule, draining Bowman's capsule. As the tubule becomes less convoluted, it is termed the pars recta, then becomes the loop of Henle. The tubule bends toward the capsule again and becomes the distal convoluted tubule. The tubule again straightens and empties into the collecting duct, which in turn drains into papillae and leaves the kidney at the pelvis of the organ.

The functions of the kidney include removal of metabolic wastes; maintenance of electrolyte balance and water balance; maintenance of acid-base balance; secretion of organic anions and cations into tubular urine; secretion of renin, erythropoietin, and 1,25-dihydroxycholecalciferol into the blood; and gluconeogenesis. The kidneys produce approximately 180 liters of glomerular filtrate in 24 hours; they reabsorb 179 liters, leaving 1 liter to be excreted as urine.

Special Considerations in Toxic Renal Injury

- The kidneys have an extremely rich blood supply: 20–25% of cardiac output for only 0.5% of body weight; therefore, this tissue is subject to a large exposure to circulating toxicants.

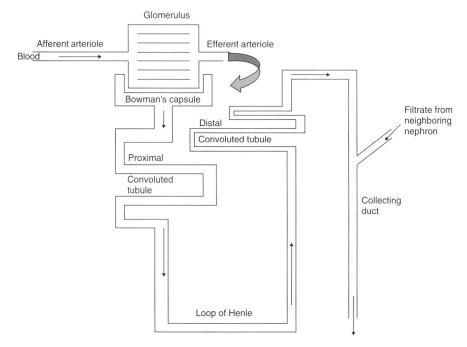

Figure 1-9 *Schematic representation of the nephron.*

- The kidneys extract substances from blood, collecting and accumulating them in tubular urine. Reabsorption of water and Na^+ back into the blood can concentrate tubular urine 5 times the plasma concentration; collecting duct urine can be 100 times the plasma concentration and 500 times if the substance is filtered and actively secreted by tubular cells.
- pH differences due to acid-base balance (H^+ secretion) at both the proximal and distal tubules cause some compounds to precipitate. The pH range of tubular filtrate is 4–8.
- Renal tissue has a high level of metabolic activity for normal endogenous compounds and for xenobiotics; it has a high rate of oxygen consumption. The kidneys contain relatively high levels of many cytochrome P450 monooxygenases, glutathione reductase, epoxide hydrase and glucuronic acid conjugating enzymes, and metallothionein.
- The kidneys have a large endothelial surface area upon which antigen-antibody complexes can deposit.
- Nephrons differ in function. Renal toxicants can shift the blood supply to different nephrons to change the chemical composition of the urine produced.

Types of Toxic Renal Injury

Glomerular Damage. The basement membrane of the glomerulus is composed of type IV collagen, proteoglycans rich in heparan sulfate, fibronectins, and laminins. It serves as a size- and charge-selective barrier to filtration from the blood into the glomerular filtrate in Bowman's capsule. The basement membrane has negative charges due to the ionization of the sulfate and carboxyl groups of the heparan sulfate and sialic acid proteins. These negative charges repel most of the plasma proteins, which are negatively charged at the pH of the blood. Thus, the membrane serves as an ionic barrier to protein filtration from the blood. Chemical agents that alter the net negative charge of the glomerular basement membrane will defeat the barrier and allow proteins to be filtered in the glomerulus; this leads to protein in the tubular filtrate and urine and the condition proteinuria.

Hemodynamic damage may reduce the glomerular filtration rate. Aminoglycoside antibiotics (streptomycin, neomycin, gentamicin, tobramycin, amikacin, netilmicin, and kanamycin, which are polycations) increase resistance in the afferent and efferent arterioles of the nephron and deplete the glomerular basement membrane of polyanions (and anionic phosphoinositides on the brush border of the proximal tubule). The result is proteinuria, oliguria, or polyuria, depending on the balance of the effects of arterioles. Increased glomerular permeability due to changes in podocytes, mesangial cells, and/or endothelial cells of afferent arterioles may cause proteinuria and nephrotic syndrome (proteinuria, hypoproteinemia, hypercholesterolemia, lipiduria, and edema). Glomerular damage may reduce filtration rate; antigen-antibody complexes clog the fenestrae of the glomerulus; doxorubicin (an antibiotic anthracycline) alters the charge and size-selective properties of the glomerulus. The result is proteinuria or oliguria.

Tubular Damage. Tubule blockage by a precipitate may reduce filtration rate, resulting in oliguria or anuria. Interstitial edema may collapse the tubular lumen and reduce filtration rate. Cephaloridine is a cationic antibiotic that is actively removed from the blood into tubular epithelial cells but not secreted by those cells into the tubular filtrate; the antibiotic accumulates in the cells, causing edema and necrosis and resulting in oliguria or anuria. There may be increased backdiffusion of the filtered solute through damaged tubular epithelia. Sulfa drugs precipitate in acidic urine of the proximal tubule to block flow. Heavy metals or parasympathomimetic drugs may cause renin-angiotensin release, which causes high pressure in peritubular capillaries; this can cause interstitial edema, which decreases fluid uptake from tubules and results in tubular occlusion.

Arsine (AsH_3) causes massive hemolysis, and much hemoglobin clears the glomerulus but stays in the lumen of proximal tubules, forming amorphous casts of methemoglobin; the tubules become blocked, and anuria and uremia result. Ethylene glycol can be metabolized to oxalic acid, which may precipitate Ca^{++} and block the flow in tubules. The proximal tubule is the site of major reabsorption of solutes and water; 80% of NaCl and water and 100% of glucose, amino acids, and small proteins filtered in the glomerulus are reabsorbed here; also, 100% of organic acid secretion takes place here. With the first concentration of tubular urine occurring here, this becomes the most frequent site for necrotic lesions produced by nephrotoxins.

Examples of Toxic Injury to the Kidney. Nonsteroidal antiinflammatory analgesics (aspirin-like compounds) can be toxic to the kidney. On an acute basis (single overdose) they produce proximal tubular necrosis, probably due in part to their active secretion via the organic acid pathway and to their metabolism to electrophilic intermediates. More often, however, they produce a chronic (exposure >5 years) toxicity characterized by renal papillary necrosis (the papillae are the projections at the end of the collecting ducts entering the pelvis of the kidney); there is also a reduction in glomerular filtration rate. Toxic or infectious damage to cells causes most cells to increase the rate of synthesis and immediate release of prostaglandins that are involved in inflammation. Nonsteroidal antiinflammatories inhibit the conversion of arachidonic acid to the unstable endoperoxide intermediate PGG2 catalyzed by cyclooxygenase. Inhibition of prostaglandin synthesis decreases the production of vasodilators by renomedullary interstitial cells to leave the vasoconstrictors (catecholamines and angiotensin II) unopposed. This results in a decrease in renal blood flow and ischemia.

The solvent bromobenzene is metabolized in the liver to a glutathione conjugate and excreted in the blood; when the conjugate reaches the proximal tubules in the kidney, an enzyme on the exterior surface of the epithelial cells (γ-glutamyltranspeptidase) hydrolyzes the glutamic acid residue from the glutathione conjugate, decreasing the water solubility of the metabolite and increasing its rate of absorption into the tubular epithelial cells. Once in the cells, the metabolite is further metabolized to produce a reactive electrophilic derivative that is cytotoxic.

Probenecid is a drug used to treat gout as it increases the urinary excretion of uric acid. Probenecid is secreted by the kidney into the tubular filtrate as an organic anion, and this competes with the tubular secretion of other organic anions such as penicillin and the herbicide 2,4,5-T, reducing their rates of excretion in the urine. This delayed excretion will provide more time in the body to allow metabolic activation of the xenobiotics and possible toxicity.

Lung as a Target Organ

Pulmonary toxicology is the study of the response of the respiratory system, especially the lung, to toxic insult, as opposed to inhalation toxicology, which is the study of the responses of all tissues in the body, but especially the lung, to inhaled substances. In most cases, the route of exposure in pulmonary toxicology is inhalation, but systemic exposure can also lead to toxic responses in the lung. The response of the lung to exposure to an airborne toxicant depends strongly on the aerodynamic properties of particles and solubilities of the gases. Particles larger than 5–10 μm do not follow the curved path of the airstream and therefore impact in the nasal passages, never entering the deep lung. Particles with diameters 1–5 μm do enter the trachea and bronchi, and those less than 1 μm reach the alveoli. Water-soluble gases dissolve in the mucus layer in the upper airways and may not reach the alveolus unless the air concentration is relatively high; lipid-soluble gases usually follow the inhaled air all the way to the deep lung, where they are absorbed rapidly into the circulating blood.

Structurally, the respiratory system is composed of a series of ducts leading from the nasal passages to the trachea, bronchi, bronchioles, terminal bronchioles, respiratory bronchioles, and alveolar ducts, ending at the alveolar sacs, where gas exchange takes place. The lungs are made up of more than 40 different cell types.

The trachea, bronchi, and bronchioles are ciliated and have goblet cells that secrete mucus. The cilia, which are frequent targets in pulmonary toxicology, move the mucus toward the mouth so that particles that deposit in these ducts do not enter the deep lung but are moved to the mouth, where they are swallowed (therefore, the absorption site is the gastrointestinal tract) or expectorated. The small bronchioles have mucus-secreting cells but not ciliated cells. Clara cells in these ducts are able to metabolize xenobiotics and therefore are often involved in toxic responses by the lung.

The alveolus contacts the blood capillary, making an air-blood barrier only two basement membranes and two cells thick: the alveolar epithelial cell (type I cell) and the capillary endothelial cell. The alveolus also contains type II cells that secrete surfactant and macrophages that engulf foreign particulate material lodged on the surface of the alveolar lumen. Type I and type II cells are also capable of metabolizing foreign compounds. The alveolar sacs provide a large surface area for gas exchange with the blood; the surface area, depending on the species, is one or two orders of magnitude larger than the area of the skin.

Other special features of the lung are (1) it receives all of the cardiac output so that any chemical in the general circulation will reach it, and (2) lung tissue has the highest concentration of oxygen in the body. The lung has a well-developed immune component, making the tissue susceptible to hypersensitivity reactions.

The respiratory system responds to toxic insult through irritation, inflammation, necrosis, fibrosis, constriction of the airways, and carcinogenicity. Usually the responses follow inhalation of the toxic agent but can follow systemic exposures to chemicals as well. Acute responses include altered reactivity of the airways and pulmonary edema. Chronic responses include fibrosis, emphysema, asthma, and cancer.

In acute toxicity inhaled chemicals can stimulate cholinergic receptors in the trachea and bronchi to release cyclic guanosine monophosphate, a transmitter that stimulates smooth muscle to contract. The muscles surrounding these airways become constricted. The large airways also contain beta-adrenergic receptors that release cyclic adenosine monophosphate, which causes dilation of the airways. Histamine, prostaglandins, leukotrienes, substance P (a polypeptide hormone), and nitric oxide also affect smooth muscle and airway reactivity. Air pollutants are important agents affecting reactivity of the airways and are especially significant in brochioconstriction.

Environmental agents can induce the synthesis of interstitial collagen in the alveolar region of the lung to result in a fibrosis; with chronic exposures, this can lead to a stiffening of the lung and obstruction of the airways. The lung produces type I and the more flexible type III forms of collagen. Certain agents such as ozone can increase the ratio of type I to type III collagen, resulting is a less compliant lung and decreased ventilation.

Chronic inhalation of agents such as tobacco smoke can lead to the destruction of lung tissue, primarily alveolar tissue, and a condition termed emphysema. This is an enlargement of the lung due to destruction of the walls of the alveolar sacs and expansion of the air spaces distal to the terminal bronchioles. The larger air spaces reduce the rates of gas exchange with the blood. The tissue destruction results from the production of reactive oxygen species and stimulation of macrophage release of proteases that digest the alveolar cells. Emphysema is characterized by dyspnea, wheezing, and coughing.

Chronic responses by the lung to inhaled toxicants include asthma, a variable change in airway resistance to flow; unlike alveolar fibrosis, asthma arises from changes in smooth muscle hypersensitivity in the larger respiratory ducts. In many cases, an inhaled allergen reacts with an immunogloblulin bound to the surface of mast cells, which in turn release multiple mediators of type I hypersensitivity including histamine, prostaglandins, and leukotrienes. These mediators stimulate the contraction of smooth muscle to produce constriction of the bronchi, increased mucus secretion and vascular permeability, and edema. Degranulation of the mast cells can also lead to increased involvement of eosinophils and neutrophils to increase the brochioconstriction and edema. If the condition continues, a delayed response arises from degranulation of eosinophils and the release of proteins that damage epithelial cells and impair the mucociliary action in the larger airways. An autonomic nerve discharge is stimulated by the damage to epithelial cells and a further increase in muscle contraction results, worsening the brochioconstriction.

Lung cancer is a leading cause of cancer deaths worldwide. Although not the only cause of lung cancer, tobacco smoking appears to be the major cause of this fatal disease. Lung cancers include squamous cell carcinomas arising in the bronchi, adenocarcinomas in the bronchioles, bronchioalveolar carcinoma, oat cell (small cell) carcinoma arising from bronchial epithelium, and mesotheliomas from the pleural lining of the lung. Radon gas may be responsible for some human lung cancers, although this has not yet been established. Mesotheliomas can be caused by exposure to high concentrations of certain asbestos fibers.

Most of the responses by the respiratory system described above are caused by inhalation exposures. Systemic exposures can and do cause toxic responses in the lung. Such environmental agents include the herbicide paraquat, the plant pyrrolizidine alkaloid monocrotaline, the anticancer agents bleomycin and cyclophosphamide, and the phytoallexin 4-ipomeanol.

Nervous System as a Target Organ

The nervous system has several features that impact on its response to toxic insult. This organ involves the entire body, has an energy requirement greater than most organs, has a high lipid requirement to maintain the excitable membranes, and relies on neurons that cannot regenerate. Circulation to the brain is controlled in part by capillary endothelial cells with tight junctions, few transport vesicles, and high electrical resistance; this limits movement of water-soluble toxicants but not lipid-soluble toxicants into the brain.

Direct chemical injury to the nervous system can result in neuronopathy, axonopathy, myelinopathy, and neurotransmission dysfunction; in addition, this system is subject to developmental injury and carcinogenesis. Neuronopathy is the result of the death of the neuron itself, which can lead to degeneration of associated nerve fibers. Since there is no regeneration of the lost neuron, this response is irreversible. In axonopathy the axon is the target for the toxicant. This loss is irreversible in the central nervous system, but axons in the peripheral nervous system can repair and regenerate distally from the point of injury. A common cause of axonopathy is induction of change in neurofilaments, such as protein cross-linking and aggregation; this results in impaired transport of nutrients in the axon and compromised impulse conduction by the neuron.

When the target cell is the oligodendrocyte of the brain or Schwann cell of the peripheral system, there is a loss of the myelin sheath protecting the axons; this is termed myelinopathy. Diphtheria toxin inhibits protein synthesis in myelinogenic cells, resulting in impaired formation of the sheath and slow conduction of impulses. Triethyltin induces edema and swelling and subsequent loss in the myelin sheath. The sites for chemical injury resulting in the interruption of neurotransmission include (1) the microtubules involved in the two-way axon transport, (2) the plasma membrane, (3) sites of synthesis and storage of transmitters, (4) sites for transmitter release and reuptake, and (5) transmitter receptors.

Inhibition of protein synthesis is a common action of toxicants. The neuron perikaryon is the principal site of the synthesis of proteins that must be transported over great distances along the neuron; the axon and cell terminal depend on this source for a continued supply of essential constituents. The nervous system, therefore, is susceptible to the many inhibitors of protein synthesis, especially those with a high lipid solubility.

The ion channels are subject to toxic attack by a variety of chemical agents. Interference of these channels results in interruption in the generation of action potentials and restoration of the resting membrane potential.

Skin as a Target Organ

The skin constitutes approximately 16% of the body weight of the average adult human, making it the largest organ of the body. It is composed of two complex layers: the epidermis (outer layer) and the dermis. Beneath the dermis is subcutaneous fat. The skin serves as an entry barrier to environmental agents such as microbes, chemicals, and physical agents, and as a barrier to the loss of water from the tissues.

Epidermis. The epidermis (see Fig. 1-5) is a continually renewing, stratified, squamous epithelium without its own blood supply; the thickness of this layer varies from 0.1 mm to more than 4 mm, depending on its location on the body, and to some extent, the nature of the environment. It is composed of five layers and has no blood circulation of its own. The deepest layer is the stratum germinativum (also basal cell layer). The other layers are derived from this layer and represent different stages of differentiation of the basal cells.

The stratum germinativum consists of a single layer of columnar epithelial cells called keratinocytes, which, upon division, push up into the stratum spinosum (prickle cell layer). The keratinocytes in the basal layer undergo frequent mitoses to generate the other layers of the epidermis. There is an extensive network of intercellular spaces in the layers above the stratum germinativum that permits nutrients from the dermal blood supply to reach these cells. In the stratum granulosum the keratinocytes lose their nuclei and die as a result of the digestive action of the lamellar bodies. The intercellular spaces fill with ceramides, which are lipids composed of palmitic acid, serine, and a second fatty acid.

The stratum lucidum is a thin, ill-defined layer of dead cells containing eleidin, an oily degradation product of the keratohyaline granules. The stratum corneum is the last layer of cellular material originating from the basal layer; it takes 26–28 days for a basal keratinocyte to migrate to the outer surface of the stratum corneum. This layer is composed of dead corneocytes flattened and overlapping each other, is coated with ceramides, and is the main component of skin that serves as the barrier function.

Specialized nonkeratinizing cells in the epidermis include melanocytes, Langerhans cells, and Merkel cells. Melanocytes are dendritic cells derived from neural crest cells and produce the dark brown to black pigments known as melanin, which absorbs ultraviolet light and protects the basal cell layer from phototoxicity. Langerhans cells are dendritic cells in the stratum spinosum. They are antigen-presenting cells of the peripheral immune system and stimulate T lymphocytes. Merkel cells are found near nerve endings from the dermis in or near the stratum germinativum. They may act as transducers; they sense pressure and activate neighboring axons of nerves innervating the dermis.

Dermis and Epidermal Appendages. The dermis is a thick fibrous network of collagen, reticulin, and elastin that supports the epidermis. It consists of two layers: (1) the outer, thinner layer called the papillary layer and (2) the inner, thicker layer called the reticular layer. Blood capillaries are found in the papillary layer. The dermis is composed of fibroblasts, fat cells, macrophages, histiocytes, mast cells, neurons, and blood vessel components; fibroblasts are the predominant cell type.

The epidermal appendages are the sweat glands, the sebaceous glands, and the hair follicles. They represent discontinuities in the stratum corneum and therefore sites where skin penetration rates are greater than other areas.

The Functions of Human Skin. In addition to serving as an interface with the environment, protecting the body from chemical, microbial, and physical invaders, the skin, especially the stratum germinativum, is biochemically active. It is an important site of xenobiotic metabolism and photoactivation of vitamin D precursors.

The skin presents two protective barriers to guard the body from toxic chemicals: the stratum corneum inhibits absorption, and the epidermal layers below the stratum corneum, especially the basal cells, are active metabolically at detoxifying chemicals. Skin (epidermis and dermis) has less than 10% of the enzyme capability that the liver has for xenobiotic metabolism; however, when considering the portion of the liver composed of hepatocytes (the xenobiotic-metabolizing cells) and the portion of the cutaneous tissues composed of basal cells, the activities of the two active cell types in xenobiotic metabolism are quantitatively about the same. There are qualitative differences in the biochemical pathways active in skin metabolism and liver metabolism; however, the similarities are striking. As in the liver, the skin contains oxidizing enzymes such as the monooxygenases (P450 system) and alcohol and aldehyde dehydrogenases. The skin also carries out carbonyl reductions, ring unsaturations, hydrolyses, and conjugations.

Percutaneous Absorption. Dermal absorption of chemicals has been reviewed by Roberts and Walters (1998). The movement of a chemical from the outermost surface of the skin into the blood and lymphatic circulatory systems occurs in two steps: first, the chemical must penetrate the stratum corneum, the most important barrier aspect of human skin; second, it must be absorbed into the living cells and pass into the circulating systems. The rate of diffusion of a toxicant through the skin is related to the lipid solubility (see Fig. 1-4) and inversely to the molecular weight of the compound. Polar toxicants diffuse through the outer surface of the keratin filaments of the hydrated stratum corneum. Nonpolar toxicants dissolve in and diffuse through the nonaqueous lipid matrix between protein filaments of the stratum corneum.

Feldman and Maibach (1967) reported on regional differences of dermal absorption in humans. Their studies demonstrated that the rates of absorption of lipid-soluble compounds (hydrocortisone and parathion) through human skin were relatively slow for the plantar region of the foot and the palm of the hand, and relatively rapid for the axilla, scalp, forehead, ear canal, angle of the jaw in men who shaved, and scrotum. Intermediate rates were reported for the ventral forearm, back, and abdomen.

The dermatotoxicologist distinguishes between skin penetration and absorption through the skin. Penetration is a measure of permeability of a chemical and the transfer of a chemical across the stratum corneum to the basement membrane of the epidermis. Absorption is the transfer through the epidermis and dermis and uptake into the circulation. Skin factors that affect penetration include thickness of the stratum corneum, hydration of the stratum corneum (more important for water-soluble toxicants), integrity of the stratum corneum, solvent, and toxicant binding to the stratum corneum. Absorption depends on these same factors and also temperature and blood flow in the dermis.

An indirect measure of skin penetration is the rate of water loss from the skin. Transdermal water loss (TDWL) is often used as an indication of skin permeability, as the skin is normally a barrier to loss of water from the dermis and the capillaries passing through the dermis. An increase in the rate at which water is lost from this tissue usually reflects, and indeed is due to, an increase in the permeability of the epidermis. An increase in TDWL is usually interpreted as an indication that a chemical will be absorbed across the skin at an increased rate.

Types of Toxic Injury to the Skin

Corrosion. The chemical (a corrosive) acts directly on the skin upon contact, causing disintegration and irreversible alteration of the epidermis and dermis; the result is ulceration and necrosis, followed by scar formation. Corrosion is a chemical burn. Chemicals that cause corrosion of the skin destroy the integrity of the stratum corneum and rupture the cells in the epidermis and dermis. These chemicals are usually acids, bases, or strong oxidants, or decompose on the skin to such active compounds. The action can be delayed, usually due to slow penetration of the stratum corneum. Inflammation follows destruction of the tissue.

Acute Irritation. The chemical (an acute irritant) causes a local, reversible inflammatory response after a single exposure and does not involve an immunologic mechanism. This is a frequent response to chemical contact with the skin and is often called (nonallergic) contact dermatitis. Determination of irritancy is clinically subjective; end points are usually erythema, eschar, and edema. Eschar is a sloughing of dead skin as a result of corrosive action. After the skin has fully returned to normal, a second exposure to the same chemical requires the same concentration as in the initial exposure to elicit irritation; this makes irritation different from sensitization and allergic contact dermatitis, in which the response can be elicited by challenge exposures at lower concentrations. Surfactants such as soaps and detergents are often acute irritants. The strongest irritants are usually classified as corrosives.

Cumulative (Repeat) Irritation. The chemical (a repeat irritant) causes a local, reversible inflammatory response after repeated exposure but not after a single exposure; an immunologic mechanism is not involved. This condition is similar to acute irritation except that it is elicited only after repeat exposure to the chemical; it is also called repeat irritation or chronic irritation. Acute and chronic irritation account for about 80% of all cases of contact dermatitis (the remainder are cases of allergic contact dermatitis). This condition is difficult to distinguish from allergic dermatitis by simple physical examination.

If no penetration of the stratum corneum occurs, the chemical does not irritate. A chemical that is usually nonirritating may elicit irritation if the skin is abraded, infected, or irritated by another chemical. Most chemicals can be irritants under the appropriate conditions. The most important intrinsic factor affecting irritation is age: the skin of infants and children is more sensitive to irritants; the skin of adults and the elderly is usually less sensitive. Sex does not seem to be important. Skin color has often been cited as important, but the problem is more clinical than scientific; it is more difficult to detect erythema in pigmented skin.

Allergic Contact Dermatitis. The chemical (a sensitizer) causes a local erythema and edema of the epidermis and perhaps also vesiculation, scaling and thickening of the epidermis. A hypersensitivity is developed via an immunologic mechanism. Clinically allergic contact dermatitis (delayed hypersensitivity) appears closely similar to irritation; mechanistically, the two types of dermatitis are different, and the difference is determined by skin testing. Allergic contact dermatitis is a cell-mediated or type IV immune reaction. There is a high degree of specificity to the sensitizing chemical, and the response is elicited by lower concentrations of the chemical once the individual is sensitized. The chemical acts as a hapten, binding to a protein to form an antigen that will bind to the surface of the Langerhans cells and/or macrophages to begin the immunologic response. Metals, plant toxins, phenylenediamines, and acrylates are among the most common sensitizers. It is estimated that more than 10% of the U.S. population is sensitized to nickel. Compounds that absorb light in the ultraviolet B region are often photosensitizers. An interesting aspect of allergic contact dermatitis is that the site of the initial reaction is "remembered," and upon exposure of the skin at a different site, the dermatitis recurs at the site of initial exposure.

Comedones and Chloracne. The chemical (a comedogen) causes a deep, disfiguring plug of keratin and sebum to accumulate in the hair follicle.

Chloracne is the most common form induced by a chemical and is often seen in workers exposed to chlorinated compounds. The compounds most frequently responsible for chloracne are 2,4,5-trichlorophenol, chloronaphthalenes, polychlorinated biphenyls (PCBs), polybrominated biphenyls, polychlorinated dibenzofurans, polychlorinated dibenzodioxins, tetrachloroazobenzene, and tetrachloroazoxybenzene. The lesions form most often in the malar crescent and behind the ear; the cheeks, forehead, neck, penis, scrotum, shoulders, chest, back, and buttocks can also be involved as the poisoning progresses. The hands, forearms, feet, legs, and thighs become involved only in the worst cases, and the nose is almost never involved.

The typical sebaceous gland in humans is located deep in the dermis and most but not all drain into hair follicles (see Fig. 1-10). The acini consist of a basal layer of undifferentiated flattened epithelial cells attached to the basal lamina. These cells continually proliferate and differentiate to fill the acini with round cells with many fat droplets in the cytoplasm. The nucleus shrinks and the cell fills with lipid, mostly triglycerides. The cells burst to produce the sebum, which contains cell debris. The sebum is composed of triglycerides, waxes, squalene, cholesterol, and cholesterol esters and is under hormonal control. Chemical agents stimulate an overgrowth of the epithelial cells. These

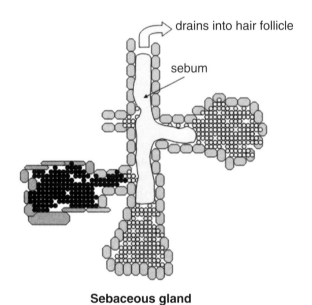

Sebaceous gland

Figure 1-10 *Development of comedones.*

hyperplastic cells eventually undergo abnormal keratinization, transforming affected follicles into keratinous sacs with vestiges of lipid-forming glandular elements. Alteration in cell growth and differentiation are essential to the formation of comedones.

Urticaria. In urticaria, the chemical targets mast cells in the dermis to release histamine and other vasoactive substances, causing the typical wheal-and-flare reaction. Usually the reaction is limited to a transient local stinging and itching of the skin, but in some cases immunologic reactions can occur, resulting in bronchial constriction and anaphylaxis.

Altered Pigmentation. Phenolic compounds, especially quinone derivatives, can inhibit melanin synthesis, thereby causing a loss of pigmentation in the skin (hypopigmentation). Alkylating agents such as psoralens, acridines, aminoquinolines, and so forth, increase melanin production, especially by macrophages, producing a hyperpigmentation.

Alopecia. Alopecia (hair loss) results from toxic injury to the papilla of the hair follicle. Antimitotic agents are especially effective in inducing alopecia as they target rapidly dividing cells.

Cancer. Skin carcinogens cause basal cell carcinomas, squamous cell carcinomas, and melanomas. Early in the development of squamous cell carcinomas, a series of progressively dysplastic changes occur, often resulting from excessive exposure to sunlight and the buildup of keratin; the changes are called actinic keratoses. These atypical cells become horny growths, like warts, often with a red or skin color. These early cancer cells produce different forms of keratin and various pigments. Keratinocytes can continue into anaplasia to become squamous cell carcinomas (moderately metastatic).

Basal cell carcinomas are slow-growing and seldom metastasize; they also arise from the stratum germinativum and maintain most of the phenotypic properties of basal cells. The tumors are usually well supplied with blood vessels and may or may not produce large amounts of melanin. If not treated, eventually they can grow large enough to locally invade bone or facial sinuses. Seborrheic keratoses are round plaques of altered keratinocytes that have increased melanin production; these dark warts are benign tumors but can go on to become the highly malignant melanoma, an aplastic melanocyte that is highly metastatic.

Light-Induced Toxic Injury to the Skin (Photodermatotoxicity). Light can injure the skin directly by interacting with constituents in the epidermal cells or indirectly by interacting with a chemical on the surface of the skin or absorbed into the epidermis (or distributed to the epidermis via the blood). The xenobiotic serves as a chromophore, absorbing light (photons) and transferring energy to, or reacting in an excited state with, cellular components; this is chemical phototoxicity. Phototoxicity is the direct action of photons on skin components. The ultraviolet light spectrum is considered in three zones: (1) UVA is light in the wavelength range of 315–400 nm. This region causes pigment darkening and much of the chemical phototoxic and photoallergic reactions; this is the region for photosynthesis in plants and the stimulus for circadian cycles and the blue end of the vision spectrum in humans. (2) UVB is light in the wavelength range of 280–315 nm. This region inhibits cell mitosis, induces sunburn and skin cancer, and activates vitamin D. (3) UVC is light in the wavelength range of 200–280 nm. This region kills cells, but light in this region is absorbed by ozone in the atmosphere and does not reach the earth's surface.

Short UV is reflected by the stratum corneum and lower layers and absorbed by urocanic acid, melanin, tryptophan, and tyrosine in the stratum corneum. About 40% of UVB light penetrates the stratum corneum. Nucleic acids serve as additional chromophores in basal cells and in the dermis. Blood passing through exposed skin is irradiated by sunlight, and this energy can be absorbed by circulating drugs and toxicants. When the chromophore absorbs photons, the molecule is transformed to a singlet electronic state (one electron pair is raised from the ground molecular orbital to an excited orbital). The electron immediately (10^{-8} to 10^{-9} sec) falls back to the ground molecular orbital, emitting energy at a lower level (wavelength longer than incident photons) or more slowly (10^{-4} to 10 sec) via a metastable triplet state, or undergoes chemical reaction (isomerization, fragmentation, ionization, rearrangement, intermolecular reaction).

Some chemicals form photoactivated products that are highly unstable, highly reactive, and short lived; these compounds must be at the cellular target site to induce phototoxic injury. Other compounds form photoactivated products that are stable and can induce phototoxicity after light exposure. The photoactivated toxicant reacts with cellular constituents such as DNA and protein to induce toxicity. Other compounds, usually those that absorb in the UVA range, produce toxicity indirectly by producing free radicals (type I photodynamic reactions) or singlet oxygen (type II photodynamic reactions), and the reactive intermediate goes on to attack the cellular target.

DNA is the most important molecular target in a cell exposed to UV light. Other molecules absorb the light, but the consequences of that absorption are not nearly as important to the survival of the involved cell. UV at wavelengths at or below 300 nm damages DNA directly; no toxic chemical intermediate is required, as the light induces the formation of cyclobutane pyrimidine dimers directly in the DNA. In other cases light activates a chromophore that subsequently binds covalently to the DNA. The formation of thymine dimers is a process greatly accelerated by sensitizers such as acetone, ethyl acetoacetate, dihydroxyacetone, acetophone, and benzophenone. The sensitizer is thought to become excited photochemically, then form a complex with the two thymines, facilitating dimerization. Cyclobutane-pyrimidine dimers can be removed from DNA by excision repair; if not removed, they can lead to mutations and/or strand breaks during DNA replication. The result may be cancer originating from the target cell.

Depigmentation (Acquired Leukoderma). Exposure to a variety of phenols and catechols causes long-term depigmentation of the skin. These chemicals appear to interfere with metabolism of tyrosine (precursor of melanin) and destroy melanocytes.

Porphyrias. Porphyrias are a class of diseases in which the production and metabolism of porphyrins (such as heme from hemoglobin) are deranged, and excessive amounts of porphyrins and porphyrin derivatives accumulate in the body. In such affected individuals, the circulating red blood cells contain unusually high levels of photosensitizing porphyrins. Light entering the capillaries of the skin can produce a protoporphyrin triplet, which in turn produces a singlet oxygen in the membrane of the erythrocyte. The singlet oxygen oxidizes unsaturated lipids in the membrane, causing hemolysis. Cholesterol in the cell membrane can also undergo photooxidation by protoporphyrin to increase the osmotic fragility and hemolysis of red blood cells passing through the dermal capillaries.

Photosensitization. Allergic contact dermatitis (delayed hypersensitivity) can be induced by some chemicals only after photoactivation. The compound does not sensitize the skin unless the skin is exposed to UVA and/or UVB after topical application.

Reproductive Systems as Target Organs

The reproductive system of the human male begins with the two testes, which produce germ cells: the sperm. The testis is composed of seminiferous tubules and connective tissue containing testosterone-producing Leydig cells and macrophages. The seminiferous tubules are made up of epithelial germ cells, which generate the sperm, and Sertoli cells, which give the tubules structure and nutrients. The testis is enclosed by a fibrous capsule, the tunica albuginea. Sperm leave the testis and are stored in the adjoining epididymus until they are fully mature. They leave the epididymus via the vas deferens and enter the seminal vesicle. Upon ejaculation, the sperm leave the seminal vesicle, enter the ejaculatory duct via the seminal vesicle duct, and pass into the urethra; secretions from the seminal vesicle and prostate gland surrounding the urethra provide nutrition and fluid transport for the sperm. As the sperm pass along the urethra, the bulbourethral gland contributes the final secretion making up the seminal fluid.

Similar to the male system, the reproductive system of the human female begins with the two ovaries that produce germ cells: the ova or oocytes. Before puberty, the ovary is composed of primordial follicles, each containing a single ovum that formed before birth. Beginning at puberty, the ovarian cycle starts with the enlargement of several of these follicles, with a cavity forming around the ovum. One follicle in one ovary grows faster than the others, which regress, and matures to a Graafian follicle. The follicle ruptures, and the ovum, which is extruded into the abdominal cavity near the external opening of the Fallopian tube, enters the end of the oviduct and passes through the uterus and out the vagina; if fertilization occurs in the uterine tube, the ovum implants in the endometrium. The ruptured follicle fills with blood, and its lining proliferates to form a corpus luteum, which secretes estrogens and progesterone. If the ovum has implanted, the corpus luteum persists; without implantation, the corpus luteum degenerates and the ovarium cycle will be repeated. In pregnancy, the endometrium undergoes hyperplasia, due to the stimulation by the estrogens and progesterone from the corpus luteum, to support the newly implanted fertile ovum.

Reproductive functions in both the male and female are under hypothalamic-pituitary control. The hypothalamus secretes a decapeptide, GnRH (gonadotropin-releasing hormone), which acts on the anterior pituitary to release two gonadotropic hormones: leutinizing hormone (LH) and follicle-stimulating hormone (FSH). LH acts on Leydig cells in the testes to produce androgenic and anabolic effects and on the theca interna of the ovary to produce estrogenic effects. FSH and testosterone act on Sertoli cells of the testis to maintain gametogenesis and secrete inhibin, which has negative feedback on FSH secretion. Testosterone inhibits the secretion of LH and GnRH. In the female, LH acts on the theca interna of the ovary to produce estrogenic effects and on the ovarian granulose to produce inhibin. FSH acts on the granulose to induce follicular growth.

Toxic injury to the reproductive system can result from chemical action of the central nervous system and/or gonads to interfere with the complex hormonal regulation by modifying ovulatory functions and altering spermatogenesis. Also, some toxic agents can act on the development of gonads in the fetus and compromise reproductive function upon sexual maturity.

Toxicity to the Male Reproductive System. Chemicals that interfere with the hypothalamic-pituitary-testicular axis, as described, have dramatic impact on male reproduction. Antagonists (e.g., diethylstilbestrol and cimetidine) to testosterone and its derivatives disrupt feedback control and normal function of the testis and accessory sex glands. Inhibitors of testosterone synthesis (e.g., dibutylphthalate) and inducers of phase I metabolism to accelerate the removal of circulating testosterone (e.g., several organochlorine compounds) decrease male fertility. Genotoxic compounds interfere with normal meiosis by damaging DNA and chromosomes and inhibiting nucleic acid synthesis; mutations in the DNA and abnormal chromosomes can generate incompetent sperm or no sperm at all.

Toxicity to the Female Reproductive System. The mechanisms of action to chemicals toxic to the female reproductive system are similar to those discussed for the male system. Interference with the hypothalamic-pituitary-ovarian axis can alter endocrine and ovarian function to affect reproduction. Competitors for estrogen receptors and modifiers of estrogen synthesis, release, and metabolism have a profound effect on ovarian and uterine function. Several halogenated insecticides and herbicides are estrogen agonists and interfere with ovulation, implantation, and embryo development.

Toxicology of the reproductive systems is a relatively undeveloped field, and little has been done on mechanisms of action of chemicals that target these tissues. There is a critical need for molecular studies on the mechanisms of action of reproductive toxins, especially widespread agents in the environment.

REFERENCES

Feldman RD, Maibach HI (1967). Regional variation in percutaneous penetration of ^{14}C cortisol in man. *J Invest Dermatol* **48**:181–183.

Klaassen CD (2001). *Toxicology the Basic Science of Poisons*, 6th ed. New York: McGraw-Hill.

Phalen RF (1984). *Inhalation Studies: Foundations and Techniques*. Boca Raton, FL: CRC Press.

Roberts MS, Walters KA, eds (1998). *Dermal Absorption and Toxicity Assessment*, New York: Marcel Dekker.

2

A Short Introduction to the Expression Profile Toolbox

**Jeffrey F. Waring, Xudong Dai, Yudong He,
Pek Yee Lum, Christopher J. Roberts, and Roger Ulrich**

INTRODUCTION

In the sixteenth century, the alchemist and physician Auroleus Phillipus Theostratus Bombastus von Hohenheim (self-immortalized as Paracelsus) stated, "Solely the dose determines that a thing is not a poison" (Borzelleca, 2000). With this dictum, expanded and restated as "All substances are poisons; there is none which is not a poison. The right dose differentiates a poison and a remedy," Paracelsus introduced the dose-response relationship as a fundamental concept in toxicology. He also developed the concept of target organ toxicity and further insisted that medicine and medical practice rely on data and facts. These principles have served toxicologists well for centuries, and in pharmaceutical research they have evolved into a defined and regulated set of safety testing criteria.

Today safety evaluation of new drug candidates, including a prediction of human risk, relies primarily on animal studies, using clinical (serum) chemistry parameters and histopathology to establish a dose-response relationship for toxic effects. Results are expressed as the numerator of the therapeutic index,

Toxicogenomics: Principles and Applications. Edited by Hamadeh and Afshari
ISBN 0-471-43417-5 Copyright © 2004 John Wiley & Sons, Inc.

$$TI = \frac{\text{median toxic dose}}{\text{median effective dose}}$$

which describes the relationship between toxicity and efficacy for new therapeutic agents. This ratio is subjective, however, as the median toxic dose depends solely on how toxicity for a compound is defined—that is, what end point or biomarker value is assigned to indicate deleterious effects. These end points can be vague or subjective. For example, while large increases in the serum levels of liver enzymes aspartic and alanine transaminases (AST and ALT) are indicative of liver damage, it is not known if small increases represent a lesser risk or no risk at all. Similarly, the median effective dose is generally based on a surrogate outcome in animal models for disease, such as glucose lowering for an antidiabetic compound in ob/ob mice, which may or may not have relevancy to efficacy in humans.

The approach to safety assessment and the biomarkers on which decisions regarding safety are made have changed little in the past several decades (discussed by Aardema and MacGregor, 2002). However, as our understanding of biological processes evolves and our technical ability to detect changes in these biological processes becomes increasingly more sensitive, the line between benefit and risk (or pharmacology and toxicity) becomes blurred and new approaches are needed. From an experimental standpoint, toxicology research is presently undergoing a major change. In particular, new technologies and experimental tools driven by or derived from genomics research are fundamentally changing our view of the toxic response and thus challenging the traditional approaches to safety assessment. It is unclear, however, if toxicogenomics will refine our ability to identify a hazard, define a toxic dose, and more accurately determine a therapeutic ratio. It is possible that analysis of chemical exposure effects at the level of the genome may redefine the relationship between dose and response. For example, while a significant induction of cytochrome P450 (CYP) activity as measured using pharmacokinetic techniques can have a profound effect on drug efficacy and in some cases may produce life-threatening toxicity, does a small induction at the transcript level carry a lesser risk or represent no risk?

It is likely that the next generation of pharmaceutical scientists will not consider the dose-response relationship for toxicity as being separate from the dose-response relationship for efficacy but will instead view beneficial and adverse effects as two points on a continuum that may even vary depending on the genetic composition of the individual patient (see Roses, 2002). All chemicals elicit a response at the molecular level, and it now appears that each compound produces a gene expressional response that is both unique and characteristic, harboring information about both intended and untoward molecular targets. The challenge is in deciphering the response into discrete cellular pathways and in recognizing the beneficial components and separating them from the deleterious ones. In the end this does not render Paracelcus' principle obsolete but rather gives it a new level of refinement.

DEFINITIONS AND EXPECTATIONS

In addition to igniting a new age of discovery, genomics has also produced a new lexicon to describe various scientific disciplines and technologies (collectively referred to as *-omics*; see Aardema and MacGregor, 2002; http://www.genomicglossaries.com/content/omes.asp for additional information) including several new terms regarding drug research. *Pharmacogenetics* studies individual polymorphisms or mutations in DNA sequence that may lead to altered drug reactions within an individual (reduced or exaggerated pharmacology or toxicity). Pharmacogenomics studies the structure and output (RNA and protein) of the entire genome across an individual or population as it relates to pharmacokinetic, pharmacologic, and pharmacodynamic responses to drugs. A genetic alteration may also follow through a population. Toxicogenomics is the study of the structure and output of the entire genome as it relates and responds to adverse xenobiotic exposure. In pharmaceutical research it is the adverse effect counterpart to pharmacogenomics. The primary new tool that enables toxicogenomics is the DNA microarray (discussed at length below), which allows for the analysis of thousands of transcripts simultaneously. Proteomics, defined as the structural and functional analysis of the complete protein complement of cells and tissues, is not discussed in this chapter.

The expectations for toxicogenomics are broad, seemingly to correct all the problems inherent in current approaches to safety evaluation. Current methods (clinical chemistry and histopathology) are largely descriptive and are used to identify and describe toxic effects, not to define mechanism. Hence it was expected that gene expression profiling using microarrays would help identify mechanisms of toxicity, providing that genes represented on the microarrays were sufficiently representative of all genes involved in toxic responses and that markers identified from expression profiling could be used to establish mechanism-based screens (Waring and Ulrich, 2000). It was also proposed that small diagnostic sets of candidate genes could be used to classify and screen for toxicity of new compounds (Thomas et al., 2001) or that toxicants could be classified based on expression patterns alone using larger gene sets compared to a database or compendium (Waring et al., 2001; Hamadeh et al., 2002; Castle et al., 2002).

Since transcriptional changes occur rapidly following xenobiotic exposure, it is hoped that transcriptional analysis using microarrays will enable the prediction of long-term effects from short-term studies. Since the chemical scale-up required to conduct in vivo studies is quite costly, it is expected that transcriptional analysis will enable the prediction of in vivo effects from in vitro models. And, since the role of the pharmaceutical toxicologist is to understand the liabilities of a new drug candidate within the context of human risk, it is further hoped that gene expression analysis will define species differences in responses (adverse and beneficial), thereby increasing the predictive value of animal models. Finally, it is hoped that toxicogenomics will allow us to identify those compounds that do not show toxicity in animals but will produce toxicity in humans (Castle et al., 2002). The latter raises questions as to how predictive toxicology can be verified; a clinical trial using a drug candidate to test the predictive value of rat gene expression results might be unethical.

Collectively the expectations for toxicogenomics are considerable, especially considering the enthusiasm for gaining new knowledge is generally counterbalanced by a concern and aversion toward generating data, which might trigger unnecessary (and costly) delays or block drug development altogether. In summary, to the toxicologist the expectation is that toxicogenomics will help convert toxicology from a largely descriptive science (what happened) to one that is predictive (what will happen and in what species) while providing insight on mechanisms. To what extent these various expectations are met remains to be demonstrated.

To some there is the hope or perhaps the concern that gene expression technologies using microarrays will replace histopathology as the primary tool for drug safety evaluation. This is not likely to occur anytime soon, if at all. Here another -*omics* term applies; economics sets limits on the use of toxicogenomics, since, as with other disruptive technologies, it can at least initially drive up the cost of pharmaceutical development. For example, a small experimental study that examines 7 basic target tissues at a single time point in 5 animals per sex per dose group, with 3 dose groups and controls, approximately 440 arrays would be used (assuming a two-color system with fluor-reverse pairs). The cost, at an estimated $500 per array, would be approximately $220,000 (excluding personnel costs) and would generate 1.1×10^7 data points to analyze. Adding positive and negative controls would increase the study size considerably. For a typical GLP-regulated study evaluating 26 tissues at a single time point in 10 animals per sex per dose, with 3 dose groups and controls, the cost escalates to $3.2 million with 7.9×10^7 data points to evaluate— more expense and effort than any laboratory is currently prepared to absorb. Economics aside, toxicogenomics provides profoundly more information regarding a given compound's effects on biological systems than any other current approach, and when used in conjunction with more traditional toxicological parameters, it can greatly increase our understanding of the consequences following xenobiotic exposure. As such, toxicogenomics is here to stay.

BACKGROUND ON MICROARRAYS

The principle enabling technology behind toxicogenomics is the microarray. DNA microarrays consist of DNA molecules arrayed at high density on a solid surface (Fodor et al., 1993; Southern et al., 1992; Schena et al., 1995). The most widely used microarray platforms consist of spotted cDNAs on glass slides (Schena et al., 1995) and Affymetrix microarrays in which oligonucleotides are synthesized in situ on silicon wafers using photolithography (Lipshutz et al.,

1999). DNA microarrays have found wide use in many areas of biological research, including (1) expression profiling, which is the parallel measurement of the steady-state levels of thousands of transcripts (Brown and Botstein, 1999; Debouck and Goodfellow, 1999; Lockhart and Winzeler, 2000); (2) genotyping, including DNA sequence determination (Chee et al., 1996), comparative genomic hybridization (Pollack et al., 1999; Hughes et al., 2000; Bruder et al., 2001), and large-scale SNP analysis (Dong et al., 2001); and (3) identification of new genes and novel splice variants of known genes (Shoemaker et al., 2001; Kapranov et al., 2002). Microarrays for gene expression profiling are now recognized as an extremely sensitive measure of cell physiology, in effect leveraging genomic sequence information to understand in depth the effects of cellular perturbations on biological systems.

Spotted cDNA Microarrays

Spotted cDNA microarray technology was developed in the laboratory of Dr. Patrick Brown at Stanford University and hinged on two key technological breakthroughs: (1) high-throughput oligonucleotide synthesis (Lashkari et al., 1995), allowing for the rapid amplification of thousands of cDNAs; and (2) a two-color fluorescence detection strategy such that the key measurement from the microarray was not the absolute intensity of the spot but rather the ratio of the fluorescence detected between two differentially labeled RNA samples (Schena et al., 1995). Thus, for each RNA species, a ratio of expression is generated between two samples, and this expression ratio is much more reproducible from one microarray to the next than the intensity values measured for each spot, thus minimizing the differences between microarrays. Through the pioneering work of Dr. Joe DeRisi in the Brown laboratory, robotic methodology was developed for mass producing these microarrays at the laboratory benchtop, thus making this technology widely accessible throughout academia and industry.

Affymetrix Oligonucleotide Microarrays

An alternative microarray technology was developed by Dr. Stephen Fodor and colleagues at Affymetrix Inc. (Lipshutz et al., 1999). Affymetrix arrays use photolithography to direct oligonucleotide synthesis in situ on silicon wafers. Because this method achieves poorer coupling efficiencies compared to phosphoramidite oligonucleotide synthesis, the length of the oligonucleotides is limited to 20mers. However, because of the precision of the photolithographic process, Affymetrix arrays contain a very large number of features compared to spotted cDNA microarrays. Thus, to achieve sufficient specificity for parallel gene expression profiling measurements, it is necessary to select many different (e.g., up to 20) "perfect-match" 20mer oligonucleotides, as well as corresponding "mismatch" oligonucleotides for each perfect-match oligonucleotide, in which a single mismatch is placed at the center of each 20mer. The mismatch oligonucleotide is used to monitor nonspecific cross-hybridization, which is subtracted from the perfect-match oligonucleotide to determine the specific intensity for each gene. Expression ratios between any two samples are then computed in silico. While more expensive than cDNA microarrays, Affymetrix arrays are also widely used in the scientific community.

Inkjet Oligonucleotide Microarrays

A third less widely used microarray technology utilizes inkjet printer heads to direct the synthesis of oligonucleotides on a glass surface (Blanchard et al., 1996). This technology was developed initially at Dr. Leroy Hood's laboratories at Cal Tech University, then the University of Washington, and later at the company Rosetta Inpharmatics, Inc. The inkjet printer head directs phosphoramidites to discreet spots on the surface of the glass slide and uses standard oligonucleotide synthesis chemistry, allowing for the synthesis of much longer oligonucleotides (e.g., 60mers) than on Affymetrix microarrays. However, the density of spots on inkjet arrays ($\sim$25,000 per $1'' \times 3''$ glass slide) is much less than Affymetrix arrays and more similar to spotted cDNA arrays, so methods for oligonucleotide selection were optimized such that single 60mers could be identified that were specific for a given gene, allowing for the simultaneous measurement of expression ratios of $\sim$25,000 different transcripts (Hughes et al., 2001). These microarrays, marketed by Agilent Technologies, Inc., have been used in only a limited number of published studies (Shoemaker et al., 2001; van't Veer et al., 2002).

Statistical Methods for Data Analysis

It was immediately apparent that the biologists using DNA microarrays were going to need powerful informatics and database tools to be able to cope with and successfully handle the large volume of data generated. While the earliest published microarray studies consisted mostly of long lists of regulated genes, clustering analysis was quickly adopted as a method where genes with similar behaviors across a set of experiments, or different experiments with similar patterns of expression changes, could rapidly be associated (Wen et al., 1998; Eisen et al., 1998; Quackenbush, 2001). Several of the most routinely used clustering methods include hierarchical agglomerative (Hartigan, 1975), K-means (Quackenbush, 2001), and self-organizing maps (Tamayo et al., 1999). Principle component analysis (PCA), or singular value decomposition, is an increasingly common method of understanding the dimensionality of the transcriptional response through the identification of principle components (Raychaudhuri et al., 2000).

Microarray Platform Error Model

To greatly accelerate the statistical analysis of microarray experiments, tools were developed at Rosetta Inpharmatics to automate many of the data-handling procedures, enabling the biologists to rapidly assess the results of experiments. A platform error model was developed for both spotted cDNA and FlexJetTM oligonucleotide microarrays that modeled the behavior of probes for all genes represented on the microarray such that statistical *P*-values are assigned for each new hybridization, indicating whether the measured red/green ratio is significantly different from the null hypothesis of no change (Roberts et al., 2000; Hughes et al., 2000). In essence, the error model describes the signal-to-noise ratio on the array, with the most significant *P*-values assigned to the spots with the greatest signal over background. Thus, very small changes in expression ratio (e.g., 30% changes) can receive significant *P*-values if the signal is sufficiently above background.

In the absence of the statistical error model describing the platform, the researcher is forced to apply fold change cuts, which gives greater statistical weight to the spots of the array that give the most trustworthy data. To perform statistical analysis of gene expression measurements from a microarray experiment, the scientist must apply statistical cuts to the data to focus the analysis on genes that show significant changes in expression. The platform error model described above allows the selection of genes with significant changes in expression at a given confidence level, and users typically will also include a fold change requirement so that genes have to show expression changes greater than or equal to a given threshold to be included in the analysis.

REFERENCES

Aardema MJ, MacGregor JT (2002). Toxicology and genetic toxicology in the new era of "toxicogenomics": impact of "-omics" technologies. *Mutat Res* **499**:13–25.

Blanchard AP, Kaiser RJ, Hood LE (1996). High-density oligonucleotide arrays. *Biosens Bioelectron* **6/7**:687–690.

Borzelleca JF (2000). Paracelsus: herald of modern toxicology. *Toxicol Sci* **53**:2–4.

Brown PO, Botstein D (1999). Exploring the new world of the genome with DNA microarrays. *Nat Genet* **21**(suppl 1):33–37.

Bruder CE, Hirvela C, Tapia-Paez I, Fransson I, Segraves R, Hamilton G, Zhang XX, Evans DG, Wallace AJ, Baser ME, Zucman-Rossi J, Hergersberg M, Boltshauser E, Papi L, Rouleau GA, Poptodorov G, Jordanova A, Rask-Andersen H, Kluwe L, Mautner V, Sainio M, Hung G, Mathiesen T, Moller C, Pulst SM, Harder H, Heiberg A, Honda M, Niimura M, Sahlen S, Blennow E, Albertson DG, Pinkel D, Dumanski JP (2001). High resolution deletion analysis of constitutional DNA from neurofibromatosis type 2 (NF2) patients using microarray-CGH. *Hum Mol Genet* **10**:271–282.

Castle AL, Carver MP, Mendrick DL (2002). Toxicogenomics: a new revolution in drug safety. *Drug Discov Today* **7**:728–736.

Chee M, Yang R, Hubbell E, Berno A, Huang XC, Stern D, Winkler J, Lockhart DJ, Morris MS, Fodor SP (1996). Accessing genetic information with high-density DNA arrays. *Science* **274**:610–614.

Debouck C, Goodfellow PN (1999). DNA microarrays in drug discovery and development. *Nat Genet* (suppl 1):48–50.

Dong S, Wang E, Hsie L, Cao Y, Chen X, Gingeras TR (2001). Flexible use of high-density oligonucleotide arrays for single-nucleotide polymorphism discovery and validation. *Genome Res* **11**:1418–1424.

Eisen MB, Spellman PT, Brown PO, Botstein D (1998). Cluster analysis and display of genome-wide expression patterns. *Proc Natl Acad Sci USA* **95**:14863–14868.

Fodor SP, Rava RP, Huang XC, Pease AC, Holmes CP, Adams CL (1993). Multiplexed biochemical assays with biological chips. *Nature* **364**:555–556.

Hamadeh HK, Bushel PR, Jayadev S, DiSorbo O, Bennett L, Li L, Tennant R, Stoll R, Barrett JC, Paules RS, Blanchard K, Afshari CA (2002). Prediction of compound signature using high density gene expression profiling. *Toxicol Sci* **67**:232–240.

Hartigan JA (1975). *Clustering Algorithms*. New York: Wiley.

Hughes TR, Marton MJ, Jones AR, Roberts CJ, Stoughton R, Armour CD, Bennett HA, Coffey E, Dai H, He YD, Kidd MJ, King AM, Meyer MR, Slade D, Lum PY, Stepaniants SB, Shoemaker DD, Gachotte D, Chakraburtty K, Simon J, Bard M, Friend SH (2000). Functional discovery via a compendium of expression profiles. *Cell* **102**:109–126.

Hughes TR, Mao M, Jones AR, Burchard J, Marton MJ, Shannon KW, Lefkowitz SM, Ziman M, Schelter JM, Meyer MR, Kobayashi S, Davis C, Dai H, He YD, Stephaniants SB, Cavet G, Walker WL, West A, Coffey E, Shoemaker DD, Stoughton R, Blanchard AP, Friend SH, Linsley PS (2001). Expression profiling using microarrays fabricated by an ink-jet oligonucleotide synthesizer. *Nat Biotechnol* **19**:342–347.

Kapranov P, Cawley E, Drenkow J, Bekiranov S, Strausberg RL, Fodor SP, Gingeras TR (2002). Large-scale transcriptional activity in chromosomes 21 and 22. *Science* **296**:916–919.

Lashkari DA, Hunicke-Smith SP, Norgren RM, Davis RW, Brennan T (1995). An automated multiplex oligonucleotide synthesizer: development of high-throughput, low-cost DNA synthesis. *Proc Natl Acad Sci USA* **92**:7912–7915.

Lipshutz RJ, Fodor SP, Gingeras TR, Lockhart DJ (1999). High density synthetic oligonucleotide arrays. *Nat Genet* (suppl 1):20–24.

Lockhart DJ, Winzeler EA (2000). Genomics, gene expression and DNA arrays. *Nature* **405**:827–836.

Pollack JR, Perou CM, Alizadeh AA, Eisen MB, Pergamenschikov A, Williams CF, Jeffrey SS, Botstein D, Brown PO (1999). Genome-wide analysis of DNA copy-number changes using cDNA microarrays. *Nat Genet* **23**:41–46.

Quackenbush J (2001). Computational analysis of microarray data. *Nat Rev Genet* **2**:418–427.

Raychaudhuri S, Stuart JM, Altman RB (2000). Principal components analysis to summarize microarray experiments: application to sporulation time series. *Pac Symp Biocomput* 455–466.

Roberts CJ, Nelson B, Marton MJ, Stoughton R, Meyer MR, Bennett HA, He YD, Dai H, Walker WL, Hughes TR, Tyers M, Boone C, Friend SH (2000). Signaling and circuitry of multiple MAPK pathways revealed by a matrix of global gene expression profiles. *Science* **287**:873–880.

Roses AD (2002). Genome-based pharmacogenetics and the pharmaceutical industry. *Nat Rev Drug Discov* **1**:541–549.

Schena M, Shalon D, Davis RW, Brown PO (1995). Quantitative monitoring of gene expression patterns with a complementary DNA microarray. *Science* **270**:467–470.

Schena M, Shalon D, Heller R, Chai A, Brown PO, Davis RW (1996). Parallel human genome analysis: microarray-based expression monitoring of 1000 genes. *Proc Natl Acad Sci USA* **93**:10614–10619.

Shoemaker DD, Schadt EE, Armour CD, He YD, Garrett-Engele P, McDonagh PD, Loerch PM, Leonardson A, Lum PY, Cavet G, Wu LF, Altschuler SJ, Edwards S, King J, Tsang JS, Schimmack G, Schelter JM, Koch J, Ziman M, Marton MJ, Li B, Cundiff P, Ward T, Castle J, Krolewski M, Meyer MR, Mao M, Burchard J, Kidd MJ, Dai H, Phillips JW, Linsley PS, Stoughton R, Scherer S, Boguski MS (2001). Experimental annotation of the human genome using microarray technology. *Nature* **409**: 922–927.

Southern EM, Maskos U, Elder JK (1992). Analyzing and comparing nucleic acid sequences by hybridization to arrays of oligonucleotides: evaluation using experimental models. *Genomics* **13**:1008–1017.

Tamayo P, Slonim D, Mesirov J, Zhu Q, Kitareewan S, Dmitrovsky E, Lander ES, Golub TR (1999). Interpreting patterns of gene expression with self-organizing maps: methods and application to hematopoietic differentiation. *Proc Natl Acad Sci USA* **96**:2907–2912.

Thomas RS, Rank DR, Penn SG, Zastrow GM, Hayes KR, Pande K, Glover E, Silander T, Craven MW, Reddy JK, Jovanovich SB, Bradfield CA (2001). Identification of toxicologically predictive gene sets using cDNA microarrays. *Mol Pharmacol* **60**:1189–1194.

van't Veer LJ, Dai H, van de Vijver MJ, He YD, Hart AA, Mao M, Peterse HL, van der Kooy K, Marton MJ, Witteveen AT, Schreiber GJ, Kerkhoven RM, Roberts C, Linsley PS, Bernards R, Friend SH (2002). Gene expression profiling predicts clinical outcome of breast cancer. *Nature* **415**:530–536.

Waring JF, Ulrich RG (2000). The impact of genomics-based technologies on drug safety evaluation. *Annu Rev Pharmacol Toxicol* **40**:335–352.

Waring JF, Ciurlionis R, Jolly RA, Heindel M, Ulrich RG (2001). Microarray analysis of hepatotoxins in vitro reveals a correlation between gene expression profiles and mechanisms of toxicity. *Toxicol Lett* **120**:359–368.

Wen X, Fuhrman S, Michaels GS, Carr DB, Smith S, Barker JL, Somogyi R (1998). Large-scale temporal gene expression mapping of central nervous system development. *Proc Natl Acad Sci USA* **95**:334–339.

3

Microarray Manufacture

Emile F. Nuwaysir

INTRODUCTION

The field of genomics is the direct result of pioneering work by Sanger and colleagues, who described a method to read the sequence of DNA (Sanger and Coulson, 1975). Technical advances to this method in speed and efficiency have resulted in the sequencing of a vast number of diverse genomes, including the publication of the first draft of the entire human genome (Lander et al., 2001; Venter et al., 2001) in 2001. To date, there are more than 125 completely sequenced prokaryotic and eukaryotic genomes in the public domain (Web site references 1 and 2) and many more sequenced organisms in the private sector. Undoubtedly, by the time this book is published (or, perhaps more accurately, by the time this sentence is finished), this number will be out of date. Suffice to say, the pace of genome sequencing will continue to quicken, and this new information will continue to play an increasing role in our understanding of organismal biology and toxicology.

Commensurate with the modern sequencing revolution of the 1990s, a new experimental method to analyze sequence information was developed: the DNA microarray. This method allowed for the parallel analysis of DNA on a genome scale and has rapidly become a mainstay technique in the modern molecular biology laboratory. DNA microarrays have proven powerful as the result of two fundamental physical characteristics: the impermeable nature of the support and the relative miniaturization of the DNA "spots" on the surface. Impermeable support allows for improved handling, hybridization,

Toxicogenomics: Principles and Applications. Edited by Hamadeh and Afshari
ISBN 0-471-43417-5 Copyright © 2004 John Wiley & Sons, Inc.

55

and washing characteristics, while miniaturization allows the entire genome to be contained on a single, relatively small surface. The combination of substrate impermeability and miniaturization permits DNA microarrays to take advantage of the fundamental nature of DNA hybridization, where every DNA molecule in the genome can find its cognate hybridization partner simultaneously in solution in a massively parallel fashion.

For traditional techniques such as Northern and Southern blotting, the solution phase is comprised of material of known composition (probe), while the material fixed to the substrate is of unknown composition (target). DNA microarrays have reversed this distinction, where the solution phase contains material of unknown composition (e.g., labeled RNA sample), and the surface-bound material is well characterized (e.g., purified PCR products). As a result, confusion has ensued on the accepted definition of the terms *probe* and *target* in the DNA microarray field. In keeping with the published definition of these terms for microarrays (*Nature Genetics*, 1999; Phimister, 1999), the term *target* will hereafter be used to define the labeled entity in solution, and *probe* will be used to define the unlabeled molecules affixed to the substrate surface.

There are many different methods for the manufacture of DNA microarrays, but these methods can be broken down into two general categories: physical deposition of presynthesized DNAs on a solid substrate, or in situ synthesis of oligonucleotides directly on solid support. Historically, the majority of microarray users have fallen into the first category, using the methods pioneered by Brown, Davis, and colleagues (Schena et al., 1995, 1996; DeRisi et al., 1997) to spot cDNA fragments of their genome on glass slides. The primary in situ method, pioneered by Affymetrix (Santa Clara, CA), involves the parallel synthesis of hundreds of thousands of 25 mer oligonucleotides on solid support using photochemistry (Fodor et al., 1991; Pease et al., 1994; Lockhart et al., 1996). This chapter first discusses the deposition-based methods, then in situ synthesis methods.

DEPOSITION METHODS: MOTION-CONTROL ROBOTICS

The common element for all deposition-based methods of microarray manufacture is motion-control robotics, which dictate where and how quickly the DNA will be deposited on the substrate surface. Brown and colleagues designed the first microarray spotter and published these designs on their Web site at Stanford University in the mid 1990s. These designs are still available today and can be used to construct a fully functioning spotter robot (Web site reference 3). Since then, many companies have designed and made available their own version of this robot. A representative list of these commercial robotic arrayer manufacturers can be found in Table 3-1.

Although all these instruments have subtle differences in design, there are fundamental and common features. These common components are (1) a stable and rigid printing platform (platen) with vibration dampening

TABLE 3-1 Commercial robotics suppliers for microarray manufacture

Company Name	Web Site	Products	Location
Affymetrix	www.affymetrix.com	417,427 Arrayers	Santa Clara, CA
Bio-Rad Laboratories	www.bio-rad.com	VersArray Writer	Hercules, CA
BioRobotics	www.biorobotics.com	Microgrid I, Microgrid II	Cambridge, UK
Cartesian	www.cartesian.com	Microsys, PixSys, ProSys	Irvine, CA
GeneMachines	www.genemachines.com	Omnigrid, Omnigrid Accent	San Carlos, CA
Genetix	www.genetix.com	Qarray, Qarray-lite, Qarray-mini	Hampshire, UK
Genomic Solutions	www.genomicsolutions.com	Arrayer	Ann Arbor, MI
Intelligent Bioinstruments	www.intelligentbio.com	IBI HT Microarrayer	Cambridge, UK
MiraiBio	www.miraibio.com	SPBIO	Alameda, CA
Packard Biosciences	www.;packardbiosciences.com	SpotArray 24, 72 and Enterprise	Meriden, CT
TeleChem	www.arrayit.com	SpotBot	Sunnyvale, CA
Virtek Vision	www.virtek.com	ChipWriter Pro, ChipWriter Compact	Waterloo, Canada

properties—that is, flat to within approximately ±100 µm across the entire printing surface; (2) X-Y-Z motion control that can be achieved by movement of an overhead robotic arm or by a combination of a moving platen and a moving overhead robotic arm; (3) an electronic servo controller element that integrates commands from the control computer and actuates the robot's control motors; (4) computer control and interface software used to configure and execute print runs; (5) a stable and rigid microtiter plate mount that can (if desired) be connected to an automated microtiter plate stacker; (6) a stable and rigid substrate mounting platter that can accommodate tens to hundreds of slides and maintain their relative position over the entire course of a multiday print run; (7) integrated wash and dry stations that eliminate sample carryover in the printing mechanism; (8) a printing head to which the printing element(s) can be mounted; and (9) rigid environmental control for temperature, humidity, and particulate matter.

The quality and integration of these elements results in a robot with measurable characteristics that can be used to judge overall instrument quality. Overall, instrument quality and usability are related to six general factors: (1) Speed. Different laboratories have different requirements for manufacturing throughput. Thus, there is no general standard requirement for overall printing speed. (2) Accuracy. The robot should be able to place spots within 10 µm or less of the intended location on the substrate. (3) Precision. Spots must be placed with an overall repetitive error of less than ±10 µm. (4) Resolution. The smallest increment of detectable movement should be within experimental requirements, typically 1–2 µm. (5) Durability. The robot should be sufficiently durable to withstand daily operation for several years. (6) Programmability. The programming interface should be simple enough to allow for quick setup of standard print runs and the relatively easy programming of custom spotting routines.

SAMPLE DELIVERY DEVICES

Contact Printing Devices

Sample delivery devices can be divided into elements that make direct contact with the substrate and those that do not contact the substrate (noncontact). The simplest and most widely used devices are the contact elements. For contact printers, the three most common types of devices are the solid-pin, split-pin, and pin-and-ring design.

Solid pins are the simplest type of arraying tool, the easiest to manufacture, and therefore the least expensive. They are also the earliest type of contact printing technology. They can provide robust and repeatable operation, and can be used to deposit DNA on many different substrate types, including glass and nylon membranes. The blunt tip is less likely to damage the substrate surface but can result in variable spot sizes and uniformity. The biggest single

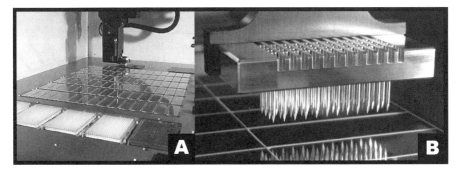

Figure 3-1 *Typical spotting robot and print head. Panel A, inside a typical spotting robot (QArraylite microarrayer, image courtesy of Genetix Ltd, Hampshire, UK). A full platen of 90 slides is shown, along with positions for microtiter plates and a vacuum wash/dry station. Panel B, a fully loaded print head with 48 split-pin microarray tips (Stealth Spotting Pins, image courtesy of Telechem/arrayit.com, Sunnyvale, CA).*

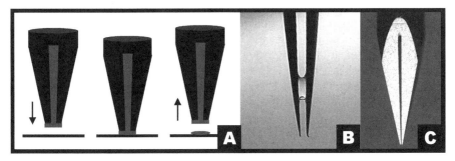

Figure 3-2 *Split-pin mechanism and example. Panel A is a cartoon of the general mechanism of split-pin function. A fully charged pin contacts the surface and leaves a small drop of liquid as a result of surface tension, adhesion, and gravity. Panel B is a photo of a standard set of laboratory tweezers acting as a sample reservoir in a similar manner to a split pin. Panel C is a high-resolution micrograph of a split pin showing the sample reservoir and pin tip that contacts the slide surface (all images courtesy of Telechem/arrayit.com, Sunnyvale, CA).*

drawback of this type of tool is that it can only deposit a single spot per loading and must revisit the source plate for repeated spotting operation. In addition, source-plate evaporation can be a significant problem because printing times are significantly increased as a result of the repeated loading of the printing pin.

The split pin or quill pin is the most widely adopted method for deposition of DNA on microarray substrates. This tool fills by capillary action and uses a combination of surface tension and gravity to deposit spots on the substrate surface. An example of a typical split pin is shown in Figure 3-2. In the late 1990s, when pins were still under rapid development, poor pin quality or clogged pins routinely resulted in missing, misshapen, or improperly aligned spots, as well as spots of variable size that often ran together with their neighbors. Improvements in pin design and manufacturing techniques have resulted in the reduction or elimination of many of these sources of error and have dramatically improved array quality.

An alternate to the split- or solid-pin design is the pin-and-ring design developed and commercialized by Genetic Microsystems (subsequently acquired by Affymetrix, Santa Clara, CA). This innovative and unique design was a combination of split-pin and solid-pin characteristics. A small ring dipped in the spotting solution picks up a volume of the reagent and serves as a reservoir for subsequent spotting. During the spotting process, a solid pin passes through the reservoir ring and picks up a small drop of reagent. This drop of reagent is then deposited on the slide surface by contact. The advantage of this approach is that the pendant drop on the pin tip produces highly uniform and reproducible spots. The disadvantage is that the ring requires a rather large volume of sample to fill the reservoir, resulting in relatively high waste volume after the full platen of slides is printed.

Noncontact Printing Devices

Inkjet or piezoelectric sample delivery devices have proven to be very useful for the delivery of oligos and cDNAs for spotted cDNA microarray manufacture (Theriault et al., 1999; Cooley et al., 2001). The principle behind the inkjet device is that certain types of materials such as ceramic deform when subject to an electrical pulse. Thus, a mechanical force can be generated and tightly controlled by the application of electric potential. The ceramic material in the device is wrapped around a small capillary pipette containing the sample of interest. When a pulse of electricity is applied to the ceramic, it constricts and generates a transient pressure wave inside the capillary, resulting in a small droplet of liquid being expelled from the end of the capillary tube. Because the frequency, strength, and duration of the electrical potential can be tightly controlled, very small sample volumes (<100 pl) can be dispensed at a rate of more than 10,000 droplets per second (Theriault et al., 1999; Cooley et al., 2001). The primary advantage of the inkjet method is the extremely rapid and tunable delivery of sample without the need for direct contact with the surface.

Bubble-jet devices have also been employed for the delivery of DNA samples to the substrate surface (Okamoto et al., 2002). The device works by reducing the viscosity of the sample by transient heating. This reduced viscosity, combined with pressurization, results in the sample being expelled from the sample reservoir.

SUBSTRATE CHEMISTRY AND COMPOSITION

Two types of substrates are the predominant choice for most microarray manufacturing laboratories: glass microscope slides and nylon filters. The majority of in-house microarray manufacturing is performed on glass microscope slides.

   Glass slides have numerous advantages including improved handling due to physical rigidity, low cost, broad availability, low porosity, high transparency, low autofluorescence, and high chemical resistivity, which allows for many different surface chemistry modifications. General problems with microscope slides are that different manufacturers and different products from the same manufacturer can vary widely in their flatness and overall geometry, as well as their surface chemistry characteristics. Minor variations in flatness can result in nonuniform scanning characteristics, and small variations in overall length and width can accumulate over the surface of the printing platen, causing serious alignment problems for the robot (Massimi et al., 2002). In addition, slide cleanliness can vary enormously between manufacturers and between manufacturing lots from the same supplier. Particulates embedded in the slide surface are one of the most common types of background noise seen in microarray experiments.

   Nylon arrays are also a common substrate material for microarray manufacture. Advantages of nylon as a substrate include low-cost, high-throughput manufacturing, simple postprocessing of manufactured arrays, compatibility with radioactive sample labeling protocols, possible reuse, possible high-throughput hybridization, easy customization and analysis similar to glass-based microarray formats (Web site reference 4). Representative commercial substrate providers are shown in Table 3.2.

SLIDE CHEMISTRIES

Initial studies with spotted microarrays employed poly-L-lysine–coated microscope slides (Schena et al., 1995, 1996), but other surface chemistry modifications have become popular alternatives, including aldehyde and aminosilane derivatized surfaces. There is no single best substrate for all microarray applications, since substrate performance is dependent on the type and purity of the DNA to be spotted, slide manufacturing protocols, and particularly sample labeling, washing, and detection protocols.

   Poly-L-lysine is the most popular substrate for the spotting of PCR products because it is inexpensive, it is easy and nontoxic to prepare, the precursor chemicals are widely available, and early protocols were established with its use. The binding of DNA to poly-L-lysine–coated slides is initially through charge interactions that are converted to covalent (or functionally covalent) bonds by UV cross-linking. Drawbacks are that poly-L-lysine slides are easily scratched by handling or by contact with the printing pins, and "aging" of the newly coated slides for up to 4 weeks is required before the substrate is ready for printing. Aminosilane-coated substrates are a popular alternative to poly-L-lysine. Binding of DNA to this substrate is initially through electrostatic interactions and is stabilized by a covalent linkage between thymidine and the substrate alkyl amine groups after UV irradiation (Schena, 2002). Aminosilane coatings tend to be more durable than poly-L-lysine. Aldehyde-

TABLE 3-2   Commercial substrate suppliers

Company Name	Product Name	Surface Chemistry	Price/100 Slides	Web Site
Asper Biotech	Genorama SA	Aminosilane	625	www.asperbiotech.com
Asper Biotech	Genorama SAL	Aminosilane/isothiocyanate	805	www.asperbiotech.com
Corning	GAPSII	Aminosilane	1,050	www.corning.com
Corning	UltraGAPS	Aminosilane	1,300	www.corning.com
Eppendorf	CreativeChip PCR	Proprietary	1,375[a]	www.eppendorf.com
Eppendorf	CreativeChip Oligo	Proprietary	1,500[a]	www.eppendorf.com
Erie	SuperChip	Aminopropylsilane	1,006	www.eriesci.com
Erie	UltraClean	No modification	506	www.eriesci.com
Erie	3Dimension APS	Proprietary	1,106	www.eriesci.com
Erie	Poly-L-Lysine	Poly-L-lysine	379	www.eriesci.com
Erie	Epoxy	Epoxy	1,006	www.eriesci.com
Genetix	Amine	Amine	134	www.genetix.com
Genetix	Aldehyde	Aldehyde	173	www.genetix.com
Genetix	Aldehyde Xcell	Aldehyde	228	www.genetix.com
Full Moon Biosystems	cDNA Slides	Proprietary	990	www.fullmoonbio.com
Full Moon Biosystems	Oligo Slides	Proprietary	1,098	www.fullmoonbio.com
Full Moon Biosystems	Xtra Clean Slide	No modification	498	www.fullmoonbio.com
Schleicher & Schuell	FAST Slides	Nitrocellulose membrane	1,250	www.schleicherandscheull.com
Quantifoil	QMT-Epoxy	Expoxy	625	www.quantifoil.com
Quantifoil	QMT-Aldehyde	Aldehyde	700	www.quantifoil.com
Quantifoil	QMT-Amino	Amino	650	www.quantifoil.com
Telechem	SuperAldehyde	Primary aldehyde	940	www.arrayit.com
Telechem	SuperAmine	Primary amine	840	www.arrayit.com
Telechem	SuperEpoxy	Organoepoxy	900	www.arrayit.com
Telechem	Superclean	No modification	640	www.arrayit.com

[a] Package of 125 slides.

coated substrates are another surface chemistry that is now widely employed. Binding of the DNA to the substrate is generally achieved through a covalent linkage between the surface aldehyde group and 5′ amino groups on the modified oligonucleotides or PCR products.

cDNA CLONE SET LIBRARIES

The primary source for DNA probe material for deposition-based microarray manufacture has been PCR product derived from cDNA clone libraries from commercial suppliers. Long oligonucleotides from 50–75 bases have become a popular alternative to the spotting of PCR products. The pros and cons of long oligo spotting are discussed later in this chapter.

cDNA clone libraries are available from a variety of sources, and detailed information can be found on the specifics of clone libraries and suppliers (Bowtell and Sambrook, 2003). These suppliers provide the cDNA clones in plate format as glycerol stocks of bacterial cultures. The primary suppliers of human, rat, and mouse clone sets licensed by the IMAGE consortium (Lennon et al., 1996; Web site reference 5) for distribution of their clone libraries are Research Genetics/Invitrogen Corporation (Web site reference 6), the German Human Genome Research Project Resource Center (Web site reference 7), and Lion Biosciences (Web site reference 8). Other suppliers of cDNA clone sets include RIKEN (Web site reference 9) and the Institute for Genome Research (Web site reference 10). In addition, Incyte Pharmaceuticals distributes purified PCR products from their own proprietary clone libraries that are suitable for direct printing on microarrays (Web site reference 11), and American Type Culture Collection (ATCC) is a good source for individual clones (Web site reference 12). Critical factors to consider when choosing clone sets for arraying are the products should be selected for optimal length (generally 500–1,500 bases), biased toward the 3′ end of the transcript, and filtered to remove redundancy.

Historically, there have been significant problems with T1 bacteriophage contamination in the cDNA stocks, as well as incorrect assignment of clone location within the plates, contamination of multiple clones within single wells, and clones that fail to grow or amplify by PCR. Estimates as high as 38% error were common when clone sets were first made available (Halgren et al., 2001). However, most clone libraries have undergone extensive resequencing and rearraying, resulting in dramatic improvements in overall accuracy and clone set quality. Current estimates are that within well-maintained clone sets 1–5% of the clones are misassigned to the wrong location or are contaminated with neighboring clones (Knight, 2001; Bowtell and Sambrook, 2003).

There are many logistical issues that must be overcome for the maintenance and safeguarding of expensive clone sets. Foremost is the validation of laboratory sterile techniques for plate replication and growth. Good descriptions of methods to test and quality control (QC) sterile techniques have been described (Holloway et al., 2002d). In addition, storage space must be considered, since storage of a single set of 96-well plates for 25,000 clones will consume more than 25% of a typical −80 freezer, and usually multiple copies of the set are maintained. Other considerations include storage location (on-site and off-site), number of replicate copies, restricted access to limited personnel, and the development of a sufficiently powerful LIMS system and relational database to allow for quick access to clone set annotation.

PCR Amplification of DNA for Spotting

The most labor-intensive, costly, and potentially error-prone step in deposition-based microarray manufacture is the large-scale preparation of cDNA fragments by PCR amplification for spotting on microarrays. The challenge is to generate a high percentage of successful PCR reactions while avoiding cross-contamination. A variety of templates can be used for this PCR reaction.

When bacterial clone libraries are used as source material, investigators have two options for template material: direct amplification from an aliquot of bacterial culture, or amplification of purified plasmid DNA inserts (Holloway, 2002a). Both methods utilize common vector primers and can be performed in plate format with 100 μl (or larger) reactions, which allows for relatively high-throughput and high-yield product. The obvious benefit of using bacterial culture as a template is that it is not necessary to purify the plasmid before the PCR reaction (Holloway, 2002b). This eliminates a very expensive and labor-intensive step, and reduces the risk of cross-contamination or pipetting error. However, direct amplification from bacterial culture media often has a higher failure rate, and can have lower-quality and lower-yield product. For this reason, many researchers choose to purify plasmid DNA before amplification, using commercially available miniprep kits in microtiter plate format (e.g., Web site reference 13).

As an alternative, genomic DNA can be used as a template for the PCR reaction (DeRisi et al., 1997; Wei, 2001). This method has the advantage that no bacterial growth, clone purification, or clone maintenance is required, thus eliminating the majority of the error and potential costs associated with the preparation of DNA for spotting. However, since each gene requires two unique primers for amplification, the cost of this method is generally not significantly less than methods that employ bacterial clone sets as templates.

Quality Control of PCR Products Before Arraying

After amplification, the overall success rate and quality of the PCR products should be determined. This is typically accomplished by high-throughput gel electrophoresis of a small aliquot of each reaction volume (Holloway, 2002c). Quality measures that can be assessed with this method include DNA yield, the presence of free primers or primer dimers, number of amplicons in each reaction, and amplicon size distribution. For this method, it is necessary to employ gel apparatus and combs that are compatible with multichannel pipettors to facilitate loading of the 1,000 or more PCR products. In addition, it is useful to employ image analysis programs that can automatically score each lane for the quality measures indicated above, although this process can be accomplished manually with significant effort. In addition, sophisticated robotics in combination with spectrophotometry or microchannel agarose gel electrophoresis (e.g., Web site reference 14) can be used to assess amplicon concentration and integrity in an automated way.

OLIGOS AS PROBES ON SPOTTED MICROARRAYS

Given the logistical difficulties in handling and amplifying large bacterial clone sets, increasing numbers of investigators have chosen to array commercially manufactured sets of oligonucleotides. This method has the obvious advantage that bacterial clone maintenance is not required, and sequence information is the only requirement for microarray design and manufacture. In addition, oligo probes are more specific and can be more easily used to detect closely related gene family members or splice variants.

There are a number of factors that are critical to consider when selecting oligonucleotide probes for spotting on microarrays. Table 3-3 reviews the major elements that must be considered for the selection of high-quality oligonucleotides. For example, it is critical that the oligo probe be the reverse complement of the labeled material in solution. This is a seemingly obvious criterion, but strandedness can be incorrectly assigned and can have enormous scientific and financial consequences.

TABLE 3-3   Critical considerations for oligonucleotide probe selection

Error Type	Result	Effect on Probe Performance
Incorrect strand	Cognate partner not present in solution	Total failure
Nonunique sequence	High nonspecific binding	Reduced specificity
High self-complementarity	Poor availability for hybridization	Reduced sensitivity
Sequence error in database	Mismatches reduce hybridization	Reduced sensitivity
Sequence polymorphism in sample	Mismatches reduce hybridization	Reduced sensitivity
Splice-site coincidence	Cognate partner not present in solution	Total failure
Improper location on transcript	Sample labeling bias can reduce signal	Reduced sensitivity
Poor synthesis characteristics	Reduced copy number on surface	Reduced sensitivity
Suboptimal melting temperature	Suboptimal hybe characteristics	Reduced specificity and/or sensitivity

Perhaps the most critical element in probe selection is uniqueness testing to ensure that the probe is specific for the gene (or splice variant) of interest. This testing process is usually the most computationally intensive aspect of probe selection. As an example, given the current estimated size of the human transcriptome at 65 Mb (Web site reference 1), approximately 4.2 quadrillion different comparisons are necessary to score all possible 60 mer oligos in the transcriptome for simple uniqueness (i.e., they occur only once in the entire data set)—a task that requires substantial bioinformatics capabilities. However, simple uniqueness is not a sufficient test to eliminate cross-hybridization properties, since a limited number of mutations can be tolerated in a 60 mer without substantial effects on overall probe performance (Hughes et al., 2001).

The most widely used method for long oligo uniqueness testing is a simple BLAST search against the genome or transcriptome of interest using a pre-defined *e*-value as a cutoff for probe selection, or simply selecting the sequence with the lowest *e*-value (Hughes et al., 2001). An example of an alternate method is to divide a potential probe into all its component *n*-mers (e.g., there are 46 possible unique 15 mers in a given 60 mer) and test the entire genome of the organism for the frequency of occurrence of each *n*-mer (Healy et al., 2003). The mer frequency score can then be summed for all the *n*-mers for a given probe, and this aggregate score can be used as a predictor of nonspecific hybridization behavior (Healy et al., 2003). Other methods require simple uniqueness from the genome, in addition to a predefined number of mismatches in specific locations (e.g., 5 mismatches in a 60 mer relatively evenly spread across the probe) as a distinction from any other 60 mer sequence in the genome.

Probes with secondary structure should be avoided by a simple algorithm to calculate self-complementarity. Database sources must be carefully curated, and the exemplar sequences used for probe selection should be of the highest quality, since a single nucleotide sequence error can reduce probe performance. In the case of short oligonucleotide arrays such as 25 mers that are designed to discriminate sequences based on a single mismatch, a single nucleotide error in a stretch of 100 bases can theoretically kill the performance of 24 of the possible 76 potential probes. Likewise, sequence polymorphisms in the target in solution can have the same effect.

A combination of all variables detailed in Table 3-3, as well as other possible issues, can effectively eliminate more than half the potential probes from a given transcript from consideration. The variability in performance that is attributable to these variables can be seen in Figure 3-3, which shows the inherent variability of oligo probe behavior for all possible probes on a given transcript.

In this case, all possible 24 mer probes from the AqpZ gene of *Escherichia coli* have been synthesized in a stepwise fashion on the glass surface using a combination of micromirror technology and photolithography (see "In Situ

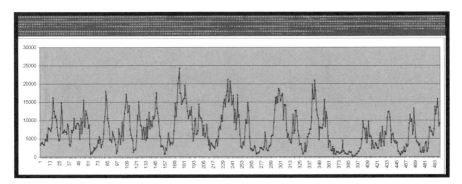

Figure 3-3 *Variability in oligonucleotide probe performance. Upper part: a section of a high-density oligonucleotide microarray. All possible 24 probes contained within the* Escherichia coli *AqpZ gene were synthesized in "tiled" sequential order on the microarray. In addition, a mismatch probe was synthesized just below the perfect match. Bottom part: hybridization intensity data for each probe on the microarray. See text for details about the experimental design.*

Synthesis Methods" section for details). Also, a control probe containing a single mismatch was synthesized adjacent to (directly below) the experimental probe in the figure. This array was hybridized with a labeled mouse liver RNA sample containing AqpZ transcript spiked in at 10 pM. For clarity, only 500 bases at the 3′ terminus are shown. As is evident, all probes do not perform equally. In fact, in this particularly severe example, many probes on the transcript have substantially reduced sensitivity and/or specificity when compared to their neighbors less than 20 nucleotides up- or downstream on the transcript. These data demonstrate the critical nature of oligo selection for microarray performance.

There are many commercial manufacturers of bulk oligonucleotides that would be suitable for the generation of custom oligo arrays. These manufacturers are listed in Table 3-4. To collect the data in Table 3-4, each supplier was asked to submit a synthesis price per base in U.S. dollars for an order of 10,000 different 60 mer oligonucleotides synthesized at 200 nmol scale. Most of these suppliers stated that they could deliver this oligo order within 1 month, while several claimed delivery within days to weeks.

In addition to custom oligo manufacturing, there are several companies that offer presynthesized and prearrayed libraries of long-mer oligonucleotides designed against recent builds of the human, rat, and mouse genome. Table 3-5 lists the suppliers of oligo libraries for these three genomes. There are many other oligo libraries available for many other genomes (e.g., *Drosophila melanogaster* from Qiagen/Operon, *Helicobacter pylori* from MWG Biotech), but the list is long, and the genomes are not as relevant for the purposes of this book.

TABLE 3-4 Commercial oligonucleotide manufacturers

Company Name	Web Site	Price per Base ($) (200 nmol scale)	Standard Purification and QC
Alpha DNA	www.alphadna.com	0.27	Desalted
BioServe	www.bioserve.com	0.75	Desalted
BioSource	www.keydna.com	0.75	Delsated, mass spec spot check
Bio-Synthesis	www.biosyn.com	0.85	Desalted
Commonwealth Biotechnologies	www.cbi-biotech.com	1.50[a]	$10 per oligo setup fee, desalted, PAGE photo
DNAgency	www.dnagency.com	1.00	Desalted, PAGE gel photo
Gene Link	www.genelink.com	2.00	Desalted, PAGE photo, min 99.5% efficiency
Illumina	www.illumina.com	0.28	Desalted
Integrated DNA Technologies	www.idtdna.com	0.50	Desalted, double quantitated[b]
Metabion	www.metabion.com	0.28[a]	Desalted, MALDI-TOF on all oligos
Midland Certified Reagent Company	www.mcrc.com	0.50	Delsalted, MALDI-TOF on all oligos
MWG Biotech	www.mwgbiotech.com	0.30	HPSF purified
Operon/Qiagen	www.operon.com	0.25	Desalted
Sigma-Genosys	www.sigma-genosys.com	0.30[a]	Desalted

[a] Inquire about volume discounts.
[b] 250 nmol scale.

TABLE 3-5 Commercial oligonucleotide libraries for human, mouse, and rat

	Compugen	Operon	MWG Biotech	Clontech	Illumina
Human Oligo Sets					
Number of oligos	18,861	21,353	9,984	12,800	22,740
Number of genes represented	18,664	21,329	9,850	12,651	20,726
Oligo length	60	70	50	68	70
Linker	5'-C6	5'-C6	5'-C6	12 base tag	n.a.
Distance from 3' end	95% within 1,309 bp	1,000 bp	n.a.	95% within 1,500 bp	n.a.
Mouse Oligo Sets					
Number of oligos	21,977	16,487	9,984	5,082	
Number of genes represented	21,825	16,463	9,850	5,002	
Oligo length	60	70	50	68	
Linker	5'-C6	5'-C6	5'-C6	12 base tag	
Distance from 3' end	95% within 1,133 bp	1,000 bp	n.a.	95% within 1,500 bp	
Rat Oligo Sets					
Number of oligos	4,854	4,283	5,760	3,959	
Number of genes represented	4,803	4,273	5,535	3,889	
Oligo length	60	70	50	68	
Linker	5'-C6	5'-C6	5'-C6	12 base tag	
Distance from 3' end	95% within 928 bp	1,000 bp	n.a.	95% within 1,500 bp	

CONTROLS USED ON SPOTTED ARRAYS

Three general categories of controls are used in the manufacture and QC of spotted microarrays: positive controls, negative controls, and spiked-in controls. Positive controls are generally used to ensure that sample labeling and hybridization have been efficient enough to achieve a standard of data quality. Common positive controls include a population of housekeeping genes, as well as other methods such as common reference pools of spotted material (Yang et al., 2002).

Negative controls are generally used to assess the degree of nonspecific hybridization and to estimate background. These controls are typically composed of DNA from distant species that are not expected to cross-hybridize with the experimental species (e.g., *Bacillus subtilis* genes as negative controls on a human microarray), or from entirely synthetic DNAs designed with no significant homology to the experimental species in question (Miki et al., 2001). Also, blank printing buffer is commonly used as a negative control.

Spiked-in controls are used to validate the efficiency of sample labeling between multiple samples, between samples with different fluorophores, and for normalization of data sets across experiments. To accomplish this, sequences of interest are transcribed and spiked in to the sample prior to labeling. These control sequences are usually derived from evolutionarily distant species and checked for cross-homology with the experimental species of interest. There are a number of commercial suppliers of control gene sets suitable for use as spike-ins (Web site references 15, 16, and 17). In addition, in pair-wise hybridizations, the reference channel or third channel can be a primer common to all sequences on the array (Dudley et al., 2002).

QUALITY CONTROL TESTS AFTER MICROARRAY MANUFACTURE

There are four general methods for quality control of microarrays after printing: visual inspection, general stain for resident DNA, hybridization to a common primer, and hybridization to a common reference RNA. Visual inspection with a stereoscope immediately after printing and after the spots have dried can give a general qualitative impression of the overall print quality for a given print run. Missing, poorly shaped, too small, too large, or wandering spots are all evident from a simple visual check. Also, a general stain that binds to DNA can be applied such as TOTO or similar compounds (Hou et al., 2002). This method can give some estimate of spot concentration, as well as the physical characteristics of the spots.

A more rigorous quality check is to perform a functional test by hybridization of a sampling of the arrays to a labeled primer complementary to a sequence within the PCR priming site or vector sequence downstream of the priming site. Alternatively, a truly functional test can be performed where a sampling of the arrays are hybridized to two different well-characterized control RNAs, and the ability of the arrays to discriminate fold gene expression changes can be evaluated. It has been recommended that 5–8% of the arrays for any given print run be used for these QC purposes (Hou et al., 2002).

IN SITU SYNTHESIS METHODS

Since DNA is a relatively simple polymer consisting of only four different building blocks, chemical synthesis of the polymer in a test tube is not diffi- cult. This simplicity allows DNA microarrays to be manufactured by in situ synthesis of the DNA molecules directly on the substrate surface. This basic method, combined with photolithographic techniques borrowed from the computer chip industry, has become a highly successful approach to array manufacture. In addition, a host of new technologies have recently appeared that promise to make high-density microarray manufacture faster, cheaper, and easier. The in situ synthesis of DNA, as opposed to the deposition of pre- viously synthesized materials, offers substantial advantages, including higher density, higher information content, lower cost per data point, highly reliable probe identification and location, and the requirement for sequence informa- tion rather than physical DNA clones for array design.

The basic concept underlying all the technologies outlined in the following section is that nucleotide monomers are added to the surface-bound growing oligos in a site-specific, spatially addressable manner. The tool used to select and activate a given site on the substrate surface can be physical methods (e.g., Agilent Technologies), electrochemical methods (e.g., Nanogen, Combima- trix), or photometric methods (e.g., Affymetrix, NimbleGen). Each of these basic approaches are discussed in more detail.

Overview of DNA Synthesis

All DNA synthesis on solid support is accomplished by the stepwise addition of nucleotide monomers. During synthesis, a reactive 3′ phosphorous group of the solution-phase monomer couples to the 5′ hydroxyl group on the surface- bound DNA chain, resulting in the addition of a single nucleotide to the surface. A general overview of this process can be seen in Figure 3-4.

These free 5′ hydroxyl groups are generated by cleavage of a protecting group by treatment with acid, or by irradiation with UV light, depending on the nature of the protecting group. Irrespective of the exact nature of the pro- tecting group and deprotection method, the acid- and light-mediated methods are mechanistically identical. After coupling of the monomer to the surface, the internucleotide linkage is converted from a phosphite to a more stable phosphate, and any unreacted 5′ hydroxyl groups on the surface that failed to couple with solution-phase monomers are capped by acetylation. This acety- lation reaction permanently inactivates the DNA strand. The remaining

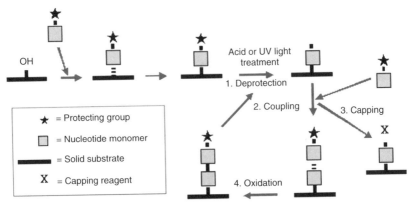

Figure 3-4 *Oligonucleotide chemistry synthesis diagram. Simplified diagram of oligonucleotide synthesis chemistry. Deprotection occurs under acidic conditions with traditional phosphoramidite chemistry or with UV light as employed in photolithographic methods.*

strands that were correctly coupled (and therefore not capped) are free to undergo the next round of synthesis.

Stepwise Efficiency

The single most important factor that determines the overall quality of the in situ synthesized DNA microarray is the stepwise efficiency of the DNA synthesis process. In general, stepwise efficiencies have been reported to be between 94% and 98% for most in situ synthesis methods—that is, for the addition of any given monomer in the oligonucleotide, 2–6% of the total sites fail to correctly complete the synthesis cycle. The percent yield of final correct product can be calculated by raising the stepwise synthesis efficiency to the nth power, where n is the length of the oligonucleotide. For example, a 25 mer synthesized with 94% efficiency will have approximately $(0.94)^{25} \times 100 =$ percent correct sequence, or 21% correct final product. That same 25 mer synthesized with a stepwise yield of 99% would yield 78% correct final product. This effect is more pronounced as the length of the oligonucleotide increases (e.g., a 70 mer synthesized at 94% efficiency yields 1% correct final product, whereas 99% synthesis efficiency would yield 49% correct final product). Thus, it is extremely important that manufacturing methods are optimized to attain the highest possible stepwise efficiency attainable for the given method.

Manufacture by Localized Monomer Delivery

One method to selectively add a monomer to site A on the microarray but not add the monomer to neighboring site B is to selectively deliver the monomer to site A and not to site B. Thus, all that is fundamentally required to manufacture the array is a precise method of substrate delivery, combined with existing DNA synthesis chemistry. One such delivery method has been adapted from inkjet printer technology (Theriault et al., 1999; Cooley et al., 2001; Hughes et al., 2001; Linsley et al., 1999) and has proven to be an effective method for microarray manufacture. The components of this manufacturing system are X-Y-Z motion control robotics, an inkjet fluidics delivery system, a flow cell, traditional phosphoramidite DNA synthesis chemistry, and an activated substrate.

The advantage of the method is that arrays can be manufactured with simple digital input to program the reagent delivery and robotics systems, and the synthesis chemistry and substrates are inexpensive and widely available. Also, inkjet-based methods can deposit solutions in an extremely rapid (greater than 10,000 spots per second) and relatively dense (up to 5,000 spots/cm^2) manner. Drawbacks are that inkjet nozzles are prone to clogging, and the system requires fairly sophisticated fluidics architecture, which does not make it immediately amenable to desktop microarray manufacture in the individual laboratory. Also, since monomer coupling is occurring in a semidry state on the substrate surface, stepwise yields are lower than with traditional phosphoramidite chemistry. Yields between 94% and 96% have been published (Hughes et al., 2001).

Manufacture by Electrochemical Control of Monomer Deprotection

Instead of selective delivery of the requisite monomer (e.g., inkjet-based methods), it is possible to deliver the monomer to all sites on the array but selectively deprotect the fraction of the synthesis sites where coupling is desired. This can be accomplished by the localizaion of high concentrations of acid at the site of interest, causing cleavage of the protecting group (Web site references 18 and 19). These acidic conditions can be generated by microelectrodes in the array substrate, and contained and localized by a zwitterionic solution phase that limits diffusion of the H+ ion. These electrodes also contribute to the hybridization as they create a local environment favorable to each step in the hybridization by offering a positively or negatively charged microenvironment. Arrays produced by this method can have up to 1,000 different oligonucleotides. The inherent strength of this approach is the flexibility of programming of the microelectrode substrate and the use of traditional phosphoramidite DNA synthesis chemistry. The primary limitation of this approach is the very low density and information content per array, and the ability of the solution phase to localize and contain the acid.

Manufacture by Photometric Control of Acid Deprotection (Xeotron)

In an interesting combination of traditional acid-labile phosphoramidite chemistry and photochemistry, investigators have utilized the Texas Instruments (TI) Digital Micromirror Device (DMD) as a means to generate acid, which can then deprotect the nucleotide (LeProust et al., 2000, 2001; Web site reference 24). This permits the use of the traditional acid-labile phosphoramidite chemistry, which is relatively inexpensive and widely available. A description of the DMD and its use for photochemistry can be found in the section on "Manufacture by Digital Light-Mediated Control of Monomer Deprotection."

Manufacture by Light-Mediated Control of Monomer Deprotection

The most common and widely adopted method for in situ synthesis of oligonucleotide microarrays is the method pioneered by Affymetrix (Fodor et al., 1991; Pease et al., 1994; Lockhart et al., 1996). In 2002, Affymetrix sold more than 400,000 individual GeneChip™ arrays, demonstrating the wide acceptance and utility of the platform (Web site reference 20). In the last several years, the price of Affymetrix arrays has dropped to within reach of most of the research community, while the information content per chip has soared and the number of species represented has increased. Currently, the standard array from Affymetrix contains almost 500,000 unique $18 \times 18\,\mu$m oligonucleotide features within a $1.28 \times 1.28\,$cm area. The company has indicated that $10\,\mu$m features are possible, which would allow more than 1.6 million data points to be collected in a single experiment.

The fundamental principle of the manufacturing method is the selective deprotection of surface-bound DNA bases by illumination with UV light through holes in a photolithographic mask. The overall mechanistic approach is the same as with standard phosphoramidite DNA synthesis chemistry, except that the acid-labile protecting group in traditional chemistry has been replaced with a photo-labile protecting group, allowing for light-directed oligonucleotide synthesis. Figure 3-5 demonstrates the basic principle of light patterning with photolithographic masks.

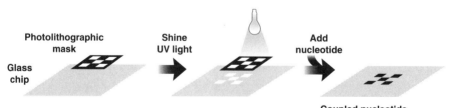

Figure 3-5 *Use of photolithographic masks to pattern light. A photoprotected glass substrate is aligned with a mask containing a chrome coating. UV light is shined on the mask, and those areas without the chrome coating allow light to be transmitted. The transmitted light strikes the glass substrate, thus cleaving or destroying the photoprotecting groups on the surface and generating an active area. The substrate is then dipped in a solution containing the DNA monomer (or the next monomer is washed through if the substrate is within a flow cell), and the monomer selectively couples to the active sites on the substrate.*

The glass substrate is first derivatized to generate active hydroxyl groups on the surface. Linker molecules containing terminal photo-labile protecting groups are then coupled to all sites on the surface. When specific regions of the substrate are exposed to UV irradiation controlled by photolithographic masks, the protecting groups are selectively removed. This generates an active site on the substrate surface that is available for subsequent coupling to the appropriate nucleotide monomer. The monomers couple to the surface through a 3′ phosphoramidite group and are protected at their 5′ position by a photo-labile protecting group. Currently, Affymetrix employs 5′-(O-methyl-6-nitropiperonyloxycarbonyl) MeNPOC as a photo-labile protecting group (Fodor et al., 1991; Pease et al., 1994). This chemistry yields an average stepwise yield as high as 95% (McGall et al., 1997).

The primary advantage to this method is that arrays with increasingly higher density and information content can be manufactured without associated increases in cost or manufacturing time. Also, the photolithographic manufacturing process is parallel in nature (49 arrays are synthesized simultaneously in a single manufacturing run) and can have very high throughput, allowing for the manufacturing of millions of arrays per year if desired. The primary drawbacks are that the MeNPOC synthesis chemistry has relatively low yield, and photolithographic masks are relatively expensive, both of which combine to make long oligo microarrays (e.g., 70mers) difficult and expensive to produce on this platform. Also, photolithographic masks are expensive to design and manufacture, so the company must charge a sizable up-front fee to recoup the cost of a custom array design. This fact, combined with the inflexibility of the mask-based process (changing a single probe requires the manufacture of an entire new set of masks) makes the tool impractical for most custom microarray applications.

Manufacture by Digital Light-Mediated Control of Monomer Deprotection

A new technology has been developed that employs photolithography to manufacture microarrays but obviates the need for photolithographic masks to pattern light (Singh-Gasson et al., 1999; Nuwaysir et al., 2002; Albert et al., 2003; Web site references 21 and 22). This method relies on the Digital Micromirror Device (DMD) developed by Texas Instruments (Web site reference 23) to pattern light. The approach promises to combine the best characteristics of photolithographic process (high density, high information content, parallel chemical synthesis) with the flexibility inherent with the use of the DMD (high flexibility, low up-front manufacturing costs) to create a new generation of high-density microarrays with increased utility. The DMD was developed for use in high-definition television and for digital movie houses, and is now in widespread use in digital projectors as well. Figure 3-6 contains micrographs provided by TI that demonstrate the DMD architecture.

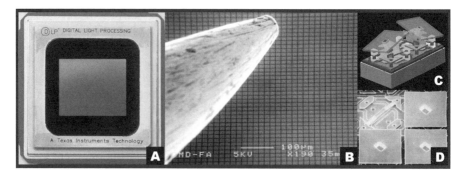

Figure 3-6 *DMD structure. Panel A is a Texas Instruments DMD. Panel B is a photomicrograph of a section of the micromirror array with the point of a straight pin shown for comparison. Each mirror is 16μm square on a 17μm pitch. Panel C is a cartoon of two micromirrors, with one mirror in the off position and one mirror in the on position. Panel D is a high-resolution photomicrograph of three micromirrors, with one micromirror removed to show the underlying micromechanical architecture.*

The DMD is an array of 786,432 aluminum mirrors contained within a 17.4 × 13.1 mm area. In this configuration, each mirror is 16μm square on a 17μm pitch. Higher-density micromirror arrays are under development that offer 1.3 million individual mirrors or more. Each mirror is mounted on a torsion hinge and can be deflected 10 degrees in the positive or negative direction from the neutral state in a voltage-dependent manner. Mirrors tilted in the +10 orientation deflect light into the light path, while mirrors deflected in the –10 orientation deflect light to an absorber. Using this method, high-resolution light patterns can be created. These patterns of light can be used to manufacture high-density oligonucleotide arrays.

The fundamental components of the manufacturing system are the Texas Instruments DMD, an optical subassembly for illumination of the DMD, a fluidics delivery system, a flow cell, an activated substrate, and DNA monomers with photo-labile protecting groups and associated ancillary synthesis chemistry. Two companies to date have developed manufacturing capabilities based on this fundamental approach: NimbleGen Systems (Madison, WI) and Febit (Mannheim, Germany).

This approach has all the fundamental advantages of conventional photolithographic manufacturing techniques that employ masks, including high density, high information content, and manufacturing scalability. However, using this approach, photolithographic masks are not required. This eliminates the fundamental cost and time barriers associated with the manufacture of custom high-density microarrays. Also, since the synthesis cycle for a given array is relatively short (3 hours for a NimbleGen array, 12 hours for a Febit array) and the design process is digital, an iterative approach to array design can be employed where experimental results can be used to improve the design of subsequent experiments. For example, it is possible with this approach to select an empirically optimized set of oligonucleotide probes for a given gene set, or perhaps the entire transcriptome, from a limited number of optimization experiments. In addition, this method employs higher-efficiency photo-labile DNA synthesis chemistry (Hasan et al., 1997) and therefore is capable of manufacturing longmer (e.g., 70 mer) arrays. Also, as TI develops higher-resolution DMDs with more micromirrors (e.g., 1.3 million instead of 786,432), this will result in a commensurate increase in the density and information content of microarrays that employ these DMDs in their manufacture.

SUMMARY

In the past several years, there has been an explosion of innovation in the field of microarray manufacture. These new techniques have made microarrays more sensitive, reproducible, accurate, easier to manufacture, versatile, cheaper, and flexible, and have made microarrays an indispensable tool for life science research. As the technology continues to improve, microarrays may provide the tangible link between the promise of the genomic revolution and our understanding of biology as a system.

REFERENCES

Albert TJ, Norton J, Ott M, Richmond T, Nuwaysir AK, Nuwaysir EF, Stengele K, Green RD (2003). Light-directed 5′–3′ synthesis of complex oligonucleotide microarrays for large-scale genotyping. *Nucleic Acids Res.*

Bowtell D, Sambrook J (2003). *DNA Microarrays, A Molecular Cloning Manual*, 1st ed. Cold Spring Harbor, New York: Cold Spring Harbor Press.

Cooley P, et al. (2001). Ink-jet deposited microarrays of DNA and other bioactive molecules. In: J Rampal (ed.), *Methods in Molecular Biology. Vol 170: DNA Arrays Methods and Protocols*. Totawa, NJ: Humana Press.

DeRisi J, Iyer VR, Brown PO (1997). Exploring the metabolic and genetic control of gene expression on a genomic scale. *Science* **278**:680–686.

Dudley AM, Aach J, Steffen MA, Church GM (2002). Measuring absolute expression with micrarrays with a calibrated reference sample and an extended signal intensity range. *Proc. Natl Acad Sci USA* **99**:7554–7559.

Fodor SP, Read JL, Pirrung MC, Stryer L, Lu AT, Solas D (1991). Light-directed, spatially addressable parallel chemical synthesis. *Science* **251**:767–773.

Halgren RG, Fielden MR, Fong CJ, Zacharewski TR (2001). Assessment of clone identity and sequence fidelity for 1189 IMAGE cDNA clones. *Nucleic Acids Res* **29**:582–588.

Hasan A, Stengele K-P, Giegrich H, Cornwell P, Isham KR, Sachleben RA, Pfleiderer W, Foote RS (1997). Photolabile protecting groups for nucleosides: synthesis and photodeprotection rates. *Tetrahedron* **53**:4247–4264.

Healy J, Thomas E, Wigler M (2003). Rapid annotation of complex genomes with counts of exact word matches.

Holloway A, Murphy M, Chandreasekharappa S (2002a). PCR amplification of probes from bacterial clone sets. In: D Bowtell, J Sambrook (eds), *DNA Microarrays, A Molecular Cloning Manual.* Cold Spring Harbor, NY: Cold Spring Harbor Press, pp 12–19.

Holloway A, Murphy M, Chandreasekharappa S (2002b). Purification of PCR products. In: D Bowtell, J Sambrook (eds), *DNA Microarrays, A Molecular Cloning Manual.* Cold Spring Harbor, NY: Cold Spring Harbor Press, pp 20–22.

Holloway A, Murphy M, Chandreasekharappa S (2002c). Visualization and assessment of PCR products. In: D Bowtell, J Sambrook (eds), *DNA Microarrays, A Molecular Cloning Manual.* Cold Spring Harbor, NY: Cold Spring Harbor Press, pp 23–27.

Holloway A, Murphy M, Chandrasekharappa S (2002d). Replication and storage of cDNA clone sets. In: D Bowtell, J Sambrook (eds), *DNA Microarrays, A Molecular Cloning Manual.* Cold Spring Harbor, NY: Cold Spring Harbor Press, pp 71–78.

Hou B, Somerville S, Holloway A, Murphy M (2002). Checking the quality of the printed slide. In: D Bowtell, J Sambrook (eds), *DNA Microarrays, A Molecular Cloning Manual.* Cold Spring Harbor, NY: Cold Spring Harbor Press, pp 71–78.

Hughes TR, Mao M, Jones AR, Burchard J, Marton MJ, Shannon KW, Lefkowitz SM, Ziman M, Schelter JM, Meyer MR, Kobayashi S, Davis C, Dai H, He YD, Stephaniants SB, Cavet G, Walker WL, West A, Coffey E, Shoemaker DD, Stoughton R, Blanchard AP, Friend SH, Linsley PS (2001). Expression profiling using microarrays fabricated by an ink-jet oligonucleotide synthesizer. *Nature Biotechnol* **19**:342–347.

Iyer V (2002). Isolation and amplification of array ready material from yeast. In: D Bowtell, J Sambrook (eds), *DNA Microarrays, A Molecular Cloning Manual.* Cold Spring Harbor, NY: Cold Spring Harbor Press, pp 30–34.

Knight J (2001). When the chips are down. *Nature* **410**:860–861.

Lander ES, Linton LM, Birren B, Nusbaum C, Zody MC, Baldwin J, Devon K, Dewar K, Doyle M, FitzHugh W, et al. (2001). Initial sequencing and analysis of the human genome. *Nature* **409**:860–921.

Lennon G, Auffray C, Polymeropoulos M, Soares MB (1996). The I.M.A.G.E. Consortium: An integrated molecular analysis of genomes and their expression. *Genomics* **33**:151–152.

LeProust E, Pellois JP, Yu P, Zhang H, Gao X (2000). Digital light-directed synthesis. A microarray platform that permits rapid reaction optimization on a combinatorial basis. *J Comb Chem* **2**:349–354.

LeProust E, Zhang H, Yu P, Zhou X, Gao X (2001). Characterization of oligodeoxyribonucleotide synthesis on glass plates. *Nucleic Acids Res* **29**:2171–2180.

Linsley P, Blanchard A, Burchard J, Coffey E, Dai H, et al. (1999). Use of FlexJet inkjet based technology to construct 50K human gene microarrays. *Microb Comparat Genomics* **4**:90–96.

Lockhart DJ, Solas D, Sullivan EJ, Cronin MT, Holmes CP, Fodor SA (1996). Expression monitoring by hybridization to high-density oligonucleotide arrays. *Nature Biotechnol* **14**:1676–1680.

Massimi A, Harris T, Childs G, Somerville S (2002). Printing on glass slides. In: D Bowtell, J Sambrook (eds), *DNA Microarrays, A Molecular Cloning Manual.* Cold Spring Harbor, NY: Cold Spring Harbor Press, pp 71–78.

McGall GH, Barone D, Diggelmann M, Fodor SPA (1997). The efficiency of light-directed synthesis of DNA arrays on glass substrates. *J Am Chem Soc* **119**: 6270–6276.

Miki R, Katoda K, Bono H, Tomaru Y, Carninci P, Itoh M, Shibata K, Kawai J, Konno H, Watanabe S, Sato K, Tokusumi Y, Kikucki N, Ishii Y, Hamaguchi Y, Nishizuka I, Goto H, Nitanda H, Satomni S, Yoshiki A, Kusakabe M, DeRisi JL, Eisen MB, Iyer VR, Brown PO, Muramatsu M, Shimada H, Okazaki Y, Hayashizaki Y (2001). Delineating developmental and metabolic pathways in vivo by expression profiling using the RIKEN set of 18,816 full-length enriched mouse cDNA arrays. *Proc Natl Acad Sci (USA)* **98**:2199–2204.

Nature Genetics (1999). The chipping forecast (supplement). 21(1). New York: Nature America Inc.

Nuwaysir EF, Huang W, Albert TJ, Pitas A, Nuwaysir AK, Singh J, Richmond T, Gorski T, Berg JP, Ballin J, McCormick M, Norton J, Pollock T, Sumwalt T, Butcher L, Porter D, Molla M, Hall C, Blattner F, Sussman MR, Wallace RL, Cerrina F, Green RD (2002). Gene expression analysis using oligonucleotide arrays produced by maskless photolithography. *Genome Res* **12**(11):1749–1755.

Okamoto T, Suzuki T, Yamamoto N (2000). Microarray fabrication with covalent attachment of DNA using bubble jet technology. *Nature Biotechnol* **18**:342–347.

Pease AC, Solas D, Sullivan EJ, Cronin MT, Holmes CP, Fodor SA (1994). Light-generated oligonucleotide arrays for rapid DNA sequence analysis. *Proc Natl Acad Sci USA* **91**:5022–5026.

Phimister B (1999). Going global. *Nature Genet* **21**:1.

Sanger E, Coulson AR (1975). A rapid method for determining sequences in DNA by primed synthesis with DNA polymerase. *J Mol Biol* **94**:441–448.

Schena M (2002). *Microarray Analysis*, 1st ed. Hoboken, NJ: John Wiley & Sons.

Schena M, Shalon D, Davis RW, Brown PO (1995). Quantitative monitoring of gene expression patterns with a complementary DNA microarray. *Science* **270**:467–470.

Schena M, Shalon D, Heller R, Chai A, Brown PO, Davis RW (1996). Parallel human genome analysis: microarray-based expression monitoring of 1000 genes. *Proc Natl Acad Sci (USA)* **93**:10614–10619.

Singh-Gasson S, Green RD, Yue Y, Nelson C, Blattner F, Sussman M, Cerrina F (1999). Maskless fabrication of light-directed oligonucleotide microarrays using a digital micromirror array. *Nature Biotechnol* **17**:974–978.

Theriault TP, Winder SC, Gamble RC (1999). Application of ink-jet printing technology to the manufacture of molecular arrays. In: M Schena (eds), *DNA Microarrays, A Practical Approach*. Oxford, UK: Oxford University Press.

Venter JC, Adams MD, Myers EW, Li PW, Mural RJ, Sutton GG, Smith HO, Yandell M, Evans CA, Holt RA, et al. (2001). The sequence of the human genome. *Science* **291**:1304–1351.

Wei Y (2001). High-density microarray-mediated gene expression profiling of *Escherichia coli. J Bacteriol* **183**:545–556.

Yang YH, Dudoit S, Luu D, Lin DM, Peng V, Ngai J, Speed TP (2002). Normalization for cDNA microarray data: a robust composite method addressing single and multiple slide systematic variation. *Nucleic Acids Res* **30**:e15.

Yue H, Eastman PS, Wang BB, Minor J, Doctolero MH, Nuttal RL, Stack R, Becker JW, Montgomery JR, Vainer M, Johnston R (2001). An evaluation of the performance of cDNA microarrays for detecting changes in global mRNA expression. *Nucleic Acids Res* **29**:E41.

WEB SITE REFERENCES

1. http://www.ncbi.nlm.nih.gov
2. http://ergo.integratedgenomics.com/GOLD
3. http://cmgm.stanford.edu/pbrown/mguide/index.html
4. http://www.grc.nia.nih.gov/branches/rrb/dna/tenreasons.pdf
5. http://image.llnl.gov/image/html/idistributors.shtml
6. http://www.resgen.com
7. http://www.rzpd.de
8. http://www.lionbiosciences.com
9. http://genome.rtc.riken.go.jp/home.html
10. http://www.tigr.org
11. http://www.incyte.com
12. http://www.atcc.org
13. http://www.qiagen.com
14. http://www.caliper.com
15. http://www.amersham.com
16. http://www.ambion.com
17. http://www.stratagene.com
18. http://www.nanogen.com
19. http://www.combimatrix.com
20. http://www.affymetrix.com
21. http://www.nimblegen.com
22. http://www.febit.de
23. http://www.dlp.com
24. http://www.xeotron.com

4

Scanners

Charles J. Tucker

INTRODUCTION

Toxicogenomic studies that are performed with microarray analysis involve monitoring the expression of thousands of genes. Typically, in-life studies are performed in which model animals are administered various doses of a chemical compound over several time points. By choosing a wide range of doses and time points, the gene expression changes in one data set can be subtle at one instance and quite noticeable at another (Hamadeh et al., 2002). Therefore, it is imperative that the sensitivity and dynamic range of the equipment collecting these microarray data be capable of capturing a wide variety of gene expression changes.

Scientists, biologists, statisticians, and bioinformaticians spend an enormous amount of time planning, executing, and mining the data from a single toxicogenomic experiment. One of the least scrutinized steps of these experiments is the actual acquisition of the signal from the microarray chip itself. The scanning step is an important one because it is here that results of the biological experiment are captured and the data are created. The most basic or raw form of the data from these types of toxicogenomic studies is the actual images created by the microarray scanners. Unfortunately, the least amount of time and resources are spent on this step of an experiment.

This chapter examines the origins of the signal from a microarray chip, then describes the various types of equipment available for acquisition. Due to the dominance of fluorescence microarrays in the genomics arena, this chapter

Toxicogenomics: Principles and Applications. Edited by Hamadeh and Afshari
ISBN 0-471-43417-5   Copyright © 2004 John Wiley & Sons, Inc.

focuses on the many technical considerations associated with this particular platform. An understanding of fluorescence image acquisition will allow the reader to fine-tune protocols and hence enhance the detection of subtle gene changes that are necessary for toxicogenomic studies.

PREMISE OF OPERATION OF SCANNERS

Toxicogenomic studies using microarrays are designed to measure the abundance of mRNAs. Most scientists simply refer to these studies as gene expression experiments, but it should be noted that mRNA transcript abundance is actually being measured. Several protocols and methods exist for isolating RNA, creating cDNA, and labeling this cDNA with a probe (Hegde et al., 2000). The one thing these protocols have in common is that the starting material is RNA and the finished product is labeled cDNA (or cRNA in some instances), which has been transcribed from this starting material. A generic schematic of labeling for a microarray experiment is presented in Figure 4-1 so that the reader will be familiar with the origins of the signal discussed in this chapter.

The complementary properties of DNA are fully exploited in a microarray experiment. A DNA microarray chip contains several thousand spots, and each of these spots contains a unique sequence of genetic material. These sequences can be either oligonucleotides or polymerase chain reaction (PCR) products and are collectively referred to as the target. Regardless of the type of material on the DNA chip, the sequences represented in the targets are complementary to those of the labeled probe. The hybridization step of a microarray experiment combines all labeled probes onto a DNA chip. This chip is then placed in either a water bath or a hybridization station where time and temperature allow the labeled probe to find and bind to its complementary target. After a washing step, the DNA chip is inserted into a microarray scanner (or reader) where the intensity of the labeled probe is measured (Schena et al., 1995).

TYPES OF SCANNERS

There are two major types of microarray platforms, which involve either labeling the probe with a radioactive or fluorescence nucleotide. Radiolabeled microarrays make up a small portion of the toxicogenomic market, so this platform is only briefly introduced in this chapter. During the reverse transcription step, the reverse transcriptase enzyme incorporates a nucleotide that contains either ^{32}P or ^{33}P. The actual substrate of the DNA chip itself used for radiolabeled microarrays can be either nylon, plastic, or most recently, glass. After hybridization, the intensity of the radiolabel is typically measured with a phosphorimager; however, X-ray film can also be used. Intensity

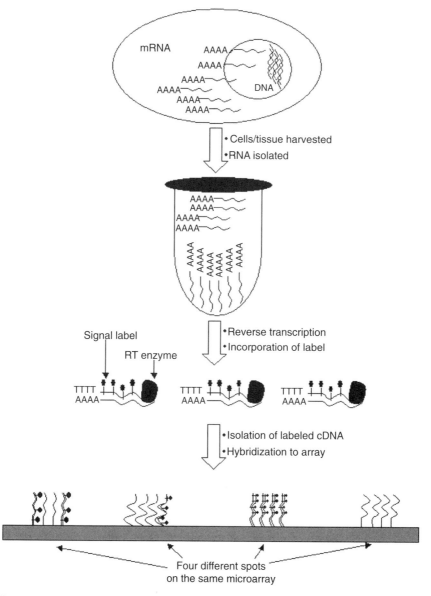

Figure 4-1 *It is important to remember the source of signal for a microarray experiment. First, cells are harvested, then the RNA is isolated. After isolation, a reverse transcription reaction is performed in which a "label" is inserted into the developing cDNA. Finally, a hybridization is performed in which the labeled cDNA binds to its complementary target on the microarray slide.*

measurements are then calculated with an image analysis software package where a background subtraction is performed, then gene annotation information is incorporated. Since radiolabeled microarrays give only a singe channel of data, if there is to be a comparison of treated versus control, then an additional microarray experiment will need to be performed with the sample of reference. The facts that most radiolabeled microarrays are low density (Bertucci et al., 1999). Only a single channel of data is obtained per array, and the obvious health risk with using ^{32}P or ^{33}P have driven most toxicogenomic research toward the fluorescence microarray platform.

When it comes to detecting the signal from fluorescence microarrays, there are two major systems: the white-light/charged-couple device (CCD) and the laser/photomultiplier tube (PMT). The white-light/CCD system utilizes wide-area, broad-wavelength sources such as a mercury or xenon arc lamp for excitation. Because the area of excitation is relatively large, a filtered CCD camera is used for detection of the emission signal. White-light/CCD systems offer several advantages over laser-based systems, such as more opportunities for excitation and less photobleaching. Despite these advantages, white-light/CCD systems have not caught on due in part to lack of demand for three-color microarrays and the complexity associated with these systems.

Laser/PMT fluorescence microarray systems are by far the most used technology in genomics. With these systems, one or two lasers are used for excitation of the microarray slide, and the emission signal is collected with a PMT. The lasers are focused on the chip, then a raster (back and forth) scan is performed down the chip. As the laser excites various regions of the array, the emission is collected with the PMT.

Within the realm of laser microarray scanners there are two varieties: confocal and nonconfocal. Confocal laser scanners are widely used because of their ability to block background signals. A pinhole is used at a conjugate focal plane to eliminate out-of-focus light. Figure 4-2 displays the optical path and shows how over- and underfocus light is removed. Confocal scanners offer the advantage of eliminating the autofluorescence signal associated with glass in the red channel. However, because lots of photons are lost with the use of the pinhole, excitation power and detector sensitivity become crucial components in this system. Nonconfocal scanners allow for the detection of out-of-focus light. However, if care is taken in choosing emission filters and nonautofluorescence glass is used, nonconfocal scanners can yield high-quality images.

THE PHYSICS OF FLUORESCENCE LASER SCANNERS

Two-color fluorescence microarray experiments are commonly used in most institutional and university genomic centers. While most genomic researchers are well trained in the areas of microbiology, toxicology, and bioinformatics, they are not as familiar with the details of fluorescence microscopy. Since the raw data of a toxicogenomic experiment are images created from a fluores-

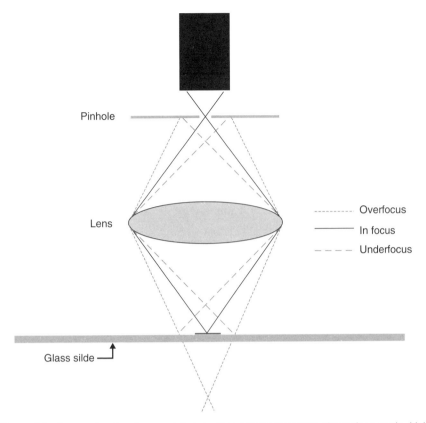

Figure 4-2 *In a confocal system, a pinhole is placed in the light path above the sample. Light that is in focus passes through the hole while out-of-focus light is blocked. Since out-of-focus light is stopped, resulting images are comprised only of a particular plane within the sample. Microarray scanners use confocality to collect signals only from the surface of a microarray while blocking signals from out-of-focus regions like the glass of the slide.*

cence scanner, it is necessary to understand the principle of fluorescence and how it applies to microarray experiments. Therefore, it is necessary to learn the specifics of fluorescence from the ground up, then apply these principles to create better microarray images.

Without diving deeply into quantum theory, let us first assume that photons are both a wave and a particle. Knowing this, we can use Einstein's theory and say that the energy of a photon is proportional to the frequency of the electromagnetic wave (EQ 1). Also, the frequency of an electromagnetic wave is the inverse of its wavelength (EQ 2). By combining these two equations, we can easily see that the energy of a photon is inversely related to the wavelength of that photon (EQ 3). For our purposes as toxicogenomic scientists, this means that a photon of blue light (~450 nm) has more energy that a photon of red light (~650 nm).

In fluorescence, an external light source is used to deliver energy in the form of photons to a molecule. If the wavelength (and hence energy) of that photon is precise enough to drive the electrons into the excited singlet state for that molecule, then the photon will be absorbed. This external energy from the photon converts the molecule from its usual ground state to an excited state. Once in the excited state, the molecule is unstable, and a photon is emitted as the compound returns to the ground state. The typical time for this transition to occur is about 10 ns, but it depends on the specific molecule excited. Because some energy is lost to heat before this transition occurs, the emitted photon has less energy and hence a longer wavelength. The difference between the excitation energy absorbed and the emitted energy is referred to as the Stokes shift. In practice, this means that the emission signal will always occur at a higher wavelength than the excitation. Figure 4-3 shows a Jablonski diagram that graphically describes electronic state transitions for fluorescence.

Not all molecules are fluorescent. Typically, only polyaromatic hydrocarbons or heterocycles are capable of absorption and fluorescence emission. Fluorescence labels are favored in microarray experiments because multiple labels that have unique absorption and emission properties can be used. Since

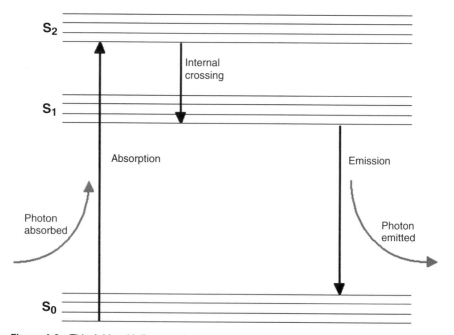

Figure 4-3 *This Jablonski diagram shows a photon being absorbed, which takes the electron orbitals of a fluorescent molecule from the ground state (S_0) to the excited state (S_2). The molecule loses some energy in the form of heat and relaxes to a lower-energy excited state (S_1). Within nanoseconds the unstable molecule emits a photon of lesser energy than what was absorbed and drops back to the ground state.*

most fluorescence microarray experiments use the cyanine dyes for labeling, we will use them as an example to explain absorption and emission properties.

Figure 4-4 shows the physical structure of Cy3 and Cy5. As you can tell, the only difference between these two compounds is the addition of two carbons in the double-bond chain linking the two indolenine moieties. This slight structural difference totally changes the absorption and emission properties (Mujumdar et al., 1993).

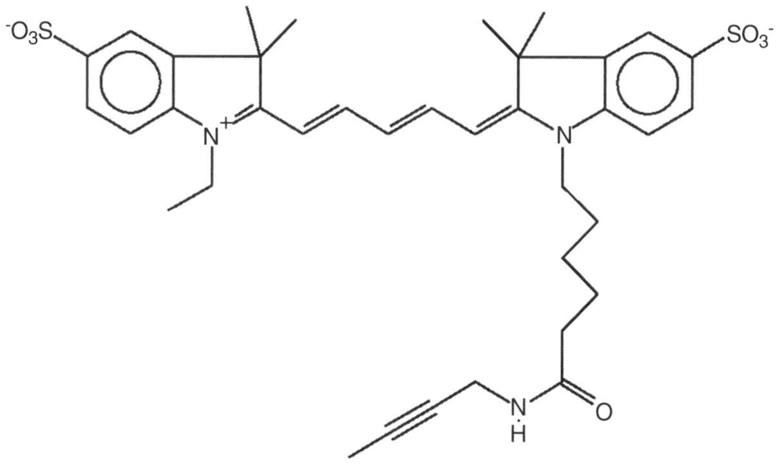

Cy5

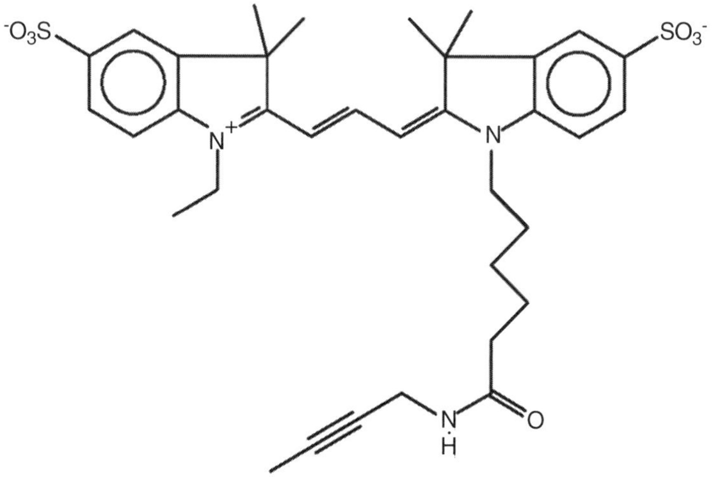

Cy3

Figure 4-4 *The physical structure of Cy3 and Cy5 is very similar. The only difference is that Cy3 has a 3-carbon chain linking the two indolenine moieties and Cy5 has a 5-carbon chain.*

For Cy3 and Cy5 the maximum absorption is 550 nm and 649 nm, respectively. Additionally, the fluorescence emission maximum for these two compounds occurs at 570 nm for Cy3 and 670 nm for Cy5. Those familiar with Cy3 and Cy5 realize that most microarray scanners do not excite exactly at the maximum absorption wavelengths. Cy3 and Cy5 both a have broad absorption spectra, meaning that absorption at wavelengths other than the peak is possible, but it will not be as efficient. The easiest way to visualize this concept is with Figure 4-5, where the absorption spectra for Cy3 and Cy5 are plotted together. According to this figure, it is possible to excite Cy3 at 500 nm, but you would be doing so at only 20% the efficiency as if you excited at 550 nm.

The precise wavelength used for excitation with a laser/PMT system is dictated by physical properties of the components in the laser. Lasers offer the benefit of delivering a directional, coherent, and monochromatic beam of light. However, the limited availability of different types of lasing media means that lasers are available only in specific wavelengths (O'Shea et al., 1977).

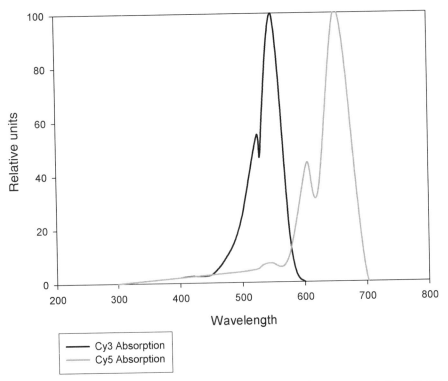

Figure 4-5 *The shapes of the absorption spectra for Cy3 and Cy5 are very similar, with the major difference being a 100 nm shift in wavelength. It is important to note the overlap region of these spectra because this zone is where it is possible to excite both fluorophores with the same excitation wavelength.*

Lasing Media	Available Lines
Argon	457 477 488 514
Helium-neon	543 594 612 633
Krypton	476 528 568 647
Neodymium:YAG	266 355 473 532 946 1064 1330
Diode	Various between 375 and 1550

Figure 4-6 *Four common lasers are listed along with the spectral lines that each is capable of producing. There are several other types of lasing media, but cost, power, and stability limit their common use.*

Figure 4-6 shows a chart with the different types of lasing media along with the wavelengths that each of them is capable of producing. Optical engineers have few options when selecting excitation laser sources in microarray scanners because of the small assortment of small, stable, and cost-efficient lasers. When it comes to exciting Cy3 and Cy5, diode lasers are the standard in microarray scanners with excitation wavelengths of 532 nm and 635 nm, respectively. While it is true that more efficient excitation of the cyanine dyes could be obtained, it would be cost-prohibitive.

The emission spectrum is made up of the collection of photons emitted from a fluorescence molecule. These photons are of a higher wavelength than the excitation signal. A large Stokes shift is desirable for a fluorophore, but in the case of two-color microarray systems it is necessary to make sure that the emission and excitation of the two fluorophores do not overlap. Such an overlap is referred to as cross-talk. In the case of Cy3 and Cy5, you must be aware of two circumstances that would yield an erroneous signal. In case 1, you want to make sure the excitation laser for Cy3 (typically 532 nm) does not overlap and excite with the excitation spectra for Cy5. In case 2, you want to make sure that the emission for Cy3 (typically 570 nm) does not overlap and excite the Cy5 fluorophore. Both of these cases would result in excitation of Cy5. As can be seen in Figure 4-7, when we overlay the emission and excitation spectra of Cy3 and Cy5, the possibility of cross-talk does exist. Specialized excitation and emission filters plus sequential scanning are the two choices to reduce cross-talk.

TECHNICAL CONSIDERATIONS FOR FLUORESCENCE ARRAYS

In toxicogenomic experiments we want to measure the abundance of specific RNA transcripts. Two-color fluorescence microarray systems give us the capa-

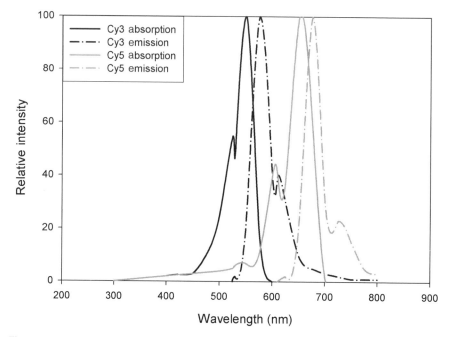

Figure 4-7   *The overlay of the absorption and emission spectra are shown for Cy3 and Cy5. Note how the emission for Cy3 overlaps the absorption for Cy5, creating the possibility for fluorescence resonance energy transfer (FRET). Additionally, there is a slight overlap of the emission spectra for these two fluorophores, creating another source of cross-talk.*

bility to measure treated populations against control populations with a single array. However, there are a few technical aspects to be considered that could potentially bias these two-color experiments. Laser power, quantum efficiency, extinction coefficients, photobleaching, and PMT wavelength sensitivity have different parameters in the context of two-channel fluorescence measurements.

Since two-color fluorescence systems use two different fluorophores for labeling, there must be two sources for excitation. Differences in the structural makeup of the fluorophores affect their optical properties in array experiments. Extinction coefficients are numbers characterizing the fluorophore's ability to absorb excitation photons of a particular wavelength. In short, some fluorophores absorb more photons than others. Additionally, the quantum efficiency is a ratio that relates the amount of emission photons to the number of absorbed photons. Therefore, when determining the correct laser power for excitation, the extinction coefficient and quantum efficiency must be balanced. Cy3 has an extinction coefficient of $150,000\,M^{-1}\,cm^{-1}$ and a quantum efficiency greater than 0.15, while the extinction coefficient for Cy5 is $250,000\,M^{-1}\,cm^{-1}$ and the quantum efficiency is greater than 0.28. These numbers mean that the

Cy5 fluorophore absorbs more photons and has a greater efficiency of converting these absorbed photons into an emission signal.

Other considerations on the emission side of two-color experiments are the photobleaching rate and the wavelength sensitivity of the PMT detector. Photobleaching is the irreversible chemical change of a dye molecule to a nonfluorescence species. Samples photobleach when the intensity of the fluorescence emission decreases as a function of laser exposure. Tests at Amersham Biosciences have shown that Cy5 emission decreases 50% in 24 hours when exposed only to ordinary laboratory strip lighting (Amersham, 2000). Figure 4-8 shows the loss of signal due to repeated scanning of Cy3 and Cy5 with an ordinary microarray slide. It should be noted that Cy5 photobleaches at twice the rate of Cy3. Increasing the sensitivity of the detector for the Cy5 channel typically compensates for differences in the photobleaching rate between these two molecules. You may think that this difference in bleaching rate could be corrected by increasing the excitation energy, but this would be counterproductive, as increasing excitation would only lead to further photobleaching.

PMTs are the standard detector used in laser-based microarray scanners. With a PMT, the emitted fluorescence signal passes through an emission filter,

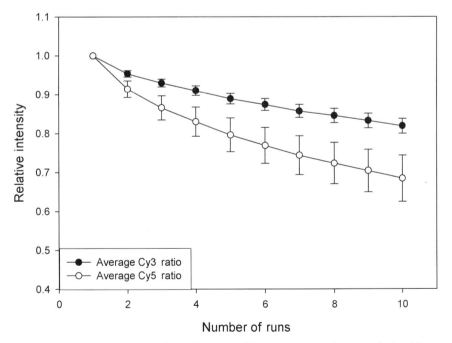

Figure 4-8 *In this experiment, four microarray chips were scanned repeatedly for 10 consecutive times. The resulting images were saved, and the mean spot intensity of each image was calculated and normalized to the first scan of each chip. Cy5 loses almost twice as much signal as Cy3.*

then strikes the photocathode of the PMT. The photocathode converts photons to electrons, which are then accelerated and amplified by multiple dynodes in the PMT. This cascade of electrons produces a current that is proportional to the number of photons that originally struck the photocathode. The material of the photocathode introduces a bias in two-color microarray experiments because it is more efficient in the blue/green region of light than the red region (Hamamatsu, Corporation). Figure 4-9 shows the typical response of a PMT versus wavelength. To compensate for this bias in wavelength with a PMT, the scanner operator can either adjust the power of the excitation laser or adjust the sensitivity of the PMT.

Signal-to-noise ratios (SNRs) are the single most important technical consideration when trying to achieve optimal sensitivity in a toxicogenomic experiment. Simply calculated, the SNR can be measured in a microarray experiment by calculating the mean intensity of the emission signal and dividing it by the standard deviation of the background. The PMTs in most micro-

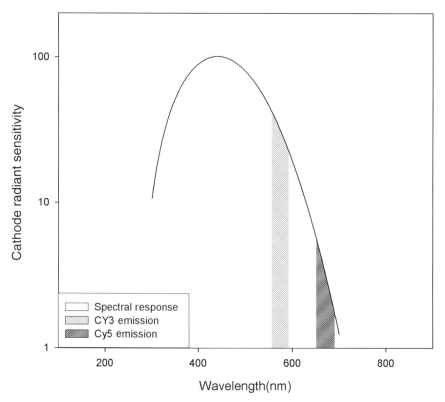

Figure 4-9 *The typical spectral response for a PMT is shown. PMTs are most efficient at detection of light in the blue and green visible light region. However, red light, like the emission of Cy5, is not detected with the same sensitivity as visible light of a lower wavelength.*

array scanners operate in the plateau region of the PMT response curve, meaning that as a person increases the voltage on a PMT, the signal and the noise increase at the same rate. So while it is possible to boost the measured signal by increasing PMT voltage, the noise equally increases.

To understand this concept, imagine you are driving in a rural section of the country and the radio station begins to fade out. The music of the radio station is considered the signal and the crackle of the fading station is the noise. If you increase the volume of the radio, you are equally boosting the signal and the noise, resulting in both louder music and static. Figure 4-10 shows the SNR as calculated for an Axon 4000B scanner for the Cy3 and Cy5 channels. Since the SNR is relatively flat over the usable PMT voltage, adjusting this PMT voltage is not beneficial to maximizing the SNR. Therefore, if you want to increase the SNR in a microarray experiment, your efforts are better spent increasing the fluorescent signal from the cDNA label as opposed to increasing the signal with PMT adjustment.

There is one trick that can be done with some scanners that will increase the SNR. Line averaging is a feature that allows longer scan times in exchange

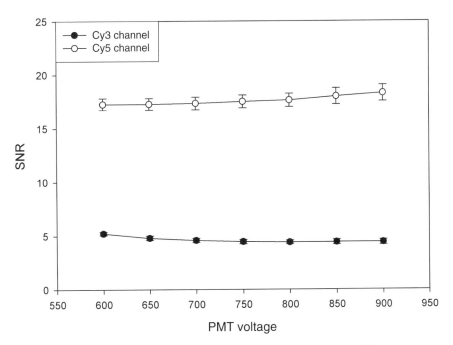

Figure 4-10 *The GP401 calibration slide provided with each Axon 4000B scanner was scanned on four different occasions. The PMTs were adjusted on each occasion from 600 to 900 volts. The mean intensity from this slide was divided by the standard deviation of the pixels to obtain a basic SNR measurement. The SNR is relatively flat in this region of the PMT. Therefore, a signal measured at 600 volts has just as much noise as the same sample at 900 volts.*

for averaging of the pixel data. The result of an image that has increased line averaging is that the intensity of the signal is smoothed and the noise is hence decreased. The downside of line averaging is that the increased scan time results in a more photobleached signal. If line averaging is available, this feature should be used for the final data scan, but any preliminary scans should be performed with minimal laser exposure to reduce photobleaching.

The final technical consideration for toxicogenomic experiments deals with the resolution and digitization of the output image. The standard output file for a microarray scanner is the 16-bit tagged image file format (tiff). With these images the value of a pixel ranges from 0 to 65,535. Since our ultimate goal is to compare the image intensities to the two channels of a scanner, we are specifically interested in calculating the ratio between Cy3 and Cy5. We should therefore plan for the mean intensity of all the spots on a microarray slide to equal 2,500. This target value allows for a 25-fold change between the two channels without achieving saturation. Saturation should be avoided because any value of 0 or 65,535 virtually contains zero information. By targeting the intensity of 2,500, we are operating in the optimal range for the two images.

In summary, Cy5 is a brighter fluorophore because of the larger extinction coefficient and the bigger quantum efficiency. However, Cy5 photobleaches more easily, and the PMTs do not measure its emission signal as effectively as Cy3. In order to compensate for this signal bias between the two channels for these fluorophores, it is necessary for the laser exciting the Cy3 signal to deliver more power than the Cy5 laser. For example, the Axon 4000B takes care of this problem by using a 17 mW 532 nm laser and a 10 mW 635 nm laser. We have also learned the limitations of PMT adjustment and the desired intensity range for signal measurement.

CONCLUSION

There are several things to be considered when trying to balance the various parameters of a two-color microarray experiment. The labeling procedure for the two fluorophores should be optimized as much as possible. We have seen that increasing PMT voltage does not increase the quality of the emission signal, so all efforts should be spent on increasing the fluorescence signal on the excitation side of a two-color experiment with either increased labeling or improved excitation efficiency.

Even though the structural differences in Cy3 and Cy5 are quite small, the differences in extinction coefficients, quantum efficiency, and photobleaching rates are significant. These differences, combined with the limitations and bias of detecting the two emission signals, need to be balanced, since our ultimate goal is to measure ratios between the two channels. Instead of adjusting PMT voltage and repeatedly scanning images until the two channels are matched, scanner operators simply need to optimize two parameters. First, they should try to increase the SNR through line averaging. Second, they should adjust the

PMTs of the scanner so that saturation (with either 0 or 65,535 pixel values) is minimized. If these two criteria are met, then an image analysis program is the best place for balancing the two channels, since PMT adjustments do not affect SNR.

The technological advances in microarray detection have drastically increased over the past few years. Most scanning systems were derived from the well-established field of confocal microscopy. As a result, microarray scanners are very advanced, containing several features that are not presently utilized. For example, there are scanners capable of detecting five different channels, and some systems are capable of 2 μm resolution. Future improvement for the detection of microarrays will probably come from two different fronts. On the biology side, labeling efficiency and hybridization kinetics will increase, enabling more signal to be obtained from smaller RNA input with less noise. On the detection side, PMTs will be replaced with CCD detectors and new technology such as avalanche photodiodes. These advances will enable the detection of subtle gene changes and further increase the sensitivity of toxicogenomics.

REFERENCES

Amersham (2000). *FluoroLink Cy3-dUTP*. Amersham Biosciences.

Bertucci F, Bernard K, Loriod B, Chang YC, Granjeaud S, Birnbaum D, et al. (1999). Sensitivity issues in DNA array–based expression measurements and performance of nylon microarrays for small samples. *Hum Mol Genet* **8**(9):1715–1722.

Hamadeh HK, Bushel PR, Jayadev S, Martin K, DiSorbo O, Sieber S, et al. (2002). Gene expression analysis reveals chemical-specific profiles. *Toxicol Sci* **67**(2):219–231.

Hamamatsu Corporation. *Photomultiplier Tubes*, 2nd ed. Hamamatsu Corporation.

Hegde P, Qi R, Abernathy K, Gay C, Dharap S, Gaspard R, et al. (2000). A concise guide to cDNA microarray analysis. *Biotechniques* **29**(3):548–550, 552–554, 556 passim.

Mujumdar RB, Ernst LA, Mujumdar SR, Lewis CJ, Waggoner AS (1993). Cyanine dye labeling reagents: sulfoindocyanine succinimidyl esters. *Bioconjug Chem* **4**(2):105–111.

O'Shea DC, Callen WR, Rhodes WT (1977). *Introduction to Lasers and Their Applications*. Reading, MA: Addison-Wesley.

Schena M, Shalon D, Davis RW, Brown PO (1995). Quantitative monitoring of gene expression patterns with a complementary DNA microarray. *Science* **270**(5235): 467–470.

5

Databases for Toxicogenomics

Pierre R. Bushel

WHAT IS A DATABASE?

In order to understand and fully appreciate the utility of databases for toxicogenomics, it is necessary to first define a database and how it has evolved into the obligatory repository mechanism for toxicogenomic information and data. To put it simply, a database is a structured collection of similar types of data.

The starting point of the utilization of database management systems is difficult to nail down, but there is reason to believe that its beginning was with the NASA APOLLO space project in the 1960s. In 1964, IBM developed a Generalized Update Access Method (GUAM) to manage the large amounts of data that NASA needed to coordinate. In 1966, IBM released the first commercially available database management system (DBMS)—the information management system (IMS)—based on the hierarchical data model. Following IBM's success, Charles Bachman headed a team at General Electric to develop an integrated data system (IDS) based on the network data model that subsequently led to Honeywell's production of IDS-2 and the formation of the Conference on Data Systems Languages (CODASYL) group, whose purpose was to tackle the issues of providing standards for DBMS. Finally, in 1970, Dr. E. F. Codd, an IBM researcher, proposed the relational data model in a theo-

Toxicogenomics: Principles and Applications. Edited by Hamadeh and Afshari
ISBN 0-471-43417-5   Copyright © 2004 John Wiley & Sons, Inc.

retical paper that was to have a major impact on the database community. Pratt and Adamski (1994) provide a more detailed overview of the history of database systems. For the purposes of this chapter and the application to toxicogenomics, the flat-file, relational, object-oriented, and extensible markup language (XML) database types are delineated.

A flat-file database is essentially the most basic type of data management entity available. It is basically an ordered collection of records in sections or in horizontal rows of the file broken up by spaces, commas, tabs, or other delimiter types into vertical columns. Flat-file databases can be indexed similar to the way a file cabinet is organized, and they are searchable. However, flat-file databases can be inefficient if large. Figure 5-1 illustrates a GenBank flat-file format used by the National Center for Biotechnology Information

```
LOCUS        AQ251347      669 bp    DNA              GSS       07-OCT-1998
DEFINITION   F25D1-T7 IGF Arabidopsis thaliana genomic clone F25D1, genomic
             survey sequence.
ACCESSION    AQ251347
NID          g3704413
VERSION      AQ251347.1  GI:3704413
KEYWORDS     GSS.
SOURCE       thale cress.
  ORGANISM   Arabidopsis thaliana
             Eukaryota; Viridiplantae; Charophyta/Embryophyta group;
             Embryophyta; Tracheophyta; euphyllophytes; Spermatophyta;
             Magnoliophyta; eudicotyledons; Rosidae; Capparales; Brassicaceae;
             Arabidopsis.
REFERENCE    1  (bases 1 to 669)
  AUTHORS    Feng,J., Dewar,K., Buehler,E., Kim,C., Li,Y., Shinn,P., Sun,H. and
             Ecker,J.
    TITLE    BAC End Sequences at ATGC
  JOURNAL    Unpublished (1997)

FEATURES              Location/Qualifiers
     source           1..669
                      /organism="Arabidopsis thaliana"
                      /strain="Columbia"
                      /note="Vector: BeloBACII; Site_1: EcoRI; Site_2: EcoRI;
                      Produced by Thomas Altmann"
                      /db_xref="taxon:3702"
                      /clone="F25D1"
                      /clone_lib="IGF"
                      /sex="hermaphrodite"
BASE COUNT      148 a     172 c     161 g     180 t        8 others
ORIGIN
        1 tgancggccg taccttttatg gtccatgtcc gattcttacc cnactttttcc cannnttacg
       61 cacaaccaac cccnctactt cttacccact attgattgat tggaaaagag ctctctgctt
      121 tatcttatct ttggtcttct atctctgctg gagcgtcaca ttctttctttt ctttccgtcg
      181 actcagttta gtggaaaatc cgtggggatc attggtctag gtagaattgg gactgccatc
      241 gcaaagaggg ctgaagcctt tagctgccca atcaattact actcaagaac cattaagcct
      301 gatgtcgcct acaagtatta tccgacggtg gttgaccttg ctcaaaactc agacatcctc
      361 gtcgtcgcat gcccgttgac cgagcagacc agacacattg tggaccggca ggtcatggat
      421 gcattaggag ctaagggcgt cctcataaac attggccgtg gaccacatgt tgatgagcaa
      481 gagcttatta aagctctaac agaaggccgc ctangtgggg ctgcccttga tgtgtttgag
      541 caggagcctc acgtgcccga ggagctcttt ggccttgaga atgtagttct cctccctcac
      601 gttgggagtg gcactgtgga aacacggaat gccatggccg atcttgncgg gggtaacttg
      661 taagcgccg
//
```

Figure 5-1   *National Center for Biotechnology Information GenBank flat-file format.*

(NCBI) to manage an annotated collection of publicly available nucleotide and amino acid sequences.

Relational databases are a collection of tables that store records, or tuples in computer science terminology, of varying data formats that are associated with one another by way of a primary or foreign key to uniquely identify and relate records. Table 5-1 shows the different data types and how they are used. Database tables are typically structured with records fields and data formats to capture the entities, relationships, and properties that are characteristic of the user business concern or experimental design. Figure 5-2 illustrates a portion of the database schema of a MicroArray Project System (MAPS) database (Bushel et al., 2001) structured with database tables according to a typical microarray experimental process. A one-to-one relationship allows a single row in one table to be associated with a single row in another table. Figure 5-3 depicts the foreign key values in one table as always being the primary key values in another table, permitting the records to be related to each other in a one-to-many type of relationship. The many-to-many relationship is used to allow one or more rows in one table to be related to one or more rows in another table. However, this kind of association is often too complex and can be more efficiently managed with multiple one-to-many relationships.

Object-oriented databases have become popular for managing data and information. In an object-oriented system, an object is some unit of data that has actions that can take place on the object. An object database management system (ODBMS) has a collection of objects representing general features of a business domain or experimental design and attributes that describe the properties of the objects. Figure 5-4 shows the object-oriented design of the array class from the European Bioinformatics Institute (EBI) ArrayExpress database (Brazma et al., 2003). The actions of the objects are methods with definitions of processes to be executed. One of the major advantages of ODBMS is that a subclass of objects can inherit or take on the structure and properties of the class of which it is a member.

XML databases are the new class of database management systems that take advantage of XML's ability to describe the logical structure of data it encapsulates and its straightforward use over the Internet. XML-enabled databases contain extensions for transferring data between XML documents and their own data structures. They are used by data-centric applications, except

TABLE 5-1   Data types and uses

Data Type	Purpose
Numeric	Quantities, numbers
Character	Strings, descriptors
Date	Date, time
Boolean	Yes, no, true, false, on, off switches
Binary	Pictures, sound, video, graphics, documents, large objects

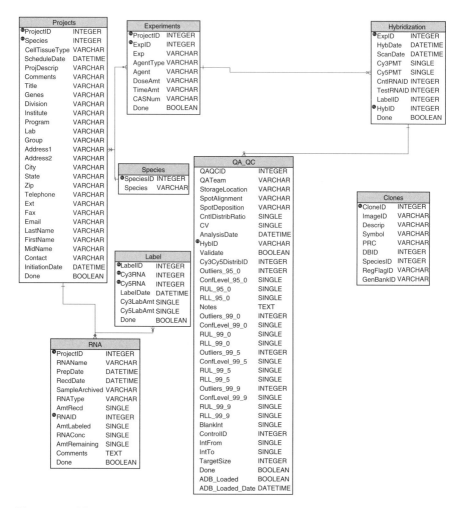

Figure 5-2   *MicroArray Project System laboratory information management system simplified database table schema.*

when the database also supports native XML storage. The difference between XML-enabled and native storage is that XML-enabled storage uses schema-specific structures that must be mapped to the XML document at design time. Native XML storage uses generic structures that can contain any XML document.

DATABASE DESIGN AND DEVELOPMENT FOR TOXICOGENOMICS

The advent and utilization of microarrays for global transcript profiling has recently fed into the realm of toxicology. Gene expression profiling promises

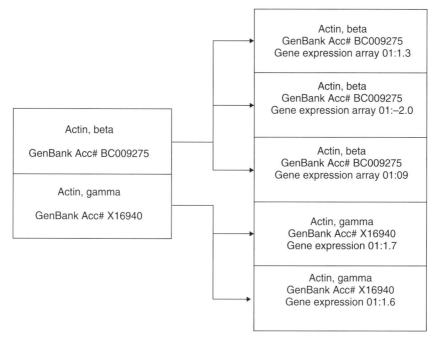

Figure 5-3 *Data key one-to-many relationship.*

to aid in chemical characterization, predictive toxicology, and risk assessment. The unification of high-scaled biology with toxicology bore the new field of toxicogenomics, where transcriptomics, proteomics, metabonomics, and functional genomics meet toxicological sciences. Although a compendium of genomics data and toxicology information regarding mechanistic effects on biological systems following environmental and toxicant perturbations will be critical for embarking upon toxicogenomics in silico, building such a knowledge base is challenging and is currently elusive.

To steer the microarray community in a common direction toward standardization of gene expression data, the Microarray Gene Expression Data (MGED) International Society has formulated Minimal Information About a Microarray Experiment (MIAME) guidelines (Brazma et al., 2001) to assist with the development of microarray repositories and data analysis tools. The guidelines encourage capturing the experimental design, sample information, hybridization, measured data, normalization method, and platform design details in accordance with the microarray process (Fig. 5-5). The ArrayExpress database developed by the European Bioinformatics Institute (EBI), the Gene Expression Omnibus (GEO) developed by the National Center for Biological Information (NCBI), and CIBEX developed by the DNA Data Bank of Japan are three public gene expression databases whereby ArrayExpress and GEO are slated as data repositories for investigators publishing results of gene expression studies in peer review journals. At this time, the ArrayExpress database is fully MIAME compliant.

Class Array {Analysis} derived from: Identifiable

Documentation
The physical substrate along with its features and their annotation

Parent Package	Array	Abstract	No
Export Control	PublicAccess	Link Class for	None
Class Kind	NormalClass	Cardinality	n
Space		Concurrency	Sequential
Persistence	No		

Assigned Components	Array

Attributes

Name	Type	Initial Value
arrayIdentifier	String	
arrayXOrigin	float	
arrayYOrigin	float	
originRelativeTo	String	

Figure 5-4 *Array class from the European Bioinformatics Institute ArrayExpress database.*

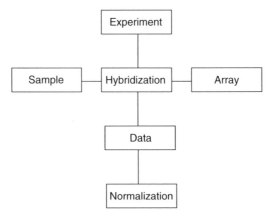

Figure 5-5 *Components of the microarray experimental process.*

To streamline the exchange of microarray information and data, MGED developed the MicroArray Gene Expression object model (MAGE-OM) and extensible markup language (MAGE-ML) data format (Spellman et al., 2002). MAGE-OM is modeled using the Unified Modeling Language (UML) and has been accepted as a standard by the Object Management Group (OMG). MAGE-ML (Fig. 5-6) is the data exchange format implemented using XML for import or export to or from databases or software adhering to MAGE-OM. To assist with the conversion between MAGE-OM and MAGE-ML, MGED released a collection of packages called MAGE Software Toolkit (MAGEstk), a loader for importing MAGE-ML validated data sets into the ArrayExpress database or a MAGE-OM data structure, and MIAMExpress, a Perl Web application and a MySQL database to annotate microarray experiments for MIAME compliant submission. Other laboratory information management systems (LIMS) and analysis information management system (AIMS) have been developed to track the microarray gene expression experimental process (Anderle et al., 2003).

The MIAME/Tox document is based on the MIAME 1.1 document produced by the MGED Society. Although MIAME concentrates on information content and should not be confused with a data format, it also tries to provide a conceptual structure for microarray experiment descriptions. Similarly,

```
<?xml version="1.0"encoding="UTF–8"?>
<!DOCTYPE BioMaterial_package (View Source for full doctype...)>
– <BioMaterial_package>
  – <BioMaterial_assnlist>
    – <BioSource identifier="BioSource:1"name="Human breast tumour">
      – <MaterialType_assn>
          <OntologyEntry category="organism part" value="breast" />
        </MaterialType_assn>
      – <Characteristics_assnlist>
          <OntologyEntry category="species" value="Homo sapiens"/>
          <OntologyEntry category="tissue" value="breast"/>
          <OntologyEntry category="disease state" value="turmor"/>
        </Characteristics_assnlist>
      + <SourceContact_assnref>
      </BioSource>
    + <BioSample identifier="BioSample:1" name="Human breast tumour">
    + <LabeledExtract identifier="T3518C">
    + <BioSource identifier="BioSource:2">
    + <BioSource identifier="BioSource:3">
    + <BioSample identifier="BioSample:2" name="PooledCellLineRef">
    +<LabeledExtract identifier="RefRNA">
    </BioMaterial_assnlist>
  </BioMaterial_package>
```

Figure 5-6 *Sample of microarray gene expression markup language (MAGE-ML).*

MIAME/Tox seeks to provide a conceptual structure in the context of toxicogenomics. Although details for particular toxicogenomic experiments as published in the literature may be different, MIAME/Tox, like MIAME, aims to define the core that is common to most toxicogenomic experiments. Here, for MIAME/Tox, the original MIAME 1.1 document has been adapted and extended to fulfill the need of toxicogenomic data capture and exchange. Like MIAME, MIAME/Tox is not a formal specification but a set of guidelines.

MIAME/Tox supports a number of different objectives—for example, linking data within a study, linking several studies from one institution, and exchanging toxicogenomic data sets among public databases. In fact, the major objective of MIAME/Tox is to guide the development of toxicogenomic databases and data management software. Efforts to build international or community-wide public toxicogenomic databases are underway at the National Institute of Environmental Health Sciences (NIEHS) National Center for Toxicogenomics (NCT), the National Center for Toxicological Research (NCTR), the Mount Desert Island Biological Laboratory (MDIBL), and EBI in conjunction with the International Life Sciences Institute's Health and Environmental Sciences Institute (ILSI HESI). The NIEHS database system will be a component of the Chemical Effects on Biological Systems (CEBS) knowledge base (Waters et al., 2003) to enable the investigation of systems toxicology and facilitate the understanding of hypothesis-driven and discovery research that contributes effectively to drug safety and the improvement of risk assessments for chemicals in the environment. MDIBL's Comparative Toxicogenomics Database (CTD) is under development as a community-supported genomic resource for understanding gene, proteins, and toxic agents of human toxicologic significance with a focus on annotating data from aquatic and mammalian organisms (Mattingly et al., 2003). The MIAME/Tox document addresses these objectives by outlining the minimum information required to unambiguously interpret and potentially reproduce and verify array-based toxicogenomic experiments.

Following the MIAME rationale, sufficient and structured information should be recorded for toxicogenomic experiments to correctly interpret and replicate the experiments or retrieve and analyze the data. Minimum information to be recorded about toxicogenomic experiments is defined in subsequent sections and should include the following data domains:

- Experimental design parameters, animal husbandry information or cell line and culture information, exposure parameters, dosing regimen, dose groups, and in-life observations
- Microarray data specifying the number and details of replicate array bioassays associated with particular samples and including PCR transcript analysis if available
- Numerical biological end-point data including necropsy weights or cell counts and doubling times, clinical chemistry and enzyme assays, hematology, urinalysis, and so on

· Textual end-point information such as gross observations, pathology, and microscopy findings

The institute and animal number, study number, and dates are sufficient identifiers to link data domains from in-life studies. In vitro experiments will establish a similar unique identifier for cultures within a study. Linking microarray data to biological end-point data will be through the individual sample assay number following MIAME conventions for microarray standard sample extracts. It should be noted that MIAME only covers array-based gene expression experiments. A Proteomics Standards Initiative (PSI) that aims to define community standards for data representation in proteomics to facilitate data comparison, exchange, and verification is currently in the developmental stage.

The Proteomics Standards Initiative was founded at the Human Proteome Organization (HUPO) meeting in Washington, D.C., April 28–29, 2002 (Kaiser, 2002). As a first step, the PSI will develop standards for two key areas of proteomics: mass spectrometry and protein-protein interaction data (Merrick, 2003). Development of a standardized general proteomics format will also be considered. At the present time, no metabonomics standards initiative is known. Therefore, this MIAME/Tox document presently outlines the requirements for array-based toxicogenomic experiments with the expectation that proteomics and metabonomics standards will be introduced very soon.

The MGED Society is developing an ontology to provide standard terms for the annotation of microarray experiments. The MGED Ontology is intended for the use of investigators in annotating their microarray experiments and for software and database developers. These terms will enable structured queries of elements of the experiments and allow unambiguous descriptions of how the experiments were performed. A core MGED Ontology was established in September 2003 and will remain unchanged to facilitate software applications development. However, as new applications of microarray technology arise, requiring new descriptive terms, a second layer of ontology will be built. The MGED Ontology also participates in the global open biological ontologies effort (GOBO) that collects freely available ontologies for use within the genomics and proteomics domains. Therefore, the MGED Ontology only includes reference to publicly available resources.

Toxicogenomic applications require specific, common terminologies that need to be identified and subsequently integrated within the MGED Ontology. MIAME/Tox content areas for gene expression experiment descriptions include sections that will be provided in a free text format, along with recommended information to be provided by maximum use of controlled vocabularies or external ontologies (such as species taxonomy, cell types, anatomy terms, histopathology, toxicology, and chemical compound nomenclature). The use of controlled vocabularies is needed to enable database queries and automated data analysis.

Since few controlled vocabularies have been fully developed, MIAME/Tox encourages the users, if necessary, to provide their own qualifiers and values identifying the source of the terminology. This is achieved through the use of (qualifier, value, source) triplets—for instance, qualifier: *cell type*, value: *epithelial*, source: *Gray's Anatomy, 38th ed.* This is recommended instead of, or in addition to, free text format descriptions wherever possible and will allow the community to build a knowledge base of the most useful controlled vocabularies for describing microarray experiments.

TOXICOGENOMIC KNOWLEDGE-BASED SYSTEMS

A knowledge-based system consists of data, information, logic, and rules that govern associations underlying complex data sets. An intelligent knowledge-based system uses inference to apply knowledge toward a particular problem-solving task (Forsyth, 1984)—that is, when provided with sufficient domain facts, an expert knowledge-based system can be designed and programmed with heuristics to bring forth further insight about a particular area of concern. The key to building useful knowledge-based systems is to supply the database with adequate information and data to represent a problem broadly without being too general or overdescriptive, then extracting the symbolic knowledge embedded in the trained system as meaningful and logical information. Figure 5-7 illustrates a typical expert system containment. Because human experts

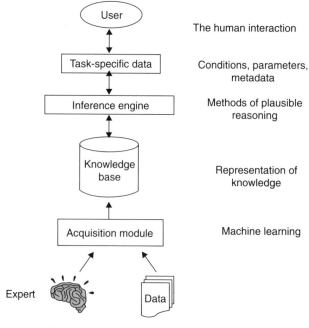

Figure 5-7 *Typical expert system containment.*

usually tend to think along the lines of a condition leading to an action or a situation leading to a conclusion, *if-then* rules are for the most part the predominant form of encoding knowledge in expert systems.

Metadata are regarded as descriptive tools of the underlying data that can be used in order to reach the information level to be provided. In this sense, metadata can be a priori defined according to the already known domains of the scientific discipline and experiment, or may be derived from the underlying data, especially in cases where a non–a priori–definable domain of discourse addressing the scientific discipline must also be considered. Therefore, metadata are used for the analysis and interpretation of scientific data aiming at the extraction of information (see Fig. 5-8a). Information in terms of already known knowledge concerning the domain of discourse of the scientific application can be conceptualized and used in order to steer the learning process (extraction of knowledge—metadata) from the underlying measurement and observation data (see Fig. 5-8b).

Constructing an expert knowledge-based system for toxicogenomics requires a database and relationships that can integrate genomic data with toxicological information and data. For instance, the toxicology domain of a microarray gene expression experiment captures the clinical chemistry and histopathology of biological samples exposed to toxicants or environmental pressures. This "tox-arm" of the microarray data is considered the metadata that describe biological end points of the toxicology experiment. In essence, the metadata can be viewed as anchoring the gene expression profile to the phenotype observed in the exposed biological sample for identification of early biomarkers of toxicity.

Relational database management systems (RDBMS) are powerful tools used to organize the storage, retrieval, and integrity of high-ordered data. A key functionality of RDBMS is the ability to link multiple files in many types of relationships for sophisticated architecture of complex data structure. Professional RDBMS products offer enhanced portability to various operating

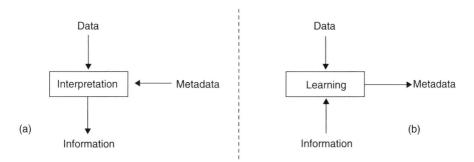

Figure 5-8 *Utilization of metadata for learning and interpretation of data and information.*

system environments, flexibility to interactive application programs, and reliability of performance speed/accuracy of data processing. Application programming interfaces (APIs) such as CORBA's (common object request broker architecture) distributed object management architecture, DBI's (database interface) pure Perl solutions, and ODBC (open database connectivity) are typically used as abstract interface layers to provide transparent and seamless integration of databases. Various Web browsers, Java, Active Server Pages, and hypertext markup language (HTML) have been used to access and query ODBC-compliant databases. In addition, customized user interfaces and generation of standard reports are readily available using a common gateway interface (CGI), object-oriented programming, and HTML to facilitate information sharing among multiple users.

A client/server computing architecture consists of one or more workstations designated as client machines running application programs with a remote computer, designated as the server machine, which services requests from the client. The RDBMS resides on the server, and the application program on the client machine interfaces with a middleware software product that serves to communicate requests and transactions between the application and the RDBMS (Fig. 5-9).

The National Institute of Environmental Health Science (NIEHS) National Toxicology Program (NTP) has developed a Toxicological Database Management System (TDMS) to monitor the conduct of subchronic and chronic in-life studies and manage data collected during the in-life and pathology portions of these studies for public peer review. TDMS consists of two components: (1) the Laboratory Data Acquisition System (LDAS), which is a microcomputer application and hardware uniquely designed to directly collect in-life and pathology study data and to transmit these data to the TDMS database facility; and (2) the TDMS database and associated software, which receives and manages the study data transmitted from the individual laboratories. In addition, specifically designed software is used to generate reports

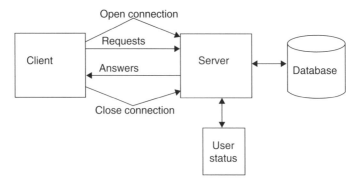

Figure 5-9   Basic client/server computer architecture.

for quality control and to help evaluate the progress and interpretation of study results.

To date, there has been a pharmacogenomics database developed to integrate large databases on gene expression and molecular pharmacology: the Pharmacogenetics Knowledge Base (PharmGKB), which links genomic, phenotypic, and clinical information collected from ongoing pharmacogenetic studies (Hewett et al., 2002); a database to manage high-density tissue microarray data associated with pathology and clinical outcomes on prostate cancer patients (Manley et al., 2001); and separate published efforts to combine results generated from gene expression analysis with data collected from toxicology (Burczynski et al., 2000; Waring et al., 2001, 2003; Boess et al., 2003) or molecular pharmacology experiments (Scherf et al., 2000; Blower et al., 2002); yet there has been no successful public effort to synergize microarray gene expression data with toxicological information and data in a database system for the development of a true learning expert knowledge base.

TOXICOGENOMIC DATABASE MINING

Analysis of microarray data can be partitioned into a number of successive steps (Bushel et al., 2002) in order to successfully capture meaningful gene expression patterns for classification and prediction of biological samples (Fig. 5-10). Without quality assessment of acquired data and selection of informative genes, classification of biological samples exposed to toxicants can be an elusive process. In SAS data mining, the process that clarifies this approach is sample, explore, modify, model, and assess (SEMMA). Sample the data by

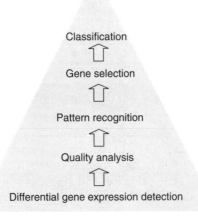

Figure 5-10 *Simplified model for the gene expression classification and prediction process.*

extracting a portion of a data set large enough to contain the significant information yet small enough to manipulate quickly. Explore the data by searching for unanticipated trends and anomalies. Modify the data by transforming the variables or imputing missing values. Model the data by automatically searching for a variable combination that reliably predicts a desired target or outcome. Assess the data by evaluating the usefulness and reliability of the findings from the data mining process(es). A database can be extended to serve as an AIMS to detect altered data, statistically validate differentially expressed genes, and perform multidimensional analysis of gene expression profiles. Typically microarray data are scrutinized for quality control and are subjected to preprocessing, normalization and transformation, postprocessing, and multivariate analyses. Visualization and inspection of data can take the form of observing the array image, surveying summary statistics and quality metrics, displaying data as a scatter plot or histogram, grouping the gene expression profiles, or reducing the dimension of the data.

Analysis of microarray data has aided investigators with interpreting the complex data structures and vast gene expression profiles extracted from biological systems. Reducing the dimension of microarray gene expression data is an effective technique to extract meaningful information pertaining to the complex alterations in gene expression from chemical, mechanistic, and environmental effects on biological systems. Several statistical procedures and computational tools have been developed to specifically analyze microarray data. In particular, clustering algorithms (Eisen et al., 1998; Tamayo et al., 1999), Bayes' methods (Baldi and Long, 2001; Newton et al., 2001; Friedman et al., 2000), statistical distributions and permutations (Callow et al., 2000), mixed linear (Wolfinger et al., 2001) and analysis of variance (ANOVA) models (Kerr et al., 2000) have been implemented in various software and applications to analyze gene expression data.

Utilization of supervised and unsupervised pattern recognition methods have been applied to microarray data to group the inherent symmetrical relationships between genes and treatments or condition of biological samples. Hierarchical cluster analysis, principal component analysis, and self-organizing maps are examples of unsupervised pattern recognition procedures previously used to predict the class of biological samples based on gene expression data. No prior knowledge or information pertaining to the disease state, chemical treatment, mechanistic pathway, gene function, or promoter region is required. Conversely, supervised methods such as support vector machines, linear discriminant analysis, and neural networks employ prior knowledge of the biological samples to guide the classification process. Recently, more sophisticated hybrid pattern recognition procedures have been developed that utilize multivariate analysis capabilities of two or more algorithms to enhance the gene selection and classification grouping of microarray gene expression data (Li et al., 2001; Ooi and Tan, 2003). The field of artificial intelligence and reasoning has for sometime addressed the area of machine learning by developing sophisticated computer algorithms that can deduce and extract additional facts from those already in place. Backpropagation, neural networks, and fuzzy logic are a few approaches to machine learning of information and data.

REFERENCES

Anderle P, Duval M, Draghici S, Kuklin A, Littlejohn TG, Medrano JF, Vilanova D, Roberts MA (2003). Gene expression databases and data mining. *Biotechniques* Suppl:36–44.

Baldi P, Long AD (2001). A Bayesian framework for the analysis of microarray expression data: regularized *t*-test and statistical inferences of gene changes. *Bioinformatics* **17**(6):509–519.

Blower PE, Yang C, Fligner MA, Verducci JS, Yu L, Richman S, Weinstein JN (2002). Pharmacogenomic analysis: correlating molecular substructure classes with microarray gene expression data. *Pharmacogenomics J* **2**:259–271.

Boess F, Kamber M, Romer S, Gasser R, Muller D, Albertini S, Suter L (2003). Gene expression in two hepatic cell lines, cultured primary hepatocytes, and liver slices compared to the in vivo liver gene expression in rats: possible implications for toxicogenomics use of in vitro systems. *Toxicol Sci* **73**:386–402.

Brazma A, Hingamp P, Quackenbush J, Sherlock G, Spellman P, Stoeckert C, Aach J, Ansorge W, Ball CA, Causton HC, Gaasterland T, Glenisson P, Holstege FC, Kim IF, Markowitz V, Matese JC, Parkinson H, Robinson A, Sarkans U, Schulze-Kremer S, Stewart J, Taylor R, Vilo J, Vingron M (2001). Minimum information about a microarray experiment (MIAME) toward standards for microarray data. *Nat Genet* **29**:365–371.

Brazma A, Parkinson H, Sarkans U, Shojatalab M, Vilo J, Abeygunawardena N, Holloway E, Kapushesky M, Kemmeren P, Lara GG, Oezcimen A, Rocca-Serra P, Sansone SA (2003). ArrayExpress—a public repository for microarray gene expression data at the EBI. *Nucleic Acids Res* **31**:68–71.

Burczynski ME, McMillian M, Ciervo J, Li L, Parker JB, Dunn RT II, Hicken S, Farr S, Johnson MD (2000). Toxicogenomics-based discrimination of toxic mechanism in HepG2 human hepatoma cells. *Toxicol Sci* **58**:399–415.

Bushel PR, Hamadeh H, Bennett L, Sieber S, Martin K, Nuwaysir EF, Johnson K, Reynolds K, Paules RS, Afshari CA (2001). MAPS: a microarray project system for gene expression experiment information and data validation. *Bioinformatics* **17**: 564–565.

Bushel PR, Hamadeh HK, Bennett L, Green J, Ableson A, Misener S, Afshari CA, Paules RS (2002). Computational selection of distinct class- and subclass-specific gene expression signatures. *J Biomed Inform* **35**:160–170.

Callow MJ, Dudoit S, Gong EL, Speed TP, Rubin EM (2000). Microarray expression profiling identifies genes with altered expression in HDL-deficient mice. *Genome Res* **10**:2022–2029.

Eisen MB, Spellman PT, Brown PO, Botstein D (1998). Cluster analysis and display of genome-wide expression patterns. *Proc Natl Acad Sci (USA)* **95**:14863–14868.

Forsyth R (1984). *Expert Systems: Principles and Case Studies.* New York: Chapman and Hall.

Friedman N, Linial M, Nachman I, Pe'er D (2000). Using Bayesian networks to analyze expression data. *J Comput Biol* **7**:601–620.

Hewett M, Oliver DE, Rubin DL, Easton KL, Stuart JM, Altman RB, Klein TE (2002). PharmGKB: the Pharmacogenetics Knowledge Base. *Nucleic Acids Res* **30**:163–165.

Kaiser J (2002). Proteomics. Public-private group maps out initiatives. *Science* **296**:827.

Kerr MK, Martin M, Churchill GA (2000). Analysis of variance for gene expression microarray data. *J Comput Biol* **7**:819–837.

Li L, Weinberg CR, Darden TA, Pedersen LG (2001). Gene selection for sample classification based on gene expression data: study of sensitivity to choice of parameters of the GA/KNN method. *Bioinformatics* **17**:1131–1142.

Manley S, Mucci NR, De Marzo AM, Rubin MA (2001). Relational database structure to manage high-density tissue microarray data and images for pathology studies focusing on clinical outcome: the prostate specialized program of research excellence model. *Am J Pathol* **159**:837–843.

Mattingly CJ, Colby GT, Forrest JN, Boyer JL (2003). The Comparative Toxicogenomics Database (CTD). *Environ Health Perspect* **111**:793–795.

Merrick BA (2003). The Human Proteome Organization (HUPO) and environmental health. *Environ Health Perspect* **111**:797–801.

Newton MA, Kendziorski CM, Richmond CS, Blattner FR, Tsui KW (2001). On differential variability of expression ratios: improving statistical inference about gene expression changes from microarray data. *J Comput Biol* **8**(1):37–52.

Ooi CH, Tan P (2003). Genetic algorithms applied to multi-class prediction for the analysis of gene expression data. *Bioinformatics* **19**:37–44.

Pratt PJ, Adamski JJ (1994). *Database Systems: Management and Design*, 3rd ed. Boston: Course Technology, Inc.

Scherf U, Ross DT, Waltham M, Smith LH, Lee JK, Tanabe L, Kohn KW, Reinhold WC, Myers TG, Andrews DT, Scudiero DA, Eisen MB, Sausville EA, Pommier Y, Botstein D, Brown PO, Weinstein JN (2000). A gene expression database for the molecular pharmacology of cancer. *Nat Genet* **24**:236–244.

Spellman PT, Miller M, Stewart J, Troup C, Sarkans U, Chervitz S, Bernhart D, Sherlock G, Ball C, Lepage M, Swiatek M, Marks WL, Goncalves J, Markel S, Iordan D, Shojatalab M, Pizarro A, White J, Hubley R, Deutsch E, Senger M, Aronow BJ, Robinson A, Bassett D, Stoeckert CJ Jr, Brazma A (2002). Design and implementation of microarray gene expression markup language (MAGE-ML). *Genome Biol* **3**(9):research0046.1–research0046.9.

Tamayo P, Slonim D, Mesirov J, Zhu Q, Kitareewan S, Dmitrovsky E, Lander ES, Golub TR (1999). Interpreting patterns of gene expression with self-organizing maps: methods and application to hematopoietic differentiation. *Proc Natl Acad Sci (USA)* **96**:2907–2912.

Waring JF, Jolly RA, Ciurlionis R, Lum PY, Praestgaard JT, Morfitt DC, Buratto B, Roberts C, Schadt E, Ulrich RG (2001). Clustering of hepatotoxins based on mechanism of toxicity using gene expression profiles. *Toxicol Appl Pharmacol* **175**:28–42.

Waring JF, Cavet G, Jolly RA, McDowell J, Dai H, Ciurlionis R, Zhang C, Stoughton R, Lum P, Ferguson A, Roberts CJ, Ulrich RG (2003). Development of a DNA microarray for toxicology based on hepatotoxin-regulated sequences. *Environ Health Perspect* **111**:863–870.

Waters M, Boorman G, Bushel P, Cunningham M, Irwin R, Merrick A, Olden K, Paules R, Selkirk J, Stasiewicz S, Weis B, Houten BV, Walker N, Tennant R (2003). Systems toxicology and the Chemical Effects in Biological Systems (CEBS) Knowledge Base. *Environ Health Perspect* **111**:811–824.

Wolfinger RD, Gibson G, Wolfinger ED, Bennett L, Hamadeh H, Bushel P, Afshari C, Paules RS (2001). Assessing gene significance from cDNA microarray expression data via mixed models. *J Comput Biol* **8**:625–637.

6

Statistics for Toxicogenomics

Thomas J. Downey Jr.

INTRODUCTION

What makes a toxicogenomic study substantially different from traditional studies in toxicology is the nature of the observed response. Rather than observing just a few response variables, today's toxicologist uses microarray technologies to measure thousands of variables per specimen. Taken as a whole, these variables can be viewed as a high-dimensional response vector, which in the ideal case offers a comprehensive snapshot of the genome. This comprehensiveness and very high dimensionality are new to toxicologists and statisticians alike and present new opportunities and dangers. The opportunity is the ability to gain a global view of the genome. But with that comes the danger inherent with such high-dimensional response data—that is, a myriad of false discoveries. With so many variables, the number of potential relationships between the variables becomes so large that seemingly compelling correlations can appear merely by chance.

The purpose of this chapter is to introduce important statistical concepts and techniques useful for designing, analyzing, and interpreting a toxicogenomic experiment. Concepts are illustrated, when possible, using a single published experiment that is typical of the types of experiments being conducted in the field of toxicogenomics today. In addition, some of the common pitfalls that lead to false discovery are addressed.

Toxicogenomics: Principles and Applications. Edited by Hamadeh and Afshari
ISBN 0-471-43417-5   Copyright © 2004 John Wiley & Sons, Inc.

EXAMPLE EXPERIMENT

An experiment was conducted to investigate the global and gene-specific expression changes resulting from toxic and nontoxic doses of a once over-the-counter medication, methapyrilene (Hamadeh et al., 2002). In the experiment, 36 rats were randomly assigned to 9 study groups (4 rats/group) and dosed for 1, 3, or 7 days with vehicle, 10 mg/kg/day methapyrilene, or 100 mg/kg/day methapyrilene. Dose selection was based on previous studies. The low dose is considered relatively nonhepatotoxic in rats, and the high dose is considered acutely hepatotoxic in rats. The time points were chosen because hepatotoxicity in rats due to methapyrilene treatment is usually evident within several days of exposure at the 100 mg/kg/day dose level (Hamadeh et al., 2002). Each group of treated animals was necropsied with a group of time-matched control animals. An RNA sample corresponding to livers derived from pooling the 4 animals at each treatment condition was used for hybridization to 4 custom National Institute of Environmental Health Sciences (NIEHS) cDNA arrays (Nuwaysir et al., 1999). A total of 24 arrays were used to measure the gene expression from the animals, 8 arrays for each time point. The pooling of animals and assignment to the arrays from the 3 doses at each time point is shown in Figure 6-1.

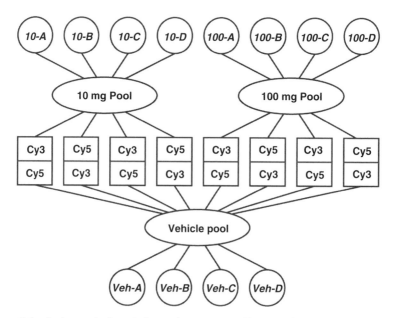

Figure 6-1 *Assignment of pooled samples to arrays. Four rats from each treatment group were pooled together, then allocated to 4 arrays (in the case of the low and high doses) or 8 arrays (in the case of the controls). This design was replicated at the day 1, 3, and 7 time points.*

The pooled sample from the low and high doses at each time-point group was hybridized to 4 separate arrays, while the pooled sample from the control animals was hybridized to 8 arrays. The low dose and high dose for each time point were cohybridized to 4 arrays along with a control sample from the same time point. On the first 2 of the 4 arrays, the control sample was labeled with a green (Cy3) dye, and the treated sample was labeled with a red (Cy5) dye. On the other 2 arrays, the remaining aliquots were labeled with the dyes reversed (control sample labeled with a red dye and treated sample labeled with a green dye). This has been referred to as a "dye-swap" design (Kerr et al., 2000).

EXPERIMENT DESIGN

Types of Variables in the Experiment

In the previous experiment, dose and time are referred to as the independent variables because the experimenter manipulates them in order to answer the research questions. The experimenter systematically chose the dose levels and time points to be administered, and the assignment of the animals to each treatment condition. The ~7,000 gene expression variables are called dependent or response variables because they represent the responses by the rats, which are believed to be at least partially dependent on the treatments administered. The extraneous variables are similar to the independent variables in that they may have an effect on the response variables, except they are not controlled during the experiment. For this reason, extraneous variables are also sometimes referred to as nuisance variables because they usually have a negative impact on the experiment. Extraneous variables are not necessarily a nuisance, as is the case when they are not confounded with the independent variables, and they are also predictive of the response variables. Likewise, nuisance variables are not always extraneous. In the methapyrilene experiment, the array and the dye are nuisance variables because the researcher is not directly interested in their effects. In this experiment, the dye and the array have been considered in the design, so they are not extraneous.

In addition to its type, each variable is measured on either a categorical or numerical scale. Categorical (or nominal) variables take on the values of named categories. Dose, time, dye, array, and animal pool are categorical variables. The ~7,000 gene expression variables are numerical.

Observational Studies Versus Controlled Experiments

In an observational study, the natural order of events is recorded to determine whether two variables are correlated. For example, a study could be conducted to determine if there is an association between methapyrilene use and the incidence of liver disease in humans. Even if the study establishes a significant correlation between methapyrilene exposure and liver disease, it can only establish the strength of the correlation; it cannot establish that methapyrilene use *causes* liver disease. For example, participants exposed to methapyrilene may have other medical or lifestyle differences that are correlated to both methapyrilene exposure and liver disease. It is possible that one of these other medical or lifestyle differences was actually responsible for the higher incidence of liver disease.

In order to establish that there is a causal relationship between exposure to methapyrilene and liver disease, all reasonable alternate explanations have to be ruled out. This illustrates an important but often overlooked fundamental of statistics: *correlation does not establish a cause and effect relationship*. This is especially true in observational toxicogenomic studies, where significant correlations between gene expression and phenotypic end points do not establish that either the gene expression caused the phenotype or vice versa.

The desire to establish causality between two variables is one of the leading motivations for conducting a designed experiment. In contrast to an observational study where the researcher passively observes events, an experiment is a planned and systematic intervention of natural events. The researcher introduces one or more treatments or conditions that have a measurable effect on the subject. Alternative explanations are systematically controlled or removed through a process called random assignment, leaving the treatment as the only reasonable cause of any significant changes in biological response. In gene expression studies, use of knockout mice models, antisense and RNAi technologies allow controlled experiments to determine causal relationships between gene expression and phenotypic end points.

Random Sampling

Experiments and studies are conducted in order to make an inference about a population based on a relatively small sample. The population is the group of all subjects to which the conclusions are generalized. In order to statistically generalize results to the target population, it is required that each subject in the population has an equal chance of being selected for inclusion in the experiment or study. Selection bias occurs in observational studies when the relationship between exposure to the risk factor and disease is different between those who participate in a study and those who are theoretically eligible to participate.

Selection bias can result in a study whose conclusions can be very large, very powerful, and very wrong. Consider the following example: It was feared that there was an increased incidence of leukemia among soldiers who were exposed to radiation during the U.S. Army's 1957 Smoky atomic bomb tests. Researchers at the Centers for Disease Control (CDC) conducted a study to determine whether there was an association between participation in the atomic bomb test and incidence of leukemia (Caldwell et al., 1983). The researchers concluded that there was a significant association between participation in the test and incidence of leukemia. Unfortunately, the investigators could not locate all of the participants of the bomb test, and many of the subjects who were included in the study contacted the CDC on their own due to illness (self-selection). As a result, there was a higher proportion of diseased subjects included in the study than those who were not included in the study. This form of selection bias is called volunteer bias. To control selection bias, researchers should take precautions to make sure that study subjects adequately represent all those who could have participated. The selection method is, however, seldom perfect, and thus a major question in any observational study is how much the results are biased by the selection process.

Due to the nature of toxicological experiments, animals are usually used instead of humans. The researcher chooses an animal that is believed to be a good model of the target population, and scientific generalization rather than statistical generalization is required to make inferences about the target population. For this reason, and because subjects are randomly assigned to treatment groups, random selection of subjects is frequently not critical in a designed experiment. The failure of random selection in observational studies, however, is one of the most common sources of erroneous conclusions from human medical research studies.

Random Assignment

The most important characteristic of a designed experiment that distinguishes it from an observational study is that all subjects are identical and identically treated except for the factor (or factors) being manipulated by the researcher. In the methapyrilene experiment, the animals were the same gender, strain, and age. Additionally, they were supplied the same diet and living environment. But even with strict selection of animals and identical treatment, significant variability between individual rats will still remain. Therefore, in order to avoid confounding treatment conditions with unforeseen extraneous variables, the animals in this experiment were randomly assigned to the nine treatment conditions. Random assignment abides by the mathematics of probability and is the mechanism by which a properly conducted experiment can establish a cause and effect relationship.

Completely Randomized Versus Randomized Block Designs

The Completely Randomized Design. There are many types of designs, but the most fundamental and important building block in experimental design is the completely randomized design. Assignment of animals to the nine treatment groups in the methapyrilene experiment is an example of a completely randomized design (Fig. 6-2) because the animals were assigned to the treatment groups completely at random.

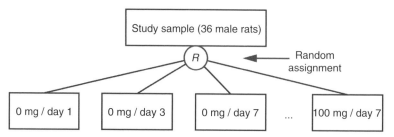

Figure 6-2 *Completely randomized design. Rats are randomly assigned to the treatment groups. In this experiment there is a constraint to place an equal number of rats in each treatment group.*

Random assignment of subjects to treatment groups guarantees that there are no significant differences in extraneous variables between treatment groups if and only if the sample size is relatively large (i.e., greater than 100 subjects). With the smaller sample sizes frequently used in genomic experiments, it is possible that an "unlucky" random assignment can result in the confounding of treatment conditions with exogenous variables. For example, suppose the methapyrilene experiment includes both male and female rats (18 of each). Since each of the 9 treatment groups contains only 4 animals, it is possible that an unlucky random assignment of rats could place 4 male rats in one treatment group and 4 female rats in another treatment group. In this case, gender becomes confounded with the treatment groups and the researcher will be unable to statistically conclude that the effects are due to the treatment and not gender.

The Randomized Block Design. In order to protect against confounding with small sample sizes, it is often advantageous to "block" the samples, then randomly assign samples from each block to the different treatment groups. Continuing with our example, gender can be used as a blocking variable by grouping the animals into a block of males and a block of females. The block of 18 male rats is randomly assigned to each of the 9 treatment groups, placing 2 males in each group. The same is done with the block of female subjects (Fig. 6-3). This ensures that 2 males and 2 females are assigned to each treatment group, and that gender is not confounded with the treatment. Since each gender occurs in equal proportion in each treatment group, what could have been a nuisance variable becomes a blocking factor. Because gender is not confounded with the treatment, if it turns out that there is a significantly different response between male and female rats, the gender effect can be quantified and removed from the statistical noise using analysis of variance.

The randomized block design is favored when blocks of biologically similar units can be formed or when specimens can be divided into aliquots. For the latter reason, most in vitro experiments can usually use randomized block designs to reduce statistical noise.

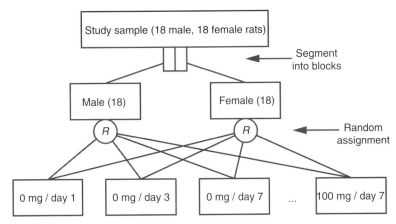

Figure 6-3 *Randomized block design. The study sample is first divided into male and female blocks. Each of the blocks is then randomly assigned in equal proportion to the nine different treatment groups.*

Since the samples were allocated in equal proportion to the Cy3 and Cy5 dyes, dye and array have been used as blocks in the design, and thus this experiment uses elements of completely randomized and randomized block designs.

DATA ANALYSIS

Image intensity values for each gene on each microarray from the methapyrilene study were quantified from both the Cy3 and Cy5 channels independently. Subsequently a ratio of the treated sample to the time-matched control sample from the same array was created. This resulted in a spreadsheet containing ~7,000 ratio measurements for each of the 24 arrays. The ratio is commonly used because the intensity for a given gene in a treated sample can only be fairly compared to the matched control on that same array (since the array-to-array variability for any particular gene is significant). At the time of this writing, it is common to analyze two-color gene arrays as ratios, although several statisticians experienced with gene expression two-color gene arrays have recommended analyzing the data without explicitly forming ratios (Kerr et al., 2000; Wolfinger et al., 2001). Specifically, the ratio suffers from at least three limitations:

1. If the proposed ratio has an intensity value equal to zero in the denominator, the ratio cannot be properly computed.
2. Important information about signal intensity is lost when the ratio is formed (e.g., 1,000/100 is the same ratio as 10/1).
3. Information is lost when the numerator and/or denominator contain negative values.

For these reasons, the data will also be analyzed using the individual intensities for the Cy3 and Cy5 channel without taking the ratio. This alternative representation of the data yields a spreadsheet containing ~7,000 intensity measurements for each of the 48 hybridized samples.

It may seem a little confusing to compare the original number of rats (36) to the total number of arrays (24) or samples (48). This is due to the fact that the control samples were hybridized once after pooling on each of the 24 arrays while the low and high doses were hybridized to only 12 of the 24 arrays each (see Fig. 6-1).

The remainder of this chapter demonstrates how a few useful statistical techniques can be used to analyze the described data. First an exploratory analysis technique is used to identify global effects across the genome and unanticipated effects in the data. After a preliminary exploratory analysis, statistical inference is used to identify genes that are differentially expressed between treatment groups. Finally, a predictive model is constructed that uses a gene expression signature to distinguish between animals exposed to high dose versus those exposed to low dose. All of the numerical and visual analyses were performed using Partek Pro Version 5.2 statistical analysis software (http://www.partek.com). Simpler versions of those analyses could be conducted with some commercially and publicly available analysis tools (see Chapter 15); however, data volume handling is limiting. Partek Pro, SAS, and R are some of the relatively more powerful high-end packages for handling larger data volumes.

Exploratory Data Analysis

Exploratory data analysis (EDA) makes use of visual and numerical analysis and presentation of data with the goal of revealing patterns that are interesting. EDA is not necessarily concerned with answering specific questions or hypotheses but is useful for finding patterns in the data that are not expected or that the researcher would have not otherwise looked for. Common benefits of exploratory analysis include discovery of outliers, significant effects due to nuisance variables, severe departures from normality, and other anomalies.

Most exploratory techniques are multivariate, meaning multiple variables are simultaneously considered. Given the high dimensionality of the response vector in a genomic experiment, techniques that examine all of the response variables together are very powerful for identifying global effects on the genome. Exploratory analysis techniques that are commonly applied to genomic data include data visualization and data reduction techniques, such as cluster analysis, principal components analysis, and multidimensional scaling. It is important that the exploratory tool is capable of analyzing all ~7,000 genes simultaneously, as certain commonly used gene prefiltering techniques can produce false discoveries even in random data (see "False Discoveries from Genomic Studies" at the end of this chapter).

Principal Components Analysis. Principal components analysis (PCA; Jolliffe, 1986) is an exploratory technique that is used to describe the structure of high-dimensional data by reducing its dimensionality. Although it is a relatively old technique (Hotelling, 1933), recent advances in computing power and interactive 3-D graphics make it very useful for today's very high-dimensional genomic data.

Put simply, PCA is a linear transformation that converts n original variables into n new principal components (PCs) that have three important properties:

1. The PCs are ordered by the amount of variance explained.
2. The PCs are uncorrelated.
3. The PCs explain all variation in the data (there is no loss of information in the transformation if all PCs are considered).

PCA of Log-Ratio Data. Figure 6-4 shows a PCA mapping of the ratio data from this experiment. In this visualization, there is one data point for each of the 24 arrays. The location of each point is determined by the first three

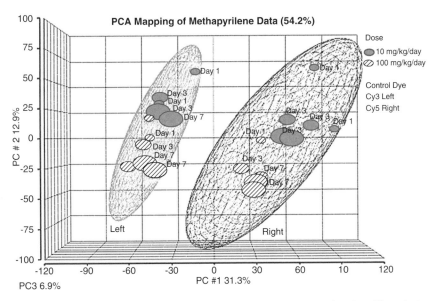

Figure 6-4 *PCA mapping of ∼7,000-dimensional log-ratio gene expression data. Three factors of the experiment are displayed together with a mapping of the expression data. The color of each data point indicates the dose (shaded = 10 mg, diagonal lines = 100 mg), the size of each point indicates the day (small = day 1, medium = day 3, and large = day 7), and the wire-mesh ellipsoids enclose points that use the same dye for the control sample (left ellipsoid = Cy3, right ellipsoid = Cy5).*

principal components of the ~7,000 log ratios of the treated samples to control which appear on the x-, y-, and z-axes, respectively. The first PC (x-axis) contains the largest amount of information among the PCs (34.3% of the total variation in the data), the second PC (y-axis) contains the next largest amount of information (12.9% of the total variation in the data), and the third PC (z-axis) contains the third largest amount of information (6.9% of the total variation in the data). Since the PCs are uncorrelated (orthogonal), the individual sources of variability can be added together, indicating that 54.1% of the total variation in the data can be explained by the first three PCs.

The data points in Figure 6-4 correspond to arrays and are colored by the dose and sized by the time. Additionally, a wire-mesh ellipsoid is used to distinguish which dye was used for the control sample (remember that the control is labeled with Cy3 on 12 arrays and with Cy5 on the remaining 12 arrays). The position of a point relative to the others indicates the array's similarity to the other arrays. Data points that are close together in the PCA scatter plot correspond to arrays that are similar in overall gene expression, and points that are far apart correspond to arrays that are relatively dissimilar in overall gene expression.

An important feature of PCA is that the positions of the points in the scatter plot are not determined using variables such as dose and time. Even though the factors manipulated by the researcher (dose, time, and control dye) are used to control the color, size, and so forth, they are not considered in the computation of the PCs. (PCA is an "unsupervised" analysis technique.) The following observations can appropriately be made from Figure 6-4:

- The distance between any two points is related to the similarity between the ~7,000 dimensional response vectors from the corresponding arrays.
- Points that are close together correspond to arrays that have relatively similar expression pattern across the ~7,000 dimensional response vector.
- Points that are far apart correspond to arrays that have relatively different expression patterns across the ~7,000 dimensional response vector.
- The total amount of variance explained by the first three PCs is 54.1% (34.3% for PC 1 + 12.9% for PC 2 + 6.9% for PC 3).
- Note that the PCs are ordered by the amount of information they convey. PC 1 has the most information and differentiates arrays by the dye used for the control sample. PC 2 has the second largest amount of information and differentiates between the low and high doses. The fact that the difference in control dyes appears on PC 1 and the difference in dose appears on PC 2 indicates that the largest effect in the data is due to the dye used for the control sample, and the second largest effect in the data is due to the difference between the low and the high doses.

In summary, the following three conclusions can be made about the data based on the PCA visualization of Figure 6-4:

1. The orientation of the dye labeling seems to be the largest effect in the data. Separated mostly by PC 1, the difference between Cy3 and Cy5 dyes appears to be the most substantial difference between the arrays.
2. The low dose and high dose are well separated mostly by PC 2, and thus this is the second largest effect in the data (the second largest difference between arrays).
3. There appears to be a trend where day 1 samples appear near the top of their respective clusters and day 7 samples appear near the bottom. This indicates that the difference in time points is also significant but less so than effects due to dye and dose.

In short, the largest global effects making one array different or similar to another are (1) dye, (2) dose, and (3) time, in that order.

PCA of Nonratio Data. Figure 6-5 shows a PCA mapping of data from the methapyrilene experiment with a single data point for each sample hybridized to an array. Since the 2 samples on an individual array were not combined as

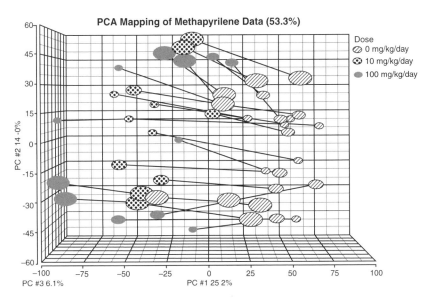

Figure 6-5 *PCA mapping of ~7,000-dimensional gene expression vectors for individual analysis of Cy3 and Cy5 channels. Three factors of the experiment are displayed together with a mapping of the expression data. The color of each data point indicates the dose (diagonal lines = 0mg (vehicle), dots = 10mg, shaded = 100mg), the size of each point indicates the day (small = day 1, medium = day3, and large = day 7), and connecting lines are drawn to identify two samples from the same array.*

a ratio, there are 48 data points corresponding to the 48 samples (24 vehicle, 12 low dose, and 12 high dose).

The most striking observation from Figure 6-5 is that the control samples are significantly different from either the low dose or the high dose, regardless of which dye is used. The difference is apparent on PC 1, so this is the most substantial overall effect that makes one sample similar or dissimilar to another. Since most of the lines that connect two samples from the same array together are relatively orthogonal to the *y*-axis, PC 2 largely explains the array-to-array variability (the second largest effect in the data). The following observations can be made from Figure 6-5:

- The total amount of variance explained by the first three PCs is just about the same as Figure 6-4: 53.3% (33.2 for PC 1 + 14.0% for PC 2 + 6.1% for PC 3). The reason for this similarity is the high degree of correlation between the two different representations of the data.
- PC 1 has the most information and shows that samples are significantly different depending on whether the sample was treated. Note that the difference between the 10 mg/kg/day and 100 mg/kg/day is much smaller than the difference between the control dose and either treated dose. This is interesting and possibly unexpected, since the 10 mg dose is supposed to represent a relatively nonhepatotoxic dose and the 100 mg is supposed to represent an acutely hepatotoxic dose.
- PC 2 has the second largest amount of information and shows a difference from one array to another.
- The difference between time points is obvious in the PCA plot; however, a pattern can be seen where the larger points (day 7) appear close to the top and bottom of the plot and the smaller points (day 1) appear near the center of the scatter plot.
- The difference between low and high doses is not evident in the PCA plot, since there is no clear pattern differentiating the orange (10 mg/kg/day) and the blue (100 mg/kg/day).

In summary, the largest global effects making one sample different or similar to another are (1) control/treated, (2) differences from one array to another, and (3) differences between time points.

Statistical Inference

Statistical inference is the process of making conclusions (inferences) about a target population based on a much smaller sample of that population. Given a research hypothesis, an experiment is performed on a sample of subjects and an inference (conclusion) is made about the hypothesis. The research hypothesis, however, cannot directly be proven. Instead, an opposite null hypothesis (H_0) is formed and is formally tested. There are only two possible outcomes of the test: either the rejection H_0 or the failure to reject H_0.

A statistical test begins by assuming that H_0 is true, then trying to prove that assumption unlikely. The methapyrilene experiment was conducted to test the research hypothesis that there is a difference in expression of one or more genes between those rats that receive a low dose and those that receive a high dose. The research hypothesis (H_A) is that there is a difference between the average expression of one or more genes in rats exposed to a low dose and those exposed to a high dose. The null hypothesis is the opposite of the research hypothesis, being that the mean expression of the low-dose subjects is equal to the mean for the high-dose animals. Formally stated,

$$H_A : \mu(control) \neq \mu(treated) \text{ (research hypothesis)}$$

$$H_0 : \mu(control) = \mu(treated) \text{ (null hypothesis)}$$

The result of a statistical test is called a p-value. The p-value gives the probability that H_0 is true. Therefore, small p-values lead to the rejection of the null hypothesis and the conclusion that the research hypothesis must be true. Normally, a threshold is set for declaring the p-value significant prior to performing the test. This threshold is called the α level and is commonly set at $\alpha = 0.05$ (The somewhat arbitrary α level of 0.05 was first suggested by Sir Ronald A. Fisher [Fisher, 1925]; however, few have found much reason to dispute his suggestion since then.)

Type I and Type II Errors

The result of a statistical test is not perfect, and incorrect conclusions are sometimes the result. There are two types of errors that can occur as a result of a statistical test: type I and type II. A type I error occurs if H_0 is rejected when it is in fact true. This is also referred to as a false positive. In the methapyrilene experiment, a false positive would occur if there was no real difference in the expression of a particular gene between the low and high doses, but the statistical test concluded that there was a difference. A type II error, also known as a false negative, occurs if H_0 is accepted when it is in fact untrue. In the example, a false negative occurs if there really is a difference in gene expression, but the statistical test fails to show the difference. When testing thousands of response variables, plenty of type I and type II errors are sure to occur. Figure 6-6 summarizes the types of errors that can result from a statistical test and the associated probabilities of making each type of error. The power of a test refers to the probability of detecting a difference when one really exists and is defined as $1 - \beta$, where β is the probability of committing a type II error.

Parametric Versus Nonparametric Tests

Statistical tests such as the t-test and ANOVA make assumptions about the underlying distribution of the response variable. If the assumptions are true,

Truth \ Decision	Fail to reject H_0	Reject H_0
H_0 is true	True negative Probability = $1 - \alpha$ (Specificity)	False positive (type I error) Probability = α
H_0 is false	False negative (type II error) Probability = β	True positive Probability = $1 - \beta$ (Power, or sensitivity)

Figure 6-6 *A decision table summarizing type I and type II errors and associated measures of sensitivity and specificity.*

then the response data can be accurately described by parameters such as the mean and variance. For these reasons, tests such as the *t*-test and ANOVA are known as parametric tests. The three assumptions are

1. *Independence.* If the subjects randomly are sampled and randomly assigned to the treatment groups, then the independence assumption is usually satisfied. Prior to pooling the animals, the assumption of independence was met. However, tissue from 4 rats in each treatment group was pooled, then 4 samples were drawn from that pool. Clearly these 4 samples are related and do not represent independent samples. This is discussed in detail in the section on analysis of variance.

2. *Normality.* Parametric tests like the *t*-test and ANOVA assume that the data come from a normal (bell-shaped) distribution. For large sample sizes, however, the central limit theorem (CLT) gives protection against nonnormality. Recall that the CLT states that the sample means will be normally distributed even if the population that the samples are drawn from is decidedly not normal. Therefore, for relatively large sample sizes, the CLT assures us that parametric statistical tests are robust against violations of this assumption.

3. *Homogeneity of variance.* Also called homoscedasticity, homogeneity of variance requires that the variance in each treatment group be the same. Although there are a number of tests for homogeneity of variance, most of them are affected to some degree by nonnormality of the data. Thus it is not necessarily recommended to run tests for homogeneity of variance as a matter of course, because the variances or standard deviations of each group can be inspected by eye. If the variances in any two groups differ by more than a factor of 4 (or standard deviations differ by more

than a factor of 2), then the researcher should be concerned that this assumption has been violated.

In many cases, the data will violate the assumptions of normality and homogeneity of variance, especially if the response variable is measured on an ordinal or categorical scale. In these cases the researcher may consider use of a nonparametric test, which makes relatively few assumptions about the nature of the distribution of the response data. Under some circumstances, particularly when the response variable is not numeric, use of nonparametric tests can detect treatment effects that cannot be detected by parametric tests such as ANOVA.

Although parametric tests make assumptions about normality and homogeneity of variance, in many cases they are very robust even when these assumptions are violated. For example, when the sample size is large, the CLT gives protection against nonnormality. Additionally, when there are an equal number of samples in each group, the tests are also relatively robust against violations of homogeneity of variance. As an example, consider a distribution that is not normal but uniform. Application of ANOVA on these data will give the same results as a nonparametric test. However, if the data are normally distributed and within-group variances are approximately equal, the parametric tests provide substantially more power than nonparametric alternatives.

Although some researchers prefer nonparametric tests because of the few restrictive assumptions, these tests have several shortcomings relative to parametric tests, particularly with the small sample sizes typically used in toxicogenomic experiments. For these relatively small sample sizes, nonparametric tests are substantially less powerful than parametric alternatives, and it would be unlikely that after multiple test correction (see below) they could produce results that would lead to rejection of the null hypothesis. Additionally, for analysis of multifactor experiments (such as the example used here), nonparametric tests may not be available. For these reasons, nonparametric tests should be used when no alternative parametric test is reasonable.

Data Transformation. If the assumptions of homogeneity of variance and normality are not met, a transformation of the data will often fix the problem. Transformations that improve homoscedasticity are often referred to as variance stabilization transformations, and those that improve normality are referred to as normalization. Luckily, transformations that stabilize variance usually also solve the normality problem (and vice versa). Common transformations include log and square root. Log transformations are commonly employed when effects are multiplicative (larger means have larger variances), when the data are a ratio, or when data are significantly skewed to the right. The square root should be considered when the data are counts, such as white blood cell counts, as it usually improves homoscedasticity for these types of data.

The ratio data for the methapyrilene experiment were log transformed. This is a standard practice with ratio data, since they are naturally from a lognormal distribution. For the purposes of this chapter, a log transformation of the nonratio version of these data were also used. Whether or not to log scale nonratio data has been the subject of great debate—usually with statisticians recommending a log transformation and biologists feeling that changing the data in this way is inappropriate. When the response vector contains thousands of variables, however, it is safe to conclude that a transformation such as log scaling will be beneficial for some variables and not for others.

Analysis of Variance

A one-way analysis of variance (ANOVA) is a parametric statistical test used to compare the means of two or more groups of continuous observations where the groups are defined by levels of a single factor. Sir Ronald A. Fisher, the father of modern statistics, first introduced ANOVA in 1925 (Fisher, 1925).

Analysis of variance detects differences between groups by partitioning the total variability in the response data into variability between treatment groups and variability within the treatment groups. The variance within a single treatment group is also called error variance or statistical noise. A ratio is formed as the variability between groups divided by the variability within the groups. This ratio is called the F-ratio (named after its inventor), and F-ratios that are significantly greater than 1 lead to the rejection of the null hypothesis:

$$H_0 : m_1 = m_2 = \ldots = m_k$$

where k is the number of levels in the factor.

Factors and Levels. A factor is an independent variable whose levels are controlled by the researcher. Dose and time are examples of factors in the methapyrilene experiment. The levels of a factor are the different values that the independent variable can take on. The factor dose has three levels in this experiment: 0 mg/kg/day, 10 mg/kg/day, and 100 mg/kg/day. The factor time also has three levels: day 1, day 3, and day 7. Experiments such as this one that investigate more than one factor simultaneously are called factorial designs. In a factorial design, the treatment groups are formed as combinations of different levels of the factors. The methapyrilene experiment is therefore a 3×3 factorial design. Factorial designs are more efficient than one-factor-at-a-time experiments, and they allow interactions (see below) between factors to be detected.

Blocking Factors. Blocking factors are used to reduce noise in an experiment by making sure that nuisance variables that can be incorporated into the design are not confounded with treatment conditions. In the methapyrilene experiment, dye and array were used as blocking factors, ensuring that the treatments of interest are not confounded with dye or array. Other common blocking variables in an experiment like this one might be gender, strain, litter, or wash batch.

Main Effects and Interactions. For a two-factor experiment, a two-way ANOVA may be used to identify sources of variability due to the two factors and also due to the interaction between the two factors. An interaction is the variation among the differences between means for different levels of one factor over different levels of the other factor. For example, in the methapyrilene experiment (and in most toxicological experiments), an interaction between dose and time is expected, since the differences between high and low doses are expected to be different at the day 1 and day 7 time points (you would expect the high dose to have an effect at an earlier time point than the low dose). For the methapyrilene experiment, dose and time are main effects and a dose * time interaction is expected.

Balanced, Crossed, and Partially Crossed Designs. The following factors will be considered in the methapyrilene experiment:

- Dose (0 mg/kg/day, 10 mg/kg/day, 100 mg/kg/day)
- Time (day 1, day 3, day 7)
- Dye (Cy3, Cy5)
- Array (array 1, array 2, . . . , array 24)
- Animal pool (pool 1, pool 2, . . . , pool 9)

One factor is crossed with another if and only if every level of one factor is administered in combination with every level of the other factor. In the methapyrilene experiment, the two factors, dose and time, are crossed, since each level of dose is administered in combination with each time point. Additionally, since there are an equal number of rats (4) allocated to each treatment combination, these two factors are also balanced (see Table 6-1). Array is only partially crossed with dose, since each array only has a control dose and either a low dose or a high dose but not both a high dose and a low dose (see Table 6-2).

Nested/Nesting Relationships Between Factors. One factor is nested in another factor if and only if all of the levels of one factor occur within only

TABLE 6-1   Dose crossed and balanced with time

Time/Dose	0 mg	10 mg	100 mg
Day 1	4 rats	4 rats	4 rats
Day 3	4 rats	4 rats	4 rats
Day 7	4 rats	4 rats	4 rats

TABLE 6-2 Dose partially crossed with array

Array/Dose	0 mg	10 mg	100 mg
Array 1	1	1	0
Array 2	1	0	1
Array 3	1	1	0
. . .	. . .	. . .	. . .
Array 24	1	0	1

TABLE 6-3 Array nested in time

Array/Time	Day 1	Day 3	Day 7
Array 1	2	0	0
Array 2	0	2	0
Array 3	0	0	2
. . .	. . .	. . .	. . .
Array 24	0	0	2

one level of the other factor. Array is nested in time (array is the nested variable and the time is the nesting variable), since each array occurs only within a single time point (see Table 6-3).

Random Versus Fixed Effects. Fixed effects are factors for which the levels included in the experiment represent all the levels of interest to the researcher. For fixed effects, conclusions can be inferred only on the levels of the factor tested. Random effects are factors that represent only a random sample of all the possible levels, not just those that are included in the experiment. For random effects, conclusions can be inferred to the population from which the random sample is drawn.

Sometimes it is difficult to determine if a factor is a fixed effect or a random effect, but it is important to know which it is. Here is an easy way to decide if a factor is a random or fixed effect: Imagine that the experiment would be performed again. If the same levels of the factor would be administered, it is a fixed effect. If different levels of the factor would be administered, it is a random effect.

The methapyrilene experiment has examples of both fixed and random effects. Dose, time, and dye are fixed effects. Array is a random effect because the 24 arrays in this experiment represent only a random sample of all the arrays that could have been used. Animal pool is also a random effect, since the pool of animals in the experiment represents only a random sample of all the pools of animals that could have been used.

Covariates. Consider an extraneous variable that is numerical such as age, weight, or scan order. Certainly age and weight may be expected to have an effect on gene expression, but will scanning order? Since the scanner likely "warms up" and has other changes during the day, it is possible that arrays scanned early in the day are different than arrays scanned late in the day. Numerical explanatory variables such as age, weight, and scan order can be considered in the analysis as covariates. This type of ANOVA is called analysis of covariance (ANCOVA).

Multiple Comparison Corrections. When an analysis of variance shows a significant F-ratio when testing for a difference in means between more than two groups, it is common to follow up by making pairwise contrasts between the different pairs of levels. Consider a simple one-way ANOVA that tests the following null hypothesis:

$$H_0 : \mu_{control} = \mu_{10\,mg/kg/day} = \mu_{100\,mg/kg/day}$$

If the F-ratio is sufficiently large, resulting in a p-value that is significant, it is common to follow up by asking, Which means are different? There are three contrasts to be tested:

$$H_0(1) : m_{control} = m_{10\,mg/kg/day}$$
$$H_0(2) : m_{control} = m_{100\,mg/kg/day}$$
$$H_0(3) : m_{10\,mg/kg/day} = m_{10\,mg/kg/day}$$

With an α level of 0.05, a 1/20 chance of making a type I error is expected if testing a single hypothesis. But when testing several hypotheses, the probability of making a type I error increases substantially. Therefore, when following up on a significant F-ratio from an ANOVA by performing multiple comparisons of pairwise contrasts, a suitable multiple comparison correction should be made to the α level. There are many different multiple comparison corrections, and most can be achieved by either adjusting the α level or by adjusting the p-value resulting from the individual contrasts.

The simplest multiple comparison correction is the Bonferroni correction. The Bonferroni correction can be applied to the p-values from the individual contrasts by multiplying each p-value by the total number of comparisons (three in the example above). Alternatively, the α level can be divided by 3. The formula for the Bonferroni adjustment for the p-value is

$$P_{Bonferroni} = p * N$$

where N is the number of comparisons.

Notice that this simple formula can easily result in corrected p-values that are greater than 1. This is because the Bonferroni correction is actually only an approximation. The actual corrected p-value is given by the Sidak correction:

$$P_{Sidak} = 1 - (1 - p)^N$$

where N is the number of comparisons.

For mathematical correctness, the Sidak correction is preferred over Bonferroni. However, most researchers are familiar with the Bonferroni correction and not familiar with Sidak. For adjusted p-values below 0.05, the Bonferroni is an excellent approximation to the Sidak correction, so the choice of which to use makes little practical difference.

Multiple Test Corrections. Besides the multiple comparisons that result when multiple contrasts are considered on a single gene, today's genomic experiments result in multiple tests in which a statistical test is performed on each of the thousands of genes in the response vector. For the methapyrilene experiment, if testing all ~7,000 genes using an α level of 0.1, you would expect ~7,000 × 0.1 = ~700 type I errors. Therefore, even if no true treatment effects occur, it might be tempting to nonetheless conclude that 700 genes have been significantly affected by the treatment. This is known as multiple testing problems, and there are several ways to address this situation.

The Bonferroni or Sidak corrections can also be used to correct for multiple tests. They are the most conservative corrections, and it can be difficult to pass this strict correction when the number of response variables is very large. The reasons for these corrections being conservative are two-fold:

1. The Bonferroni and Sidak corrections assume that the ~7,000 tests are independent, meaning that the ~7,000 genes are uncorrelated. Since it is not possible to measure thousands of response variables on a biological sample without a large degree of correlation, the actual number that the p-value should be adjusted by is smaller than N, where N is the number of genes or tests.

2. The second reason the Bonferroni and Sidak corrections are conservative is because they produce a family-wise error rate (FWER). The FWER is the probability of at least one false positive among all of the genes being tested. When testing thousands of response variables, the researcher may be willing to accept some percentage of false positives, and this correction does not allow for control over the proportion of false positives.

Bootstrap (or resampling) methods randomly reassign the classification of each sample to a category in such a way that the number of samples in each category remains unchanged. After random reassignment of categories, the statistical test is performed and the resulting p-values are tabulated. This procedure is repeated many times, and for each original p-value, the fraction of times that a p-value that is as small or smaller occurs in the randomly reassigned data is the bootstrap estimate for the adjusted p-values. It empirically measures the likelihood of obtaining the uncorrected p-value by chance.

Resampling methods do not suffer from the assumption that the tests are independent, but they are still conservative in that they also produce a FWER that does not allow control over the proportion of false positives. Benjamini and Hochberg (Benjamini et al., 1995) proposed an alternative to a FWER that allows the researcher to control the false positive rate, or false discovery rate (FDR). FDR provides the most lenient multiple test correction of those described so far. Because of the very large number of statistical tests resulting in a genomic experiment and the relatively small number of samples, it is sometimes difficult to overcome the relatively conservative multiple test corrections such as Bonferroni or resampling methods. For this reason, FDR has become a popular choice for analysis of genomic experiments.

Analysis of Log-Ratio Data. The methapyrilene experiment has examples of fully crossed, partially crossed, balanced, unbalanced, and nested relationships between factors. Additionally, it has examples of random and fixed effects, so it makes a great example of the power and flexibility that ANOVA offers for the analysis of a designed experiment.

Simple tests such as the *t*-test (which is actually a special case of a one-way ANOVA) also test for a difference in means between groups; however, the *t*-test is unable to handle a comparison of more than two groups and is also unable to deal with the multiple factors in this experiment, including dose, time, and dye. Nonparametric tests such as Mann-Whitney and Kruskall-Wallis also cannot handle the multiple factors in this experiment.

Because the experiment was designed to investigate the effects of two factors, dose and time, these should be included as main effects in the analysis of variance. As discussed, you normally expect to find an interaction between dose and time in a toxicological experiment, so we would like to include the dose * time interaction in our analysis as well. In addition, the PCA analysis summarized in Figure 6-4 indicated that the dye used for control was the most substantial source of variability in the data, so it should be included in the analysis as well.

Our ANOVA model should include the following effects:

- Dose
- Time
- Dose * time
- Control dye

However, there is a problem. Remember the assumption of independent samples required for a parametric test like ANOVA? The four samples taken from each pool are clearly not independent samples, so the animal pool needs to be included as a blocking factor in the ANOVA. But the animal pool is confounded with the dose * time interaction (each treatment group used only one pool of animals), meaning that without some additional assumptions and a lot of work, the dose * time interaction cannot be estimated because the animal pool is confounded with the treatment groups. In addition, assuming that a dose * time interaction really exists, the error variance is contaminated with the variance due to the dose * time interaction, which will make the F-ratio smaller for the main effects. In short, pooling the animals changed a replicated design into a nonreplicated design, denying the ability to estimate the dose * time interaction and reducing power to detect main effects.

Since the dose * time interaction cannot be included in the ANOVA model, a four-way mixed-model ANOVA was run individually on all ~7,000 genes to estimate the effects due to dose, time, control dye, and animal pool, with animal pool as a random effect. (Analysis of variance that includes both fixed and random effects is referred to as a mixed-model ANOVA.) For each of these four effects, histograms for the resulting p-values are shown in Figure 6-7.

Inspecting these histograms is a form of FDR and can be interpreted as follows. For dose, there are 2,185 genes with p-values below 0.1. By chance you would expect ~7,000/10 = ~700 genes with p-values below 0.1. There are therefore 2,185 − ~700 = ~1,485 more genes below 0.1 than are due to chance.

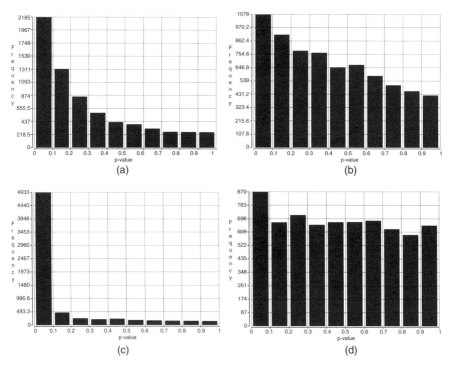

Figure 6-7   Histograms of ~7,000 p-values for (a) dose, (b) time, (c) control dye, and (d) animal pool.

In terms of a false discovery rate, if you consider those genes with p-values below 0.1, $\sim$700/2,185 are false positives, giving a false discovery rate of about one-third for those 2,185 genes with p-values below 0.1. Note that using the false discovery rate of Benjamini and Hochberg provides only a threshold value for a specified false discovery rate. The histograms in Figure 6-7 give more information than a single threshold. Histograms with taller peaks to the left (indicating more small p-values than due to chance) indicate effects that are seen across many genes. Compare the results of the ANOVA in Figure 6-7 to the results of PCA in Figure 6-4. Both analyses are consistent, showing the largest effect is due to the dye used for the control, followed by dose, and third by time. Variability from animal pool to animal pool (Figure 6-7d) is evident but does not affect a substantial number of genes. Also note that the analyses summarized in Figures 6-4 and 6-7 both represent global effects across the studied portion of the genome. PCA, however, uses a multivariate approach, and the histograms from the $\sim$7,000 ANOVAs represent a summary of $\sim$7,000 univariate (one gene-at-a-time) tests.

None of the genes passed a Bonferroni correction at the 0.05 level or the FDR (step-up) correction at 0.05 for the main effects of dose and time. Although the global effects can be seen in Figures 6-4 and 6-7, the design did not provide the statistical power to identify that individual genes changed due to either dose or time with a sufficient degree of confidence. The reason is partly due to the fact that the animals were pooled and partly due to the fact that an analysis using ratio data is less powerful than an alternative analysis using nonratio data.

Analysis of Nonratio Data. To analyze the data as nonratio, the dye and the array need to be included as blocking factors in the ANOVA. So the ANOVA model for the nonratio data includes the following factors:

- Dose (0 mg/kg/day, 10 mg/kg/day, 100 mg/kg/day)
- Time (day 1, day 3, day 7)
- Dye (Cy3, Cy5)
- Array (array 1, array 2, . . . , array 24)
- Animal pool (pool 1, pool 2, . . . , pool 9)

A five-way mixed-model ANOVA was run individually on all $\sim$7,000 genes to estimate the effects due to dose, time, dye, array, and animal pool. Array and animal pool are random effects, and array is nested in time. For each of these five effects, histograms for the resulting p-values are shown in Figure 6-8.

Analyzing the nonratio data produced a dramatic increase in power to detect differences due to dose. For dose, there are 6,360 genes with p-values below 0.1. By chance you would expect $\sim$7,000/10 = $\sim$700 genes with p-values below 0.1. We have 6,360 $-$ $\sim$700 = $\sim$5,660 more genes below 0.1 than are due

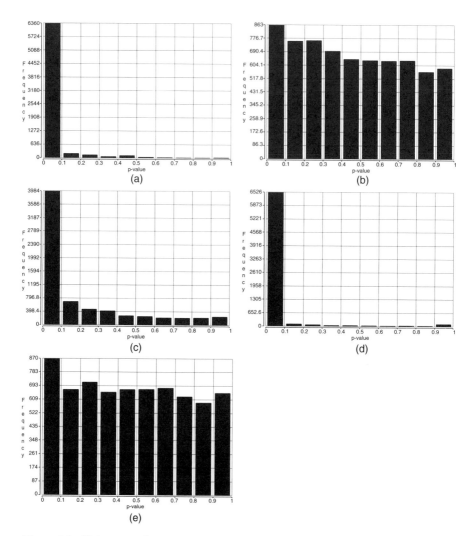

Figure 6-8   *Histograms of ~7,000 p-values for (a) dose, (b) time, (c) dye, (d) array, and (e) animal pool.*

to chance. In terms of a false discovery rate, we can consider that of the genes with *p*-values below 0.1, ~700/6,360 are false positives, giving a false discovery rate of about one-ninth.

Individual Contrasts Between Different Doses. A specific contrast between the 10 mg/kg/day dose and the 100 mg/kg/day dose is an example of preplanned contrast, since the investigator planned to make that comparison prior to running the overall ANOVA. Figure 6-9 shows the individual pairwise contrasts between the three doses. It is clear from Figure 6-9 that there is a sub-

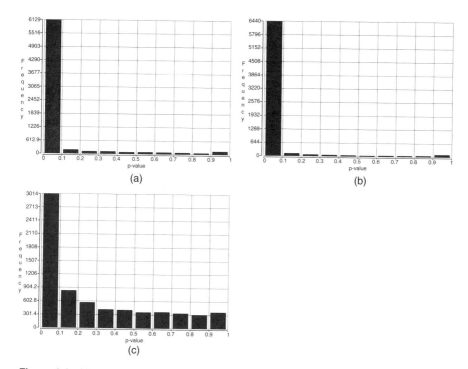

Figure 6-9 Histograms of ~7,000 p-values for the individual contrasts between doses (a) vehicle vs. 10 mg/kg/day, (b) vehicle vs. 100 mg/kg/day, and (c) 10 mg/kg/day vs. 100 mg/kg/day. Significant differences are seen in each contrast but less so in the contrast between the low and high doses.

stantial number of genes differentially expressed between all three pairwise contrasts, but there are relatively fewer between 10 mg/kg/day and 100 mg/kg/day than between the control group and either of the other dose levels.

Compare the results of the ANOVA in Figures 6-8 and 6-9 to the results of PCA in Figure 6-5. Again, both analyses are consistent, showing the largest effect due to control versus treated and array-to-array variability.

Summary of Analysis of Variance. The methapyrilene experiment was used as an example to demonstrate many facets of a many-way mixed-model ANOVA. In addition, the analysis was performed using ratio data and non-ratio data. Analyzing the data as ratios had significantly less power to detect main effects due to dose and moderately less power to detect changes due to time. The design of the experiment and the subsequent analysis is certainly made more challenging when using a two-color array such as was used for this experiment. By using the two-color platform, the experiment design becomes more complicated, and dye and array become additional factors in the subsequent ANOVA.

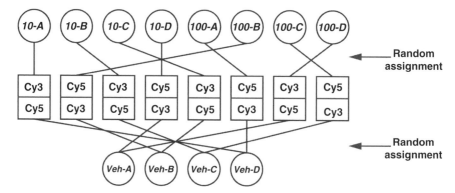

Figure 6-10 *Proposed design that omits pooling. The 4 rats from each of the low and high treatment groups are randomly assigned to the 8 arrays for this time point. Two samples were taken from each of the 4 rats in the control group for this time point and randomly assigned by blocks, such that each animal is blocked and crossed with both dye and 10 mg/kg/day versus 100 mg/kg/day. This design is replicated at the day 1, 3, and 7 time points.*

Pooling of Specimens. Equal amounts of RNA from each of the four animals in each treatment group were pooled together prior to hybridization on the arrays. Pooling of specimens from different animals is usually not a good idea and should only be done if there is insufficient RNA from each individual animal for hybridization. The methapyrilene experiment is no exception, and unfortunately pooling has reduced what was a very robust replicated design into a nonreplicated design. By pooling the samples, the effective sample size has been reduced from 36 to 9—one for each treatment group. The result is substantially reduced power to detect differences due to time and dose, and also it becomes just about impossible to detect the interactions between dose and time that are usually seen in toxicological experiments.

Figure 6-10 shows an alternative design that uses the same number of animals and arrays. Note, however, that there is no pooling of the RNA from the animals. The result is a simpler design, resulting in a simpler ANOVA that has more power to detect main effects due to dose and time, and can also estimate the effect of a dose * time interaction.

FALSE DISCOVERIES FROM GENOMIC STUDIES

The opportunity to gain a global view of the genome has been demonstrated as promised. Not only were individual genes significantly altered by the treatments in the experiment and identified using ANOVA but global patterns of gene expression affected by the treatments were identified using both PCA and ANOVA. But what about the myriad of false discoveries that can be the result of analyzing a very high-dimensional response vector?

In this section, we use randomly generated synthetic data to demonstrate just how easy it is to make false discoveries using genomic data. The reason for using random data is that in order to know that a discovery is indeed false, you must be sure that there do not happen to be any real effects in the data. Just like random assignment is required to infer cause and effect in a designed experiment, the use of purely random data is required to safely infer that seemingly compelling patterns are purely due to chance.

For the purpose of this section, we created a spreadsheet containing exactly 7,000 genes and 20 samples. Random numbers drawn from a normal distribution with a mean of 0 and a standard deviation of 1, $N(0,1)$, were used to generate simulated gene expression values. Subsequently, the 20 samples were randomly assigned to groups named "Diseased" and "Normal," 10 per group. Figure 6-11 shows a PCA plot of the 20 samples mapped from 7,000 dimensions to 3-D for viewing.

Next, a two-sample t-test was performed to identify those genes that are differentially expressed between the two groups. The distribution of those p-values is shown in Figure 6-12 and does not show any overall difference between the two groups.

The next step is one of the most dangerous practices in the analysis of genomic (or any very high-dimensional) data. I filtered out those genes that

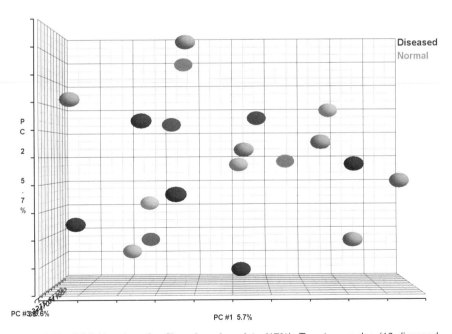

Figure 6-11 *PCA Mapping of unfiltered random data (17%). Twenty samples (10 diseased, 10 normal) are mapped from 7,000 dimensions to 3-dimensions using PCA. Since the data are randomly generated from a single normal distribution, it is not surprising that no patterns differentiating the samples are evident.*

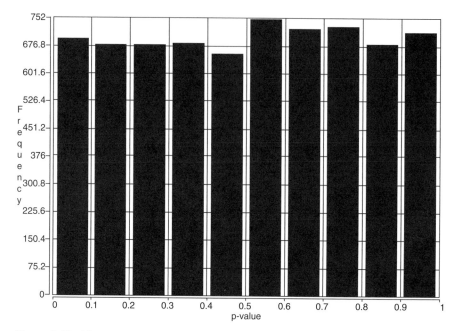

Figure 6-12   *Histogram of* p*-values from 7,000 tested genes. There are 696 genes with* p*-values below 0.1—just about what we expect (7,000/10 = 700).*

did not show much difference between the two groups by removing 1,000 genes with the largest *p*-values in the above *t*-test. This is the first amazing false discovery. Figure 6-13 shows a PCA mapping of the 6,000 remaining genes. The samples now look perfectly separable by PCA. Other exploratory analysis methods such as cluster analysis will yield similar misleading results, separating the two classes into two distinct clusters.

The results on these random data will likely surprise scientists and statisticians alike who are not trained to properly analyze such high-dimensional data. Following are just three false discoveries that can be the result of the prefiltering step:

1. *False clusters.* Cluster analysis of these filtered data would result in two clusters that perfectly agree with the two predefined classes. Of course, this clustering is a false discovery, since the only differences between the two predefined categories are due purely to random chance.

2. *Overly optimistic adjusted* p*-values.* If further statistical tests are performed on the filtered data, the multiple test corrections will be invalid, because they will be based on the number of tests being 6,000 when it should actually be 7,000. The more genes filtered out, the larger the bias in the corrected *p*-values.

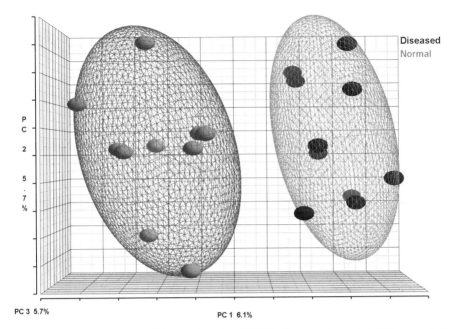

PC 3 5.7%

PC 1 6.1%

Figure 6-13 *PCA mapping of filter random data (17.5%). By removing only 1,000 genes that do not show a difference between the groups, a very distinct difference appears when the remaining data are analyzed.*

3. *Biased estimates of prediction accuracy.* Using only the filtered data, subsequent classification of these data, if only validated by reuse of the training samples (i.e., cross-validation), will indicate that the samples can be classified with a high degree of accuracy.

Unfortunately, these types of false discoveries in genomic data are state of the art in the field today. The last of the false discoveries is the trickiest. Many recent publications have claimed the ability to diagnose or predict clinical outcomes using only a few genes. Prediction accuracies are frequently estimated using a technique called cross-validation, where an entire data set is iteratively partitioned into training and test sets. However, if a gene selection process is applied external to the cross-validation, the data held out for testing do not represent independent test samples, since they were used to select genes that are predictive of the classes. During cross-validation, any gene filtering steps need to be repeated inside or internal to the cross-validation step in order for the estimates of prediction accuracy to be valid (Ambroise and McLachlan, 2002).

Many methods used to analyze genomic data cannot accommodate the full, high-dimensional response vector. Examples include hierarchical clustering and many implementations of PCA. Researchers should be very clear on why data are being filtered prior to analysis. In many cases they are filtered because ratios are not easily formed. However, ratios do not need to be created from microarray data and in many ways cause more trouble than they are worth. If researchers feel that genes expressed at low levels are not believable, they should consider waiting until after the analysis to filter those genes out.

Failure to properly design experiments, inappropriate use of statistics during analysis, and prefiltering of genes prior to analysis account for most of the false discoveries resulting from genomic studies today. The design and analysis of a genomic study is a challenging undertaking. Due to the complexity of the experiment design, statistical analysis, and the many opportunities for false discovery, scientists should seek to work closely with statisticians who are experienced with such studies.

REFERENCES

Alizadeh AA, Eisen MB, Davis RE, Ma C, Lossos IS, Rosenwald A, Boldrick JC, Sabet H, Tran T, Yu X, Powell JI, Yang L, Marti GE, Moore T, Hudson J Jr, Lu L, Lewis DB, Tibshirani R, Sherlock G, Chan WC, Greiner TC, Weisenburger DD, Armitage JO, Warnke R, Levy R, Wilson W, Grever MR, Byrd JC, Botstein D, Brown PO, Staudt LM (2000). Distinct types of diffuse large B-cell lymphoma identified by gene expression profiling. *Nature* **403**(6769):503–511.

Ambroise C, McLachlan GJ (2002). Selection bias in gene extraction on the basis of microarray gene-expression data. *Proc Natl Acad Sci (USA)* **99**(10):6562–6566.

Caldwell GG, Kelley DB, Heath CW Jr (1980). Leukemia among participants in military maneuvers at a nuclear bomb test. A preliminary report. *JAMA* **244**(14):1575–1578.

Caldwell GG, Kelley D, Zack M, Falk H, Heath CW Jr (1983). Mortality and cancer frequency among military nuclear test (Smoky) participants, 1957 through 1979. *JAMA* **250**(5):620–624.

DaSilva L, Cote D, Roy C, Martinez M, Duniho S, Pitt ML, Downey T, Dertzbaugh M (2003). Pulmonary gene expression profiling of inhaled ricin. *Toxicon* **41**(7):813–822.

Farmer PB (1995). Monitoring of human exposure to carcinogens through DNA and protein adduct determination. *Toxicol Lett* **82–83**:757–762.

Fisher RA (1925). *Statistical Methods for Research Workers.* Edinburgh and London: Oliver & Boyd.

Golub TR, Slonim DK, Tamayo P, Huard C, Gaasenbeek M, Mesirov JP, Coller H, Loh ML, Downing JR, Caligiuri MA, Bloomfield CD, Lander ES (1999). Molecular classification of cancer: class discovery and class prediction by gene expression monitoring. *Science* **286**(5439):531–537.

Gunther EC, Stone DJ, Gerwien RW, Bento P, Heyes MP (2003). Prediction of clinical drug efficacy by classification of drug-induced genomic expression profiles in vitro. *Proc Natl Acad Sci (USA)* **100**(16):9608–9613.

Hamadeh HK, Knight BL, Haugen AC, Sieber S, Amin RP, Bushel PR, Stoll R, Blanchard K, Jayadev S, Tennant RW, Cunningham ML, Afshari CA, Paules RS

(2002). Methapyrilene toxicity: anchorage of pathologic observations to gene expression alterations. *Toxicol Pathol* **30**(4):470–482.

Hotelling H (1933). Analysis of a complex of statistical variables into principal components, *J Educ Psychol.*

Jolliffe IT (1986). *Principal Component Analysis.* New York: Springer-Verlag.

Kerr MK, Martin M, Churchill GA (2000). Analysis of variance for gene expression microarray data. *J Computat Biol* **7**:819–837.

Mao R, Zielke CL, Ronald Zielke H, Pevsner J (2003). Global up-regulation of chromosome 21 gene expression in the developing Down syndrome brain. *Genomics* **81**(5):457–467.

Mead R (1988). *The Design of Experiments: Statistical Principles for Practical Applications.* Cambridge, UK: Cambridge University Press.

Montgomery DC (2000). *Design and Analysis of Experiments*, 5th ed. New York: John Wiley & Sons.

Nuwaysir EF, Bittner M, Trent J, Barrett JC, Afshari CA (1999). Microarrays and toxicology: the advent of toxicogenomics. *Mol Carcinogen* **24**(3):153–159. Review.

Perou CM et al. (1999). Distinctive gene expression patterns in human mammary epithelial cells and breast cancers. *Proc Natl Acad Sci USA* **96**:9212–9217.

Rothman KJ, Greenland S (1998). *Modern Epidemiology.* Philadelphia: Lippincott-Raven.

van't Veer LJ, Dai H, van de Vijver MJ, D. He YD, Hart AAM, Mao M, Peterse HL, van der Kooy K, Marton MJ, Witteveen AT, Schreiber GJ, Kerkhoven RM, Roberts C, Linsley PS, René Bernards R, Friend SH (2002). Gene expression profiling predicts clinical outcome of breast cancer. *Nature* **415**:530–536.

Wolfinger RD, Gibson G, Wolfinger ED, Bennett L, Hamadeh H, Bushel P, Afshari C, Paules RS (2001). Assessing gene significance from cDNA microarray expression data via mixed models. *J Computat Biol* **8**(6):625–637.

7

Real-Time and Quantitative PCR

Jeanelle M. Martinez and Nigel J. Walker

QUANTITATIVE REAL-TIME RT-PCR AS A TOOL FOR MICROARRAY VALIDATION

History and Background of PCR

In 1993 the Nobel Prize for Chemistry was awarded to Kary Mullis for developing PCR (Mullis, 1990). PCR is an extremely sensitive and invaluable tool used widely in clinical medicine, molecular biology, and forensics. Simply by knowing the sequence of a specific region of a gene, short single-stranded oligonucleotides (primers) can be designed to amplify a single gene to over a billion copies by using a thermostable DNA polymerase. DNA polymerase elongates a preexisting DNA strand, targeting an area that is specified by where the primers bind on the DNA. PCR is composed of a repeated series of cycles of varying temperature. In the first step a DNA sample is heat denatured (95°C), forcing double-stranded DNA into single strands, followed by steps at lower temperatures (50–75°C) that allow primer annealing and primer extension using the thermostable DNA polymerase. For every cycle that occurs, the amount of DNA doubles. These cycles are repeated typically anywhere from 20 to 40 times. DNA synthesis increases in an exponential manner until one of the reagents becomes limiting and the amount of DNA synthesized reaches a plateau; hence the name polymerase chain reaction.

Toxicogenomics: Principles and Applications. Edited by Hamadeh and Afshari
ISBN 0-471-43417-5   Copyright © 2004 John Wiley & Sons, Inc.

Reverse Transcription PCR

While RNA per se is not a substrate for PCR, cDNA can be prepared from the RNA using viral reverse transcriptase, which can then be amplified by PCR. The combination of PCR with reverse transcription (RT) has increased the sensitivity of mRNA phenotyping from as little as one to a few thousand cells (Rappolee et al., 1988). Quantitation of transcripts from a single cell with less than 100 copies of RNA became possible. The sensitivity for detection of genes by RT-PCR is based on the elimination of noncoding DNA by amplification of the expressed DNA. This technique allows detection and quantitation of low-abundance mRNAs between treated and nontreated samples, or between normal and diseased tissues. RT-PCR is the most common method for detection of mRNA species, since it provides a minimal input with a maximal output—that is, little tissue is needed; the number of gene products screened can be increased and allows the stabilization of material for long-term storage.

Components of the RT reaction consist of reverse transcriptase, primer, dNTPs, Mg^{2+}, and buffer system. There are three types of primers that can be used for the RT step, each with their own pros and cons: (1) Oligo (dT) primers target the poly(A) tails on the majority of mRNAs, which constitute 1–2% of the total RNA population but can fail to produce full-length cDNAs, and therefore may not be the best choice for target sites at the 5′ end of a gene. (2) Random hexamers are nonspecific and produce cDNA from the entire RNA pool, but they prime at random locations within an RNA, so they may not be suitable for very short RNAs or synthetic RNA standards. (3) Gene-specific primers are 12mers located immediately 3′ of the 3′ end of the amplicon site. If a synthetic RNA standard is used, gene-specific primer sites need to be included within the synthetic RNA standard. Gene-specific primers alleviate potential problems in the use of synthetic RNAs that result from differences in reverse transcription efficiency caused by the location of the amplicon site with a cDNA or the failure of random hexamers to fully reverse transcribe full-length cDNAs from the standards. If a gene-specific primer is used, the annealing temperature needs to be considered, and a reverse transcriptase that has thermostability may be required. Routinely random hexamers are usually the first choice and work reasonably well for most situations.

Quantitative PCR

Typical PCR amplification of a target gene occurs in an exponential manner, and quantitation is theoretically based on the following mathematical equation:

$$Y = X \cdot (1 + E)^{n}$$

where Y is the total number of PCR product, X is the starting number of copies of target segment of DNA of interest, E represents the efficiency of a PCR amplification that ranges from 0 to 1, and n is the number of PCR cycles (Raey-maekers, 1999). Two commonly used methods for quantitative PCR are competitive RT-PCR and kinetic real-time RT-PCR (discussed below).

In competitive PCR, typically a series of tubes containing a constant amount of RNA with the target mRNA are coamplified with a serially diluted amount of an identical internal standard synthetic RNA. Since the internal standard shares the same primer recognition and internal sequences with the primary target, it will "compete" for reagents when it is coamplified in a tube with the same reaction mixture and conditions (Walker et al., 1999). At the completion of the PCR cycle, analysis of the ratio of the amplified target and synthetic RNA–derived products gives an indication of the amount of target RNA. While competitive RT-PCR remains a gold standard in quantitative RT-PCR due to its inherent ability to give absolute RNA copy numbers, it has never gained widespread use because of the effort needed to create a synthetic RNA standard for each target gene of interest.

Real-Time PCR

There has been an explosion in the use of quantitative PCR due to the development of kinetic PCR, more commonly referred to as real-time PCR. In essence, real-time PCR refers to the ability to monitor DNA amplification at every cycle in the PCR reaction. In order to perform real-time PCR, the very first step is to choose an instrument suitable for studies to be investigated. Thermocyclers for real-time PCR combine the amplification, detection, and quantification of product formation all within a closed-tube system.

Fluorescence is induced during the PCR by distributing a laser light to thin-walled reaction tubes with optical fibers. Fluorescence emission returns with the fibers and is directed to a spectrograph with a charge-coupled device camera. The entire PCR cycle is monitored and thus is the basis of real-time PCR. In some instruments, amplification can be observed in true real time, whereas in others the PCR reaction is recreated after the PCR is completed. There are basically two types of instruments to choose from: one that is flexible in changing the cycle parameters each time and can run small batches of samples, or a high-throughput instrument for running numerous samples with similar parameters (Table 7-1).

Threshold Cycle

The threshold cycle (C_T) value is the primary measure in real-time PCR and the basis for all forms of quantitation, regardless of the method of fluorescence detection (discussed later in this chapter; Fig. 7-1). In real-time quantitative PCR techniques, fluorescent signals released from laser stimulation are monitored concurrently as they rise above background levels, and measurements occur before the reaction reaches a plateau. Initial template levels can be

TABLE 7-1   Real-time PCR instruments

Instrument	Company	Web Site
Flexible Systems		
Smart Cycler® System	Cepheid	http://www.eurogentec.com
ABI PRISM® 3100	Applied Biosystems	http://www.appliedbiosystems.com
Corbett Rotor-Gene	Corbett Research	http:// corbettresearch.com/
High-Throughput Systems		
ABI PRISM® 7000, 7700, 7900HT	Applied Biosystems	
Lightcycler	Roche Molecular	http://www.roche-applied-
	Biochemicals	science.com/lightcycler-online/
Icycler	BioRad	
MX-4000 Multiplex Quantitative	Stratagene	www.mx4000.com
DNA Engine Opticon	MJR	http://www.mjr.com/

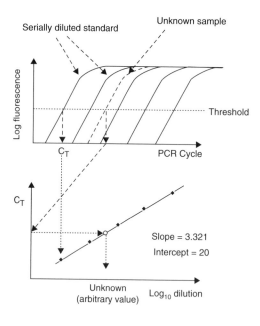

Example calculation

Dilution	Log10	Ct
1	0	20
4	0.60	22
16	1.20	24
64	1.81	26
256	2.41	28

Unknown $C_T = 25$

By linear regression:

Log10 diln = (Ct–intercept)/slope

= (25 – 20)/3.321 = 1.51

Dilution = $10^{1.51}$ = 32

Arbitrary value = 1000/32 = 31.25 units/per unit input RNA

(assuming undiluted RNA contains 1000 units of target RNA per unit total RNA)

Efficiency = $10^{(1/\text{slope})-1}$
= $10^{(1/3.321)-1}$
= $10^{0.301-1}$
= 2^{-1}
= 1

Figure 7-1 *Quantitation by real-time PCR using a standard curve. A serial dilution series of standard cDNAs are amplified along with unknown cDNAs by real-time PCR. C_T values for the standards are fit by linear regression relative to input copy number/dilution. Arbitrary values for unknown cDNAs are determined by interpolation against the standard curve. cDNAs can be derived from a reference total RNA sample containing the target of interest, a reference DNA sample of the target amplicon, or a synthetic RNA corresponding to the target of interest prepared by in vitro RNA transcription. If the copy number of the target amplicon in the reference is known, values for the unknown can be determined as copy number per unit input RNA. Otherwise, values are determined in arbitrary values.*

calculated by analyzing the shape of the curve or by determining when the signal rises above some threshold value.

The amount of fluorescence emitted is plotted against the number of amplification cycles and is proportional to the amount of product generated in each cycle. The more template present at the beginning of the reaction, the fewer cycles it takes to reach the exponential phase of amplification. The fluorescence is measured during every cycle, and the point at which the signal is statistically significantly above background to the specific fluorescence threshold is recorded as the threshold cycle (C_T) or crossing point (CP).

The C_T value assigned to a particular well thus reflects the point during the reaction at which a sufficient number of target amplicons have accumulated in that specific well at a statistically significant point above the baseline. Quantification using real-time PCR is based on the threshold cycle (C_T) that occurs during the exponential phase of amplification. The C_T value is determined early in the PCR reaction at the point of most efficient amplification where reagents are not limiting. Hence it is less sensitive to differences between samples in PCR amplification efficiency that are often observed with post-PCR end point–based methods.

Choices for Real-Time Amplification Detection

Real-time PCR can detect a specific fragment of DNA, or nonspecifically any double stranded DNA, depending on the type of fluorescence detection system used. Three types of chemistries are used in real-time PCR: DNA-binding dyes, hybridization probes, and hydrolysis probes. The least expensive and most widely used is SYBR® Green detection. SYBR green is a DNA-binding dye that binds during elongation to the newly formed double-stranded DNA. When using SYBR green, all double-stranded DNA generated during the reaction will emit fluorescence. Specificity is therefore determined solely by choice of PCR primers. Amplification specificity can only be truly validated by sequence analysis of the amplified product.

Quality control checks routinely include size validation by agarose gel electrophoresis coupled with validation of the expected melting temperature (Tm) of the amplified product. Analysis of the melting curve can show if a single product is produced and if it is likely to be the product specified (Fig. 7-2). Advantages of using SYBR green are that it is inexpensive, easy to use, and sensitive. The disadvantage is that SYBR green will bind to any double-stranded DNA in the reaction such as primer dimers or other nonspecific reaction products. For single PCR product reactions with well-designed primers, SYBR green can work extremely well, with nonspecific background only showing up past the exponential phase of amplification, a point where C_T values are not calculated.

Other formats for real-time PCR detection are based on fluorescence resonance energy transfer (FRET) technology, which involves the transfer of energy from an electronically excited molecule (the donor fluorophore or

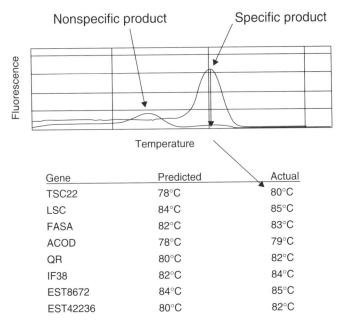

Nonspecific product   Specific product

Gene	Predicted	Actual
TSC22	78°C	80°C
LSC	84°C	85°C
FASA	82°C	83°C
ACOD	78°C	79°C
QR	80°C	82°C
IF38	82°C	84°C
EST8672	84°C	85°C
EST42236	80°C	82°C

Figure 7-2 Melting curve analysis. Analyses were conducted by adding the following temperature profile to the end of the SYBR green–based real-time PCR reaction: 95°C, 15 sec; 60°C, 20 sec; 20 min ramp to 95°C. Temperature-dependent fluorescence values were recorded during the 20 min ramp. The melting curves were obtained using ABI dissociation curve analysis software. The observed peaks represent two separate reactions containing either the specific amplified target or nonspecific products, respectively. Formation of a single peak in a reaction is indicative of a single specific product. Comparison of the peak melting temperature with the predicted melting temperature gives an indication of specificity. C_T values from reactions showing multiple peaks with melting temperatures that are highly different from expected should be interpreted with caution, since they do not likely represent values from the desired target. In general, actual melting temperatures tend to be several degrees higher than predicted.

reporter) to a neighboring molecule (the acceptor or quencher), returning the donor molecule to its ground state without fluorescence emission. If the donor and the acceptor are very close together, excitation of the donor results in energy transfer to the acceptor or a quencher that emits light at wavelengths that can be measured. If light is absorbed by a dye, it will lead to an excited state. FRET-based technology is used in a number of different systems including Taqman hydrolysis probes, molecular beacons, scorpions, and other hybridization-type probes (Walker, 2001, 2002).

Molecular beacons are dual-labeled hybridization probes that have a stem and loop structure where the target nucleic acid probe portion is located on the loop portion. The stems are designed to bind to each other by having complimentary sequences with a fluorescent marker covalently attached to one end and a quencher covalently attached to the other end. While in solution, molecular beacons adopt a hairpin conformation, but upon binding of the probe to its target sequence, a conformational change occurs, forcing the arm sequence apart and separating the reporter from the quencher to release fluorescence upon irradiation.

The Taqman assay is based on hydrolysis of a dual-labeled hybridization probe by the inherent 5′ nuclease activity of DNA polymerase. The Taqman probe contains a fluorescent reporter dye at the 5′ end that is quenched by a second fluorescent dye located at its 3′ end. During PCR the polymerase will cleave the probe that is bound to the target DNA to separate the reporter dye from the quencher dye. Separation of the quencher results in an increase in fluorescence for every cycle that is proportional to the rate of probe cleavage. This system requires a probe that has a higher melting temperature than the two primers and generates a sequence-specific fluorescent signal. Multiplexing is possible by using different fluorescent dyes to monitor different PCR products in the same reaction.

Hybridization probes used in the LightCycler system are designed as a pair to hybridize to adjacent regions of the template DNA (separated by 1–5 nucleotides). As they hybridize, the two dyes are brought together, and FRET to the acceptor results in a signal that can be measured. Fluorescence is lost when the temperature is increased above melting temperature of the oligos, the probes melt away from the template strand, and the distance between the dyes is increased. If you are simply interested in measuring relative fold changes between treated and nontreated samples, then SYBR green is a very practical choice, since it is less expensive to use during initial validations, and it allows you to measure the melting peak profile of a product to ascertain that only one product is produced.

Absolute Quantitation

PCR amplification of a standard with a known molarity that is serially diluted to produce a standard curve is required to determine the exact molar quantity of a target gene (Fig. 7-1). To assess for efficiency and allow for variation of the reverse transcription step, a standard RNA is recommended. Standard RNAs can be synthesized by subcloning an amplicon behind a SP6, T4, or T7 RNA polymerase-binding site in a plasmid vector, or by amplification by PCR of the amplicon with a forward primer that bears the polymerase-binding site. The in vitro transcribed sense RNA transcript is then digested with RNase-free DNase and accurately quantitated. Alternatively, you can isolate a stock of external RNA that has a high expression for the target gene and has been previously quantitated. The standard calibration curve can be made using either recombinant DNA or RNA (Fronhoffs et al., 2002; Pfaffl et al., 2002a), a synthetic oligonucleotide, or a purified RT-PCR product.

Once the standard is designed and made, it needs to be accurately quantified with a spectrophotometer or a fluorescent dye that binds specifically to RNA, or by competitive PCR or a ribonuclease protection assay (RPA). Using mathematical conversion of absorbance, the amount can be converted to a target copy number per microgram of RNA. Absolute quantitation is sometimes necessary in order to determine a change in gene expression and to compare the value with the one from another laboratory. This method allows an exact copy number of a gene defined per cell, mass, or total amount of RNA. Quantitation can be performed by comparing target gene expression using an external calibration curve whose absolute numbers are not known; hence final values are obtained as arbitrary numbers and are relative to this standard (Fig. 7-1).

Relative Quantitation

Relative quantitation is simply the ratio of the amount of target genes between two different samples. If a gene is induced, the ratio is greater than one; if it is repressed, it is less than one. The simplest method for relative quantitation is comparison of the C_T values for two sets of samples. Under the assumption that during the exponential phase there is a doubling of copy number, the ratio of the starting number of copies between two samples is given by the equation ratio = $2^{\Delta C_T}$, where ΔC_T is the difference in C_T values between the two samples. Similarly the relative expression ratio of a target gene to a reference gene in a sample can also be determined, which is then compared to another sample (Table 7-2). This delta C_T method assumes that the efficiency for amplification of the target and reference genes is equal. When amplicons are designed to be less than 150 base pairs and the concentrations of primer and magnesium are optimal, the efficiency will be close to 100%.

As shown in Figure 7-1, the slope of a linear regression of C_Ts versus a serially diluted sample can be used to determine the efficiency of amplification. Amplifying cDNA serial dilutions for target and housekeeping genes can test this by subtracting the reference C_T from the target gene C_T and plotting this ratio against the concentration or dilution of cDNA. You can obtain the absolute value of the slope, and if it is close to zero, then efficiencies are similar. If efficiencies are very different, there are mathematical equations to correct for them (Pfaffl, 2001; Table 7-2). For more information on the derivation and assumptions used for the delta C_T method, the reader is referred to an excellent explanation by Livak and Schmittgen (2001).

VALIDATION OF GENES FROM MICROARRAY ANALYSIS

Factors to Consider

Throughput. Methods available to measure RNA include RT-PCR, the RNAse protection assay (RPA), in situ hybridization, and Northern blots (Bustin, 2000). Advantages of real-time RT-PCR are that adaptation to high-throughput validation analysis is possible, and very little RNA sample is

TABLE 7-2 Example of relative quantitation using real-time PCR

Gene	Treatment	C_T	Average C_T	ΔC_T[a]	Fold[b] (Range)	Normalizing to the Reference Gene		
						ΔC_T[c,d]	$\Delta\Delta C_T$	Fold[e] (Range)
Target	Control	32.65	32.97 ± 0.36	0 ± 0.36	1.00 0.78–1.28	14.50 ± 0.48	0.00 ± 0.48	1.00 (0.71–1.40)
	Control	32.84						
	Control	33.48						
	Control	32.91						
	Treated	30.4	30.27 ± 0.25	−2.70 ± 0.25	6.49 5.47–7.69	12.08 ± 0.26	−2.42 ± 0.26	5.36 (4.48–6.42)
	Treated	30.55						
	Treated	30.02						
	Treated	30.12						
Reference	Control	18.13	18.47 ± 0.33	0 ± 0.33	1.00 0.80–1.25			
	Control	18.25						
	Control	18.75						
	Control	18.75						
	Treated	18.12	18.20 ± 0.09	−0.27 ± 0.09	1.21 1.14–1.28			
	Treated	18.12						
	Treated	18.27						
	Treated	18.27						

[a] Average treated gene C_T − average control C_T.

[b] Assuming 100% efficiency, the fold change relative to control is calculated as $2^{-\Delta C_T}$. The upper and lower ranges are defined using $\Delta C_T + 1$ s.d. and $\Delta C_T - 1$ s.d.

[c] If amplification occurs in the same well, the reference gene should be subtracted from the respective target gene C_T.

[d] Standard deviation is calculated based on the following equation: S.D. = (Target s.d.^{2} + reference s.d.^{2})$^{0.5}$ (PRISM, 2001).

[e] If the efficiencies for each primer set are known, then the following equation can be used (Pfaffl, 2001): $R = E^{\Delta C_T\ \text{target(control-treated)}}_{\text{target}} / E^{\Delta C_T\ \text{reference (control-treated)}}_{\text{ref}}$.

155

required. For example, most real-time PCR instruments accommodate 96-well formats, and 384-well formats are also available. Even using only a 96-well plate and analyzing triplicate control versus treated samples, on one plate you can rapidly validate 15 different genes in several hours using as little as 1 ng total RNA per gene per sample.

An advantage to using Northern blots over RT-PCR is the ability to detect the size of mRNA species. While Northern blots are not nearly as sensitive as RT-PCR, you can increase the sensitivity 20–50 times by using RPAs. RPAs are highly specific and can correctly identify and quantify mRNA species with a high degree of sequence homology. In situ hybridization has the distinct advantage of cellular localization of target genes. The disadvantage to Northern blots, RPAs, and in situ hybridization is that they are all time-consuming and cumbersome.

The number of genes to validate can also dictate the choice of the real-time PCR system. FRET-based systems generally require more assay development than dye-based systems, so unless the system is already developed, a dye-based system would likely be the first choice for validation studies.

Amount of RNA Available. Reverse transcriptase itself can inhibit real-time PCR when extremely low template levels are found—for example, in measurement of gene expression from individual cells that have been isolated using the patch clamp method (Chandler et al., 1998; Liss, 2002). In studies that target specific cell populations, laser capture microdissection from 200 to 300 cells will typically generate 1–6 ng total RNA (Wittliff and Erlander, 2002). By cleaning up the cDNA using a basic ethanol precipitation protocol, Liss (2002) showed that the inhibition problem could be overcome.

In addition, the amount of starting RNA may need to be reduced from the amount used in traditional gel electrophoresis PCR. We have found that typically 1–10 ng of total RNA is sufficient for most studies where you wish to detect small changes (<2-fold) in gene expression for most genes. For genes that are expressed at low abundance, 100 ng total RNA may be needed per assay. For highly abundant genes, pg levels of RNA are all that is required. If you are expecting a two-fold level of change, when setting up conditions for amplification of a target gene, a standard curve with a two-fold series of dilutions may be needed to test the capability of the specific assays for detecting these changes. In addition, this allows for the determination of amplification efficiency of the developed assay.

RNA Abundance. One of the advantages of real-time PCR is the ability to give a measure abundance of a gene transcript. This is an important consideration for providing a biological framework for further analyses. Using relative measures such as two-color microarray analyses or relative C_T values, ratios of expression between two samples provide an indication of an effect but without context. Absolute quantitation can provide this biological context, since RNA levels are provided as copy number per unit total RNA—for example, very low-abundance RNAs are expressed at the level of approximately 10^2–10^4 copies/μg total RNA and may be undetectable by other methods such as Northern blot analysis or protein analysis. Conversely, highly abundant transcripts expressed at levels $>10^7$/μg total RNA are easily detected by such methods. High fold changes of low-abundance messages may in some cases be reflective of effects in isolated cell populations within a tissue that is comprised of multiple cell types and thereby may direct future approaches to investigations of that particular gene. In the absence of absolute quantitation, the raw C_T value can give some indication of abundance—for example, highly abundant transcripts routinely give C_T values of <20, moderately abundant transcripts have C_Ts of 20–30, and low-abundance transcripts have C_Ts of 30–40.

Housekeeping Genes. With PCR analysis, gene transcript quantitation is often normalized to the expression of a known housekeeping gene. When using such normalization, an inherent assumption is that the expression of the housekeeping gene does not change at all. In many cases in which such normalization is used, this situation is biologically implausible—for example, genes such as actin or GAPDH can vary under different conditions, between different tissues, and especially between individuals. One approach to eliminate such errors from variable housekeeping genes is to take a geometric averaging of multiple internal control genes (Vandesompele et al., 2002). Given that housekeeping genes can and do vary, it is often more useful to consider them as null hypothesis control genes that should not be affected by the toxicological stressor under study. Quantitation relative to 18s ribosomal RNA can be used as an alternative to normalization relative to total RNA measurement in situations where sample size is limiting—for example, laser capture microdissected samples. Also when choosing an internal control gene, the transcript should have similar abundance in the same relative amount as the target gene.

Carryover. Carryover prevention should be routinely practiced when conducting PCR, since even tiny aerosols can contain thousands of copies of carried-over target molecules, thus making a true negative into a false positive. A general rule of thumb is to never open a plate that contains PCR-amplified DNA in an area where you perform pre-PCR work. Even better is to designate separate laboratories for pre-PCR and post-PCR work. It is important to use negative controls—for example, no template controls (no sample) and no amplification controls (everything but the DNA polymerase added). Addition of dUTP (instead of dTTP) and uracil N-glycosylase (UNG) eliminates DNA carryover contamination by destroying any PCR product containing uracil that may have been carried over from previous reactions. PCR workstations that use UV light in a hood-like area are available and can also help to reduce or eliminate PCR contamination.

Primers/Probes

Designing Primers. No matter what type of real-time PCR is used, primer design is a necessary step. Besides the literature, there are also Web resources for predesigned primers (Table 7-3). Typically primers for real-time PCR range from 16 to 30 base pairs long and have a G/C content of near 50%. Several real-time PCR systems require new primer designs to allow for the design of matching FRET-based probes for fluorescent detection of amplification. Dye-binding systems such as SYBR green can accommodate the use of existing primer pairs. Primer pairs should have matching Tms, but small differences are acceptable. A minimum amount of base-pair repeats (not more than 3) should be found. Avoiding Gs and Cs at the 3′ ends of primers can reduce primer-dimer artifacts.

To avoid false positive signals, you can design PCR primers that span intron-exon boundaries. There are several computer programs available that may help eliminate the use of a poorly designed pair of primers (Table 7-3). Once the primer/probe design is completed, it is useful to check the specificity of the predicted target by accessing the National Center for Biotechnology Information (http://ncbi.nlm.nih.gov/BLAST/). Commercial resources are available, some of which have over 17,000 gene expression primer/probe mixtures (Applied Biosystems) that have already been validated and optimized.

Validation of Primer Sets. After obtaining primers for genes of interest, it is important to confirm the primer specificity, especially if the design does not permit intron-exon spanning. There are three levels of validation: The first is by traditional gel electrophoresis to visualize product size. Gel electrophoresis with a 3% agarose gel is required because real-time PCR products are typically less than 150 base pairs. The second is by using SYBR green dye, which allows the melting peak profile to be visualized to ascertain that only one

---

TABLE 7-3   Real-time PCR: useful Internet sites

Predeveloped Primer Sets Freely Available

http://medgen.ugent.be/rtprimerdb/
http://www.realtimeprimers.org

Software Available to Help Design Primers (or Probes)

http://www-genome.wi.mit.edu/cgi-bin/primer/primer3_www.cgi
http://www.cybergehe.se/primerdesign/
http://bibiserv.techfak.uni-bielefeld.de/genefisher/

Quantitation Techniques

http://www.wzw.tum.de/gene-quantification/

---

product is made and that the melting temperatures match the expected design. The third and ultimate validation is to have the PCR product sequenced.

Once the product is validated, the primer efficiency needs to be determined by running serial dilution curves and examining the slope. We have found in our laboratory that concentrations of $0.2\,\mu M$ generally work well for a number of primer sets; however, if the standard deviations from triplicates run in the serial dilution curve are greater than $0.5\,C_T$s, then optimization of primer concentration needs to be performed with different combinations of forward and reverse primers (two-by-two combinations of 100, 300, or $900\,\mu M$) in tubes with aliquots of the same sample. Primer concentration is chosen from the pair that has the most amount of product formed (ΔRn) with the lowest C_T value.

Statistics. While small differences in C_T values can often be observed when comparing samples, statistical methods are needed to determine if such small gene changes, as analyzed by real-time PCR, are significant. For routine analyses, parametric tests such as the student t-test, ANOVA, and Dunnett's test (when comparing multiple groups to a control) are appropriate and assume that the control and treated groups have a normal C_T distribution and that the C_Ts differ in their mean. Randomization tests (or permutation tests) make no distributional assumptions and assume that samples are randomly allocated to two treatment groups, answering the question of how likely it is that a randomization could result in a difference as large as that obtained. Alternatively, nonparametric methods such as Mann Whitney U-tests may also be used. There are a number of mathematical algorithms that can compute an expression ratio based on real-time PCR efficiency and the crossing point deviation of an unknown sample versus a control. For example, a new software tool established by Pfaffl et al. (2002b) named REST (relative expression software tool) compares two groups and is based on the PCR efficiencies and the mean crossing point deviation between the sample and the control group. Subsequently, the expression ratio results of the four investigated transcripts are tested for significance by a randomization test.

Typically the selection of genes induced or repressed is often based on a two-fold difference in expression between samples. However, when statistical analysis using a distribution analysis is used, the level of difference in expression for genes determined to be statistically different from the control can be as low as 12%. For the subsequent validation of gene expression changes by PCR, it is seldom appreciated that small differences can be seen by random chance.

Therefore, the statistical power to detect a specific difference needs to be considered prior to validation analyses. For simple statistical pairwise group mean comparisons, depending on the variability of the real-time PCR analysis, triplicate samples may only be a sufficient sample size to detect a two-fold difference in expression (Table 7-4). If a 50% difference in expression is being

TABLE 7-4   Replicates required for 90% power to detect a significant ($P < 0.05$) expression difference by real-time PCR by pairwise group mean comparison

Mean ΔC_T	Mean Fold Change[a]	Standard Deviation of C_T Across Replicates					
		0.1	0.2	0.3	0.4	0.5	0.6
0.4	1.32	3	5	12	21	33	47
0.6	1.52	3	3	5	9	15	21
0.8	1.74	3	3	3	5	8	12
1	2.00	3	3	3	3	5	8
1.2	2.30	3	3	3	3	4	5
1.4	2.64	3	3	3	3	3	4

[a] Assuming 100% efficiency of amplification, where fold change $= 2^{\Delta C_T}$, where C_T is the difference in mean C_T Values between groups.

validated, as is often seen in a microarray experiment, as many as five or more replicates may be needed. Another approach for the statistical validation of very small changes is to use more rigorous statistical approaches that examine amplification cycle–dependent fluorescence data at multiple cycles, rather than simply the C_T values.

APPLICATIONS IN THE ANALYSIS OF GENE EXPRESSION

Validation of Microarray Analysis

The impact of microarray analysis is enormous and has definitive advantages in the ability to evaluate the expression of thousands of genes simultaneously. Limitations of microarray analysis, however, include potential incorrect gene annotation, data are often only semiquantitative, quenching of signals can result in loss of sensitivity to detect changes, and intra- and intermicroarray variations can markedly skew the interpretation of expression data. Difficulties arise in comparing results across different platforms, especially since expression measurements made across microarray technologies are not directly comparable (Kuo et al., 2002).

Gene annotation problems can occur during the process of probe development and spotting, where the potential for error in a specific gene that is plated on the microchip is inherent. To overcome this weakness, you can sequence verify every cDNA probe on the chip and/or validate differences of gene expression by using an alternate method of gene expression detection. There are several different methods to measure gene expression, and each technique has advantages and disadvantages. Northern analysis can provide useful information on mRNA size, alternate splicing, and the integrity of RNA sample. RPA assays map the transcript initiation and termination sites as well as the exon/intron boundaries, and is very sensitive in discriminating between related mRNAs of similar size. In situ hybridizations allow the specific cellular localization of a transcript within a tissue. Real-time PCR is by far the most sensitive and flexible of RNA quantitation methods. Unlike other gene expression analysis systems, it is quick and fairly simple.

In some cases the insensitivity of microarray analysis is a downfall that can easily be overcome using real-time PCR. We previously reported an increasing dose-dependent induction in the HPL1A cell line by using real-time PCR analysis. When these cells were treated with 0.1 nM TCDD, we found a two-fold induction of cytochrome P450 1A1 (a TCDD hallmark of biochemical alteration) that was not detected by microarray analysis (Martinez et al., 2002). This may be due to a quenching of fluorgenic signal or a cutoff of signal intensity values.

Predictive Toxicology and Screening

The progression of disease from exposure to a toxicant is widely known by the paradigm put forth by the National Academy of Science (NAS, 1983): *Exposure* → Internal dose → Biological effective dose → Early biological *effect* → Altered structure/function → Clinical disease → *Prognosis*. Currently, an early biological effect can be linked to the prognosis of disease. By linking exposure to early biological effect, the effect/prognosis link could be eliminated, thus enabling early prevention or treatment of a disease by defining susceptibility. Identification of genes relevant to biological effects leading to altered structure or function of a tissue or organ as a result of exposure to chemicals is a powerful tool currently being employed by major governmental research institutions and pharmaceutical companies worldwide.

Toxicogenomics is a field of science that examines many different genes that determine toxicant behavior. Predictive toxicology applies the knowledge gained from databases of gene expression profiles that link an early biological effect(s) to a disease as a result of exposure to a toxicant. Different classes of chemicals with similar biological responses that work through a common mechanism of action can provide early identifiable markings of toxicity (Hamadeh et al., 2002a; Waring et al., 2001b).

The capability to classify an unknown toxicant in rats based on in vivo toxicant altered gene expression profiles has already been effectively demonstrated by Hamadeh and coworkers (2002a). These studies and others (Hamadeh et al., 2002b; Thomas et al., 2001; Waring et al., 2001a) demonstrate that microarray analyses can be used to discriminate an optimal number of genes induced by a toxicant, which typically ranges from 12 to 50 genes. The coupling of predictive toxicity microarray analyses with high-throughput, sensitive, and quantitative real-time PCR is an obvious toxicogenomic strategy in hazard identification. Real-time PCR of defined sets of toxicant signature gene sets offers a suitable approach for toxicogenomic analysis across multiple samples, times, and doses that may be too costly for current microarray-based approaches.

SUMMARY

The impact of microarrays on the discovery of toxicologically relevant genes has helped to bring forth a new era of science: toxicogenomics. It can be used to understand toxic mechanisms of action by identifying signaling pathways and networks regulated by chemicals. Real-time PCR is a powerful tool for validating genes detected by oligonucleotide arrays. Given its high-throughput capacity where a large number of samples can be processed and analyzed in a timely, specific, and quantitative manner, toxicological hypotheses derived from microarray data can quickly be investigated.

REFERENCES

Bustin SA (2000). Absolute quantification of mRNA using real-time reverse transcription polymerase chain reaction assays. *J Mol Endocrinol* **25**:169–193.

Chandler DP, Wagnon CA, Bolton H Jr (1998). Reverse transcriptase (RT) inhibition of PCR at low concentrations of template and its implications for quantitative RT-PCR. *Appl Environ Microbiol* **64**:669–677.

Fronhoffs S, Totzke G, Stier S, Wernert N, Rothe M, Bruning T, Koch B, Sachinidis A, Vetter H, Ko Y (2002). A method for the rapid construction of cRNA standard curves in quantitative real-time reverse transcription polymerase chain reaction. *Mol Cell Probes* **16**:99–110.

Hamadeh HK, Bushel PR, Jayadev S, DiSorbo O, Bennett L, Li L, Tennant R, Stoll R, Barrett JC, Paules RS, Blanchard K, Afshari CA (2002a). Prediction of compound signature using high density gene expression profiling. *Toxicol Sci* **67**:232–240.

Hamadeh HK, Bushel PR, Jayadev S, Martin K, DiSorbo O, Sieber S, Bennett L, Tennant R, Stoll R, Barrett JC, Blanchard K, Paules RS, Afshari CA (2002b). Gene expression analysis reveals chemical-specific profiles. *Toxicol Sci* **67**:219–231.

Kuo WP, Jenssen TK, Butte AJ, Ohno-Machado L, Kohane IS (2002). Analysis of matched mRNA measurements from two different microarray technologies. *Bioinformatics* **18**:405–412.

Liss B (2002). Improved quantitative real-time RT-PCR for expression profiling of individual cells. *Nucleic Acids Res* **30**:e89.

Livak KJ, Schmittgen TD (2001). Analysis of relative gene expression data using real-time quantitative PCR and the 2(-delta delta C(T)) method. *Methods* **25**:402–408.

Martinez JM, Afshari CA, Bushel PR, Masuda A, Takahashi T, Walker NJ (2002). Differential toxicogenomic responses to 2,3,7,8-tetrachlorodibenzo-p-dioxin in malignant and nonmalignant human airway epithelial cells. *Toxicol Sci* **69**:409–423.

Mullis KB (1990). The unusual origin of the polymerase chain reaction. *Sci Am* **262**: 56–61, 64–55.

NAS (1983). *Risk Assessment in the Federal Government: Managing the Process,* Washington, DC.

Pfaffl MW (2001). A new mathematical model for relative quantification in real-time RT-PCR. *Nucleic Acids Res* **29**:2002–2007.

Pfaffl MW, Georgieva TM, Georgiev IP, Ontsouka E, Hageleit M, Blum JW (2002a). Real-time RT-PCR quantification of insulin-like growth factor (IGF)-1, IGF-1 receptor, IGF-2, IGF-2 receptor, insulin receptor, growth hormone receptor, IGF-binding proteins 1, 2 and 3 in the bovine species. *Domest Anim Endocrinol* **22**: 91–102.

Pfaffl MW, Horgan GW, Dempfle L (2002b). Relative expression software tool (REST) for group-wise comparison and statistical analysis of relative expression results in real-time PCR. *Nucleic Acids Res* **30**:e36.

PRISM (2001). *Applied Biosystems 7700 Sequence Detection System User Bulletin No. 2.*

Raeymaekers L (1999). General principles of quantitative PCR. In: U Reischl, B Kochanowski (eds), *Quantitative PCR Protocols.* Totowa, NJ: Humana Press, pp. 31–41.

Rappolee DA, Mark D, Banda MJ, Werb Z (1988). Wound macrophages express TGF-alpha and other growth factors in vivo: analysis by mRNA phenotyping. *Science* **241**: 708–712.

Thomas RS, Rank DR, Penn SG, Zastrow GM, Hayes KR, Pande K, Glover E, Silander T, Craven MW, Reddy JK, Jovanovich SB, Bradfield CA (2001). Identification of toxicologically predictive gene sets using cDNA microarrays. *Mol Pharmacol* **60**:1189–1194.

Vandesompele J, De Preter K, Pattyn F, Poppe B, Van Roy N, De Paepe A, Speleman F (2002). Accurate normalization of real-time quantitative RT-PCR data by geometric averaging of multiple internal control genes. *Genome Biol* **3**: RESEARCH0034.

Walker NJ (2001). Real-time and quantitative PCR: applications to mechanism-based toxicology. *J Biochem Mol Toxicol* **15**:121–127.

Walker NJ (2002). Tech.Sight.Featuring Real-time PCR. A technique whose time has come. *Science* **296**:557–559.

Walker NJ, Portier CJ, Lax SF, Crofts FG, Li Y, Lucier GW, Sutter TR (1999). Characterization of the dose-response of CYP1B1, CYP1A1, and CYP1A2 in the liver of female Sprague-Dawley rats following chronic exposure to 2,3,7,8-tetra-chlorodibenzo-p-dioxin. *Toxicol Appl Pharmacol* **154**:279–286.

Waring JF, Ciurlionis R, Jolly RA, Heindel M, Ulrich RG (2001a). Microarray analysis of hepatotoxins in vitro reveals a correlation between gene expression profiles and mechanisms of toxicity. *Toxicol Lett* **120**:359–368.

Waring JF, Jolly RA, Ciurlionis R, Lum PY, Praestgaard JT, Morfitt DC, Buratto B, Roberts C, Schadt E, Ulrich RG (2001b). Clustering of hepatotoxins based on mechanism of toxicity using gene expression profiles. *Toxicol Appl Pharmacol* **175**:28–42.

Wittliff JL, Erlander MG (2002). Laser capture microdissection and its applications in genomics and proteomics. *Methods Enzymol* **356**:12–25.

8

Confounding Variables and Data Interpretation

Kevin T. Morgan, Gary A. Boorman, H. Roger Brown, Zaid Jayyosi, and Lynn M. Crosby

INTRODUCTION

Toxicogenomics, a relatively new field of investigation (Castle et al., 2002), has been defined as the application of functional genomic technologies to toxicology and clinical risk assessment (Quinn-Senger et al., 2002). Toxicogenomics is currently focused on the application of large-scale differential gene expression technologies to toxicology. These rapidly evolving technologies permit the simultaneous determination of expression levels of thousands of messenger ribonucleic acid (mRNA) transcripts in cells or tissues. The entire population of such transcripts in a cell population at a moment in time is known as the transcriptome (see below). This chapter provides a brief introduction to toxicogenomics with a focus on important confounding variables and data interpretation.

Recognition of the potential power of toxicogenomics has resulted in a number of legal (Marchant, 2002), ethical (Olden and Guthrie, 2001), business (Rastan, 2001), and regulatory (Petricoin III et al., 2002) concerns. In addition to providing considerable promise for progress in health safety assessment (Castle et al., 2002; Golub et al., 1999), toxicogenomics presents a significant challenge to toxicologists (Smith, 2001). It is also important to realize that

Toxicogenomics: Principles and Applications. Edited by Hamadeh and Afshari
ISBN 0-471-43417-5 Copyright © 2004 John Wiley & Sons, Inc.

responses by the transcriptome represent only one component of an unfolding biological picture. The ability to simultaneously measure thousands of proteins (proteomics; Chevalier et al., 2000; Fountoulakis et al., 2002; Idekar et al., 2001; Ruepp et al., 2002) and metabolites (metabonomics; Oliver et al., 2002) is also providing contributions to toxicology. Parallel progress is being made in the understanding of cell-wide protein phosphorylation and glycosylation states, DNA methylation patterns, RNA expression control circuitry (Mathews et al., 2000), and interactions between these systems (Walhout et al., 2002). The scope of toxicogenomics will grow as these fields merge, but here we shall confine our attention to the transcriptome. Before attempting to understand the behavior of the transcriptome, it is necessary to learn about the nature and functions of mRNA, which entails some enjoyable reading. A small selection of citations are provided as a starting point (Aardema and MacGregor, 2002; Alberts et al., 2001; Boorman et al., 2002a; Castle et al., 2002; Farr and Dunn, 1999; Ho, 1993; Murray et al., 2000; Quinn-Senger et al., 2002).

WHAT IS A TRANSCRIPTOME?

In a simple sense, a transcriptome is considered to be the entire population of mRNA transcripts in a cell or cell population at a given moment in time. This definition begs two questions.

1. How many transcripts are there?
2. How long a moment in time is significant with respect to mRNA dynamics?

Until recently it was believed that there are about 100,000 genes in the human genome, coding for about 100,000 proteins, in addition to multiple protein products from certain genes due to recombination events, as occurs in the immune system (Abbas et al., 2000). Recent sequencing of the human genome led to the conclusion that this number should be revised downward in humans, toward 30,000 genes. Recent studies of splice variance, which can generate families of proteins from a single transcript, suggested that the number of proteins coded for by these genes may be 100,000 or more (Roberts and Smith, 2002). While the actual total remains to be determined, the behavior and interactions of these transcripts is probably a more important issue.

The transcriptome is highly dynamic, with many mRNA species having relatively short half-lives, often less than 20 minutes, while interacting with an array of regulatory proteins that influence their longevity and translational activity (Mathews et al., 2000). Knowledge of these dynamics is critical when selecting time points for toxicogenomics studies (Morgan et al., 2003). The latter point is demonstrated by gene expression responses in HepG2 cells following routine media replacement (Fig. 8-1). An effective mental picture of the transcriptome should also include consideration of interactions with other

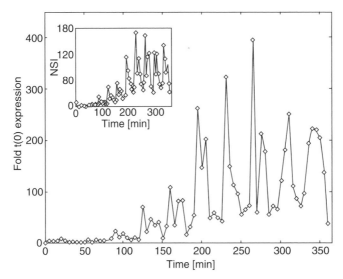

Figure 8-1 *Time course results for HepG2 cells through 6 hr following nutrient replacement at t_0 using standard cell culture conditions and media containing fetal bovine serum. CYP1A1 (Genbank Acc. # K03191) expression measured using mean of duplicates, real time reverse transcriptase polymerase chain reaction (TaqMan™), and the data normalized against 18s ribosomal RNA using the manufacturer's recommended procedures. There are steadily increasing peaks of mRNA signal, with evidence of a progressive oscillatory change superimposed upon a slowly increasing baseline. Inset: The same RNA samples examined using Clontech Nylon Gene Expression Array™ to show concordance of the two technologies for this transcript. NSI = normalized signal intensity. (Source: Published with permission from Morgan et al., 2003.)*

cellular components and functions, such as the equally dynamic proteome (Kettman et al., 2001), signal transduction pathways, and hormonal control circuitry. In fact, it is best to think of such systems as interacting networks (Cornish-Bowden and Cardenas, 2000). Your ability to effectively apply studies of the transcriptome to toxicology will be strictly limited by the quality of your mental picture of these cellular components and processes.

APPLYING TRANSCRIPTOMICS TO TOXICOLOGY

Interestingly, much of what we know about the transcriptome has been derived from a small number of divergent organisms, including watercress, yeast, soil nematode, fruit fly, and mouse (Alberts et al., 2001). Extrapolating this understanding to humans through the generation of meaningful interpretations for human disease risk assessment is one of the key challenges of toxicology today (Smith, 2001). Since the 1950s, molecular biologists have been studying small numbers of mRNA transcripts and correlating changes in their expression levels with other end points, from specific biochemical pathways to cancer.

Only with the recent advent of large-scale gene expression technology has a significant proportion of the transcriptome become accessible as an end point for toxicology.

When considering the incorporation of toxicogenomics into toxicology research, it is recommended that you begin by examining transcriptional data from normal tissues before moving on to the effects of toxins or infectious agents. Each tissue has its own transcriptional profile, which can be related to the normal functions of that tissue. For instance, compare the 50 most abundant gene expression tags reported in mice for the liver (Kurachi et al., 2002) versus the heart (Anisimov et al., 2002). In the former case, the role of the liver in the production of albumin, lipoproteins, complement, and clotting factors, combined with its functions in iron storage, processing of vitamin A and xenobiotics, and its critical activities in relation to whole-body energy metabolism, were reflected in the list of most highly expressed transcripts. In contrast, the most highly expressed transcripts in the heart reflected a dominant need for energy along with maintenance of cardiac muscle structure. Thus, organ-specific functional priorities are apparent in the structure of the transcriptome. Before attempting to apply toxicogenomics to the study of pathophysiology, we strongly recommend that you familiarize yourself with (a) the normal physiology and function of the organ of interest, and (b) the structure, and where possible the dynamics, of the normal transcriptome of the organ.

The literature provides an increasingly rich source of toxicogenomics materials (Burczynski et al., 2000; Corton and Stauber, 2000; Crosby et al., 2000; Dam et al., 2003; Fountoulakis et al., 2002; Golub et al., 1999; Morgan et al., 2002b; Reynolds and Richards, 2001; Russell et al., 2002; Tsangaris et al., 2002), and many other articles can be found via the Internet. Useful search terms include *toxicogenomics*, *bioinformatics*, *genomics*, *DNA chips*, and *cDNA microarrays*. Many institutions, such as the National Center for Biotechnology Information (NCBI), have excellent Web sites that will lead you to a wealth of toxicogenomics publications and data, analytical resources, and skilled people. An acquaintance with bioinformatics (Baxevanis and Ouellette, 2001) and the support from those trained in this field are essential for an effective toxicogenomics program. A little time spent downloading gene expression data via the Internet such as that of Iyer et al. (1999), combined with their investigation using readily available tools such as Eisen clustering (Eisen et al., 1998), will take you a long way toward developing the confidence needed to explore responses of the transcriptome to toxic chemicals.

Fragments of two toxicogenomics data sets, one from an in vivo study (rat kidney) and one from an in vitro study (primary rat hepatocytes), are provided in Tables 8-1 and 8-2, respectively. In each case the genes have been ranked by fold change, and thus the tables contain the most highly changed expression levels. The first data set is derived from a large gene expression array that detects more than 8,000 transcripts (Affymetrix Chip Technology), with both up- and downregulated transcripts shown. The second data set is from a

TABLE 8-1 Gene expression changes in the outer stripe of the outer medulla of the kidney associated with aging of rats from a range of 5–6 weeks to a range of 7–8 weeks[a]

Ratio 8 wk/6 wk	Gene
4.4816891	SP:SYUA_RAT-alpha-synuclein. P37377 rat. 10/2001-AF007758_G_AT
3.0956565	GB:AF035951-AF035951 Rattus norvegicus kinesin-related protein KRP1 (KRP1) mRNA partial cd-AF035951_AT
2.829217	AFFX_GB:AA892762-EST196565 R.norv. cDNA 3 end-RC_AA892762_AT
2.6247877	SP:HYES_RAT-soluble epoxide hydrolase (seh) (ec 3.3.2.3) (epoxide hydratase) (cytosolic epo-X65083CDS_AT)
2.5753875	GB:AF416730-AF416730 Rattus norvegicus CORO1A protein mRNA complete cds. 10/2001-RC_AA892506_AT
2.4744051	UG:Rn.36106-Maxp1 protein interacting with guanine nucleotide exchange factor-AF002251_AT
2.316367	SP:ROH2_RAT-retinol dehydrogenase type ii (ec 1.1.1.105) (rodh ii). P50170 rat. 5/2000-U33500_G_AT
2.300209	AFFX_GB:AA893273-EST197076 R.norv. cDNA 3 end-RC_AA893273_AT
2.29791	SP:VTDB_RAT-witamin d-binding protein precursor (dbp) (group-specific component) (gc-globul-M12450_AT)
2.2910266	AFFX_GB:M36151-Rat MHC class II A-beta RT1.B-b-beta gene partial cds. -M36151CDS_I_AT
2.1705921	SP:OLF5_RAT-olfactory receptor-like protein f5. P23266 rat. 7/1993-X89701CDS_F_AT
2.0959355	AFFX_GB:AA891695-EST195498 R.norv. cDNA 3 end-RC_AA891695_I_AT
2.0017064	AFFX_GB:U16359-R.norv. nitric oxide synthase gene complete cds.-U16359CDS_AT
1.9289967	AFFX_GB:X99338-R.norv. mRNA for glycoprotein 65.-X99338CDS_I_AT
1.905987	UG:Rn. 16629-Col1a2 procollagen type I alpha 2-RC_AA891828_G_AT
1.8813696	SP:PDK4_RAT-[pyruvate dehydrogenase [lipoamide]] kinase isozyme 4 mitochondrial precursor-AF034577_AT
1.7789085	UG:Rn.3979-Bcdo beta-carotene 15 15′-dioxygenase-RC_AIO14135_G_AT
1.6686251	AFFX_GB:AF023087-R.norv. nerve growth factor induced factor A mRNA partial 3 UTR.-AF023087_S_AT
1.652022	SP:SYD1_RAT-fractalkine precursor (neurotactin) (cx3c membrane-anchored chemokine) (small -RC_AA800602_S_AT)
1.6503708	UG:Rn.76652-Pcmt1 Protein-L-isoaspartate (D-aspartate) O-methyltransferase-JO1435CDS#8_S_AT
1.6470734	UG:Rn.9096-Egr1 Early growth response 1-M18416_AT
1.6258	SP:SM30_RAT-senescence marker protein-30 (smp-30) (regucalcin) (rc). Q03336 rat. 12/1998-D31662EXON#4_S_AT
1.5967974	GB:RATLY6A-M30692 Rat Ly6-A antigen gene exon 2. 4/1993-RC_AA891695_F_AT
1.5920142	GB:AF065438-AF065438 Rattus norvegicus mama mRNA complete cds. 5/1998-AF065438_AT
1.5761734	UG:Rn.9792-Tnfrsf11b tumor necrosis factor receptor superfamily member 11b (osteoprotegeri-U94330_AT)
1.5683122	SP:FSA_RAT-follistain precursor (fs) (activin-binding protein). P21674 rat. 10/2001-RC_AA858520_AT
1.5527072	AFFX-GB:RC_AI639504_AT-Rat mixed-tissue library R.norv. cDNA clone rx04791 3 mRNA sequence [R.norv.]-RC_AI639504_AT
1.5465088	AFFX-GB:X60212-R.norv. ASI mRNA for mammalian equivalent of bacterial large ribosomal subunit p-X60212_I_AT

TABLE 8-1 *Continued*

Ratio 8 wk/6 wk	Gene
1.535721	SP:LYC1_RAT_lysozyme c type 1 precursor (ec 3.2.1.17) (1 4-beta-n-acetylmuramidase c). P-RC_AA892775_AT
-1.5234843	GB:S83269-S83269 HSP70.2=heat shock protein 70 {3′ region 3.05 kb transcript} [rats PC1-Z75029_S_AT]
-1.5589305	SP:ALDR_RAT-aldose reductase (ec 1.1.1.21) (ar) (aldehyde reductase). P07943 rat. 5/2000-M60322_G_AT
-1.5761734	GB:AB019693-AB019693 Rattus norvegicus HP33 mRNA complete cds. 11/1998-RC_AA893035_S_AT
-1.590423	SP:ECHP_RAT-peroxisomal bifunctional enzyme (pbe) (pbfe) [includes: enoyl-coa hydratase (ec-K03249_AT)]
-1.6323162	SP:S6A6_RAT-sodium- and chloride-dependent taurine transporter. P31643 rat. 10/2001-M96601_AT
-1.6803465	SP:UCP1_RAT-mitochondrial brown fat uncoupling protein 1 (ucp 1) (thermogenin). P04633 rat.-X03894_AT
-1.6955378	SP:HMCM-RAT-hydroxymethylglutaryl-coa synthase mitochondrial precursor (ec 4.1.3.5) (hmg-c-M33648_G_AT)
-1.7177237	SP:CPE1_RAT-cytochrome p450 2e1 (ec 1.14.14.1) (cypiie1) (p450-j) (p450rlm6). P05182 rat. 7-S48325_S_AT
-1.7985846	AFFX_GB:X65036-R.norv. mRNA for H36-alpha7 integrin alpha chain-X65036_AT
-1.8057933	GB:AF063102-AF063102 Rattus norvegicus calcium-independent alpha-latrotoxin receptor homolo-AF063102_G_AT
-1.8112188	UG:Rn.3142-Cd36l1 CD36 antigen (collagen type I receptor thrombospondin receptor)-like 1 (-RC_AA874843_S_AT)
-1.8148449	SP:PIX3_RAT-pituitary homeobox 3 (homeobox protein ptx3). P81062 rat. 10/2001-AJ011005_AT
-1.870115	GB:RNRRNA-V01270 Rattus norvegicus genes for 18S 5.8S and 28S ribosomal RNAs. 3/2001-Rc_AA893870_AT
-2.0792349	AFFX_GB:AA874803-UI-R-EO-bw-g-08-0-UI.s1 R.norv. cDNA 3 end-RC_AA874803_G_AT
-2.1619271	UG:Rn.11372-Lox Lysyl oxidase-S66184_S_AT
-2.1902156	SP:ATF3_RAT-cyclic-amp-dependent transcription factor atf-3 (activating transcription fact-m63282_AT)
-2.2344609	SP:LOX2_RAT-arachidonate 12-lipoxygenase (ec 1.13.11.31) (12-lox). Q02759 rat. 2/1996-S69383_AT
-2.3490242	UG:Rn.1143-Kap Kidney androgen-regulated protein-U25808_AT
-2.4473358	AFFX_GB:S56464-HKII=hexokinase II [rats epididymal fat pad mRNA Partial 1456 nt segment 2 o-S56564MRNA_AT
-2.5779642	GB:RATRGB-K01592 rat 8s rna. 4/1993-RC_AI176460_S_AT
-2.6591327	AFFX_GB:AA859966-UI-R-E0-ca-g-03-0-UI.s1 R.norv. cDNA 3 end-RC_AA859966_I_AT
-2.745601	GB:MMETSB2-X56974 M.musculus mRNA for external transcribed spacer (partial) B2 element (Rn-RC_AA859372_S_AT)

[a] These data were the most statistically significantly dysregulated transcripts in a set of over 8,000 genes in the outer stripe of the outer medulla of the kidney, showing effects of age between these two control groups. The ratio of gene expression signals between 6 and 8 weeks of age is shown as up or down (negative sign) in the left column with the gene name in the right-hand column. Try your hand at interpreting the significance of these changes.

TABLE 8-2 One page of a gene expression report with changes of importance to fatty acid metabolism[a]

414	2.25	3.17	2	9.09E-03	C02h: long chain acyl-CoA synthetase 2 (LACS2); liver long chain fatty acid-CoA
415	-0.159	5.22	2	7.44E-05	C02i: aldolase C
416	0.438	2.83	2	2.55E-02	C02j: testis fructose-6-phosphate 2-kinase/fructose 2,6-biphosphate (testis 6PF)
422	1.98	2.5	2	3.82E-02	C03b: medium chain acyl-CoA dehydrogenase precursor (MCAD; ACADMA)
425	2.64	3.44	2	5.39E-03	C03e: mitochondrial hydroxymethylglutaryl-CoA synthase precursor (HMG-CoA synth)
432	0.555	2.56	2	4.20E-02	C03l: cytochrome c oxidase, subunit VIIh
438	-0.302	3.98	1	1.10E-03	C04e: cytosolic acyl-CoA thioester hydrolase (ACT); long chain acyl-CoA hydrola
441	0.645	3.46	2	6.78E-02	C04g: creatine kinase, ubiquitous, mitochondrial
445	0.22	6.98	2	2.30E-05	C04k: hormone sensitive lipase (EC 3.1.1.-; HSL)
446	-0.693	2.82	2	2.67E-03	C04l: triacylglycerol lipase precursor (hepatic)
447	-0.693	2.82	2	2.67E-02	C04m: triacylglycerol lipase precursor (pancreatic)
453	-1.23	2.83	1	7.59E-03	C05e: mitochondrial muscle carnitine O-palmitoyltransferase I(CTPI-M); carnitin
461	-0.741	3.01	2	1.81E-02	C05m: cytochrome P450 XIA1 mitochondrial precursor (CYP11A1); P450scc; choleste
464	3	2.82	2	2.08E-02	C06b: lecithin: cholesterol acyltransferase (EC 2.3.1.43; LCAT); phosphatidylcho
465	1.97	3.15	2	9.51E-03	C06c: fatty acid amide hydrolase
466	2.03	4.44	2	7.82E-04	C06d: 3-beta hydroxy-5-ene steroid dehydrogenase type III (3beta-HSD III; EC 1)
469	-0.974	4.45	1	2.07E-04	C06g: phosphatidate phosphohydrolase type 2
471	0.519	13.6	1	2.45E-08	C06i: annexin III (ANX3); lipocortin 3; placental anticoagulant protein III (PA)
476	-8.96E-02	2.81	1	1.47E-02	C06n: ytochrome P-450 19; aromatase
477	1.76	4.77	2	3.60E-04	C07a: 3-beta-hydroxysteroid dehydrogenase/delta-5→-4 isomerase, type 1; 3BETA
478	2.88	7.44	2	7.89E-06	C07b: cytochrome P450 2C11 (CYP2C11); P450(M-1); P450H; P450-UT-A; UT2
482	2	7.22	2	8.85E-06	C07f: cytochrome P450 4A3 (CYP4A3); lauric acid omega-hydroxylase; P450-LA-omeg
486	0.385	2.85	2	2.46E-02	C07j: cytochrome P-450 4F4
490	1.58	4.19	2	9.25E-04	C07n: 2-arylpropionyl-CoA epimerase; alpha-methylacyl-CoA racemase
493	2.79	2.74	2	2.34E-02	C08c: cytochrome P-450 2C23, arachidonic acid epoxygenase
500	1.77	4.63	2	4.66E-04	C08j: thymidylate (TYMS; TS)
501	3.41	6.33	2	4.93E-05	C08k: cytosolic thymidine kinase (TK1)
504	0.794	4.85	2	4.81E-04	C08n: DOPA decarboxylase (DDC); armoatic-L-amino-acid decarboxylase (EC 4.1.1.2)
505	2.26	2.82	2	1.92E-02	C09a: liver arginase 1 (ARG1)
508	1.99	3.81	1	2.51E-03	C09d: glutamate-cysteine ligase regulatory subunit (GLCLR); gamma-glutamylcyste
509	1.42	3.43	2	4.39E-03	C09e: glutathione synthetase (GSH synthetase; GSH-S; GSS); glutathione synthase
511	1.54	3.86	2	1.78E-03	C09g: dopa-tyrosine sulfotransferase

[a] Column 3 (from left) represents the fold change, column 4 is the upregulated group, and column 5 is the associated *p*-value. Output for NLR software (Kepler et al., 2001).

smaller nylon array, with 1,200 transcripts (Clontech Nylon cDNA Technology), and the data are confined to upregulated genes. In spite of marked differences in the cell system, expression platform, statistical approaches, and normalization procedures, you will see a number of common features between these data sets:

- The fold change overall is in the range of 1 to 6 times, though this is not always the case.
- Both data sets show clear indications of changes in energy metabolism; to confirm this conclusion, examine the functions of some of these genes in *Harper's Biochemistry* (Murray et al., 2000).
- Many genes, especially those in Table 8-1, are probably somewhat unfamiliar.

It is not unusual to have as many as 1,000 significantly altered genes when using very large-scale expression array platforms. With practice and a complete data set, you will be able to draw many conclusions (and hypotheses) from such data if your experiments are well designed. Optimization of the design of toxicogenomics studies is still under investigation. One critical issue to consider is whether to employ in vivo or in vitro approaches.

EXPERIMENTAL DESIGN: IN VIVO VERSUS
IN VITRO TRANSCRIPTOMICS

Despite the power of large-scale gene expression array technology, the basic rules of good science in the selection of appropriate experimental models (Ballatori and Villalobos, 2002) still apply. Toxicogenomics studies must be constructed to address the potential impact of such variables as circadian rhythm and feeding schedules (Kita et al., 2002), and meet evolving standards of study design and record keeping (Brazma et al., 2001). Toxicogenomics has been successfully applied using both in vitro and in vivo mammalian models, while bacteria, yeast, nematodes, plants, and an array of bacteria are yielding a wealth of large-scale transcriptional information (Alberts et al., 2001), much of which is directly relevant to toxicology. When working with in vitro systems, it is essential to remain cognizant of the fact that cells in culture are bereft of many signals that they normally receive from other tissues, such as hormones, nutrient changes, and neural input. In contrast, while benefiting from their intact nature, in vivo studies have the drawback that the RNA under investigation is derived from a mixture of a wide range of cell types (Figs. 8-2 and 8-3). Furthermore, the population of transcripts will be influenced by immigrations of inflammatory cells or changes in blood perfusion altering levels of blood-borne transcripts in the sample.

Cells respond to changes in their local environments by protein modifications and gene transcription, which upon translation affect lipid, phospholipid,

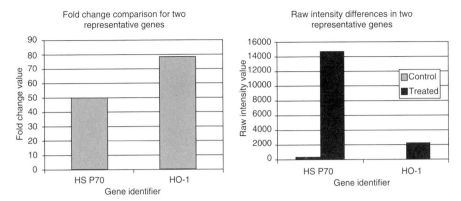

Figure 8-2 Fold change (top left) versus absolute signal intensity (top right): note that HO-1 has a greater fold change than HSP-70 but is not more highly expressed in terms of raw signal intensity. (Source: From Crosby et al., 2000, with permission.)

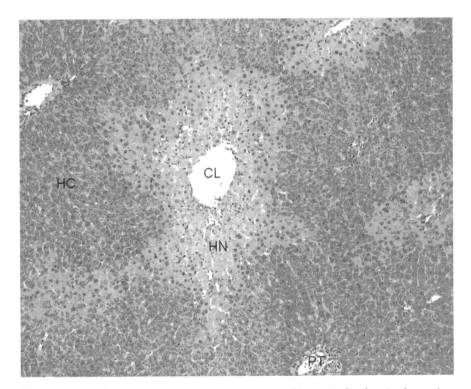

Figure 8-3 Light micrograph of the liver of a rat exposed to acetaminophen to show exten-sive areas of hepatocellular necrosis around a centrilobular vein (CL), while hepatocytes in the midzonal region (HC) and adjacent to a portal triad (PT) are apparently unaffected. When you take a slice of such a liver for gene expression studies, all regions are mixed into a homoge-neous sample, thus masking the regional nature of the responses.

protein, and carbohydrate constituents and the internal energetics and ionic milieu (Alberts et al., 2001). The architecture of a typical vertebrate necessitates that colonies of related or interacting cells maintain homeostasis of their local environments and interact cooperatively with other colonies (organs) via shared nutrient pools, external chemical messengers (hormones and cytokines), and a network of neural communication (nervous system) and protective systems (integument and immune systems).

Individual organs can modify their environments only to a limited extent. The results of a catastrophic change in the internal organ environment can be death of the organ, and, depending on the function of the organ, death of the organism. The survival strategies of cells in culture may vary radically from those they employed in their organ of origin. In organs, cells closest to the nutrient blood supply and highest level of oxygen saturation may be at an advantage or disadvantage, depending on the nature of the toxicant and the demands of the tissue. This has been well described with respect to hepatocytes in different regions of the liver lobule (Vidal-Vanaclocha, 1997).

The organ colony is in a dynamic state of flux with environmental modifications by distant colonies to adapt blood flow, blood pressure, nutrient supply, osmolality, osmolarity, pH, oxygen availability, hormonal, neuroendocrine, and neural inputs to fit the current demands. There may also be a hierarchy for cell death in organs, preservation of precursor cells, cells with critical and specialized functions, and vascular supply, all being essential for survival. In cell cultures, the conditions of a single petri dish or flask are also in flux, but the environment is much more limited, the cell type uniform, the nutrients more stable (generally in excess), the hormonal status defined by the researcher (or not), and the availability of oxygen nearly always a limiting factor. Cells in culture, even primary cell cultures, are also in extreme circumstances, either of unparalleled cell division or of progressive and often advanced senescence. Neither of these conditions may reflect the conditions of their organ of origin.

Transcriptomics of organs is by necessity based on the average message of all of the cells of the colony. Furthermore, in typical toxicology studies animals are dosed until limiting toxicities are reached, and these toxicities typically are manifested by morphologic alterations of target tissues. These changes are not static as cell populations evolve over time as the tissue attempts to recover from the insult. It is probable that there are multiple critical periods in gene transcription that should be monitored in any study where a mechanistic interpretation is desired, and each critical period will be compound and dose specific and may be species specific. The initial environmental modification that causes adaptive changes within the colony is probably the most telling.

Monitoring this environment over time for hundreds of parameters would be ideal but is impractical if not impossible in vivo. Therefore, a time course that concentrates on early first-exposure genes at progressively toxic doses is generally the most logical choice, allowing the adaptive genes to guide the investigator to the source. Typically, these time points do not result in altered morphology but may be associated with marked alterations in physiology such as blood pressure, oxygenation, and kidney perfusion, altering the elimination of metabolic by-products and resulting in clear or more subtle responses in the transcriptome for us to find and interpret. In contrast, where overt toxicity is detected at high doses or exposure concentrations, the transcriptome can reveal important underlying physiological changes at lower exposure levels that do not exhibit overt effects using standard toxicology or pathology approaches. The transcriptome is sensitive to many environmental and other variables that need to be considered when undertaking toxicogenomics studies.

THE ROLE OF ANIMAL HUSBANDRY AND PATHOLOGY IN TOXICOGENOMICS

When applying clinical pathology and histopathology to toxicogenomics, it is critical to remember in the former case that you only see what you measure (e.g., liver enzymes), and in the latter case, morphology provides only a brief snapshot of a dynamic process that is unfolding in time. It is dangerous to label findings (e.g., necrosis), then make simplistic correlations with gene expression data by claiming that you now have a pattern for a specific morphologic or pathophysiologic entity. The entire data set should be considered in the context of events that develop over time, then resolve over time.

Layered upon this dynamic aspect of pathology interpretations are a number of sources of variance or "noise." Churchill (2002) mentioned three major sources of such variation in microarray experiments, which he categorized as biological, technical, and measurement errors. By providing a morphologic or phenotypic point of reference, pathologists can often identify or explain the nature of some of the biological variation. Great care needs to be taken, however, to assure that pathologists do not inadvertently generate additional sources of variation by their procedures or diagnostic terminology. Biological variation due to the animal itself needs to be limited through the use of disease-free animals housed under defined conditions, and with an experimental design that accounts for circadian rhythm and other daily variables.

For example, in studies using cDNA arrays (Akhtar et al., 2002) or oligonucleotide arrays (Storch et al., 2002), 8% to 10% of the up to 12,000 genes examined showed robust circadian cycling in the mouse liver. In another microarray study in rats, nearly 600 genes in the liver and kidney were affected both by the time of day and the feeding state of the rats (Kita et al., 2002). Age is another critical variable (see Table 8-1; Figures 8-4, 8-5, and 8-6), as suckling rats, for example, have about 4% to 5% of hepatocytes in S-phase, compared to about 0.5% in adults (Nadal, 2000). For female rodents, it is recommended that the animals be monitored for the reproductive cycle, since many genes are under hormonal influence. In a microarray survey of approximately 10,000 genes for estrogen regulation, the kidney was third only to the uterus and pituitary in the number of estrogen-regulated genes (Jelinsky et al., 2003). An

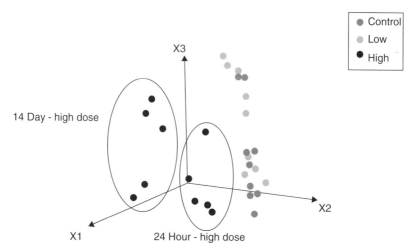

Figure 8-4   *Principal component analysis of gene expression in the outer stripe of the outer medulla of the kidney of rats treated with a toxicant. Male rats were treated with control article and low- and high-dose toxicant for 24 hours and 14 consecutive days (n = 5). Total RNA was isolated from the kidneys, processed, and hybridized onto Affymetrix RG U34A chips. PCA criteria = top 900 (intensity) present genes.*

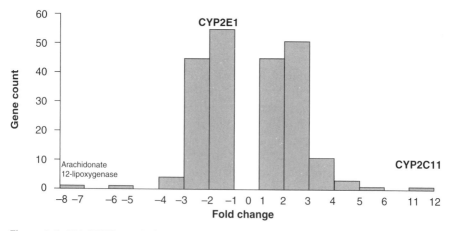

Figure 8-5   *Total RNA was isolated from the outer stripe of the outer medulla of kidneys of 5 control male rats at 5–6 weeks of age and 5 control male rats at 7–8 weeks of age. The RNA was processed and hybridized onto Affymetrix RG U34A Chips™. Statistical analysis was performed by NLR method (Kepler et al., 2001). These data indicate significant transcriptomic differences between rats due to an increase of 2 weeks in age at a stage when they are essentially rapidly maturing "adolescents." Remember that such variables can be critical when designing toxicogenomics studies.*

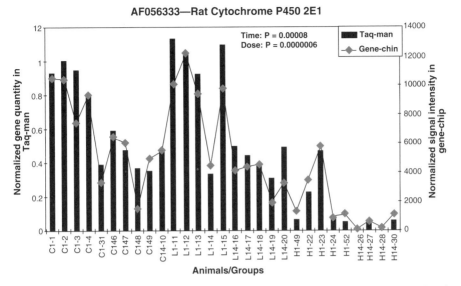

Figure 8-6 *Total RNA isolated from control and treated outer stripe of the outer medulla of the rat kidney hybridized onto Affymetrix RG U34A chips. Total RNA was also evaluated by RT-PCR (TaqMan™) for gene expression of CYP 2E1. Data were normalized for gene expression according to the respective method. Statistical analysis was performed by two-way ANOVA for the RT-PCR data. Note excellent correlation between these different methods, which increases confidence in the changes observed.*

example of a gene list demonstrating differences in expression level solely attributable to a 2-week age difference between the two untreated control groups is shown in Table 8-1.

Unless undertaken with care, the necropsy and collection of tissues can also negatively impact toxicogenomics studies. Often the animals are removed from food prior to necropsy. If this is done during the period when the animals are not eating, it may have little effect, but they should be subject to necropsy promptly, since the feeding state of the animal will affect gene expression (Kita et al., 2002). Furthermore, since immobilization (Lencesova et al., 2002) or anoxia (Seta et al., 2002) may induce gene expression alterations, methods and procedures for euthanasia need to minimize stress. The tissues should be frozen immediately or treated to inactivate RNAases that will rapidly degrade the sample (Evans and Kamdar, 1990; Heller et al., 2000). The portion of tissue collected is also important. For instance, gene expression varies by region of the brain (Meador-Woodruff et al., 1994) or location within a liver lobule (Vidal-Vanaclocha, 1997), and human liver has been shown to exhibit differences between regions for the uptake of pharmaceuticals (Jacobsson et al., 1999) and disease progression (Matsuzaki et al., 1997). Thus, the authors recommend that wherever possible, morphology and gene expression data be derived from the same region of the organ under study.

Morphological and clinical chemistry evaluations provide essential functional reference points for differential gene expression studies (Boorman et al., 2002a). A routine hematoxylin- and eosin-stained section of the tissue being analyzed for gene expression can provide information on normal structure and the nature of toxic responses (Figs. 8-2 and 8-3). Morphology will reveal unexpected "background" lesions, the extent and distribution of treatment-induced responses, the type of processes occurring (hyperplasia, degeneration, necrosis), cell types affected, and whether there are inflammatory infiltrates that can dramatically influence the population of mRNA species present in the samples. These sections will also provide some information on the extent of cell proliferation and whether apoptosis, necrosis, or other degenerative processes are present. Furthermore, the application of special stains (of which there are many thousands) can be used to reveal deposition of lipids, glycogen, or specific proteins, or provide a more quantitative measure of proliferation or apoptosis. A similar morphologic correlate is considered essential for in vitro studies (Crosby et al., 2000; Morgan et al., 2002b). Ultrastructural studies can be used to extend light microscopic investigations, to identify cell types involved, and to indicate specific organelles affected. For example, ultrastructural evaluation has shown that acetaminophen targets the hepatic mitochondria (Ruepp et al., 2002). Gene expression data can also provide clues as to the subcellular compartments involved (Drawid et al., 2000).

Throughout such studies, consistency of pathology is improved by using standard terminology, keeping a dictionary of terms used, and subjecting the study to peer review when appropriate (Boorman et al., 2002b). Grading or scoring schemes for lesion severity must also be consistent if comparisons are made across studies. In addition to the standard pathology diagnosis, it may be useful to provide some quantitative information on the tissue that is used for gene expression studies. For example, if a liver sample contains an inflammatory cell infiltrate or exhibits cellular degeneration, these changes should be carefully defined (cell type, type of degeneration) and at least crudely quantitated (e.g., % tissue affected). This is especially useful if other study samples, such as untreated controls, do not contain these changes. There are ways to address such differences. For instance, in a recently reported study, smooth muscle and connective tissue genes were downregulated in colorectal carcinomas compared to normal colon (Nottermam et al., 2001). Morphology revealed that the normal tissue samples contained smooth muscle and connective tissue while the carcinomas were comprised entirely of epithelial cells. In this case the authors censored the data for genes associated with smooth muscle and connective tissue, giving a more accurate representation of the differential gene expression of the colorectal tumors.

All members of the toxicogenomics team should become thoroughly familiar with the organ or tissue under study. This knowledge is useful during the interpretation of the differential gene expression results. For example, in a recent study of a hepatotoxicant that caused acute necrosis around the terminal hepatic venule, differential gene expression showed marked downregulation of glutamine synthetase in the treated animals. A review of the hepatic literature revealed that glutamine synthetase is expressed in the hepatic terminal plate, the two-hepatocyte-deep layer directly adjacent to the terminal hepatic venule (Gumucio et al., 1994; Watford et al., 2002). Thus, what was seen was not a downregulation in the treated rats but simply a specific loss of cells that express glutamine synthetase as a consequence of localized necrosis.

It is too easy to assume that gene expression patterns reflect the predominant cell type in the tissue under investigation. However, only 60% of the cells in the liver are hepatocytes. Hepatic stellate cells, for instance, have been reported to account for 5–8% of all liver cells, with about 13 stellate cells per 100 hepatocytes in the rat liver (Geertz, 2001; Geertz et al., 1994). Under pathological conditions, these stellate cells can differentiate into myofibroblast-like cells. Increased expression of neural cell adhesion molecules, synaptophysin, neurotrophins, and alpha B-crystallin in a damaged liver may also be related to the presence of hepatic stellate cells (Cassiman et al., 2002). Hepatic stellate cells also express glial fibrillary acidic protein (GFAP) normally discussed in the context of astrocytes in the brain. This might cause you to wonder why this intermediate filament protein is present in the liver (or is the array correctly annotated?). The study of toxicogenomics will slowly erode your fixed ideas as to which proteins occur where, and what they do there. Endothelial cells, Kupffer cells, biliary epithelial cells, and pit cells all have various unique genes and also genes that they share with hepatocytes. The study of toxicogenomics will provide you with the opportunity to rebuild your picture of basic biology!

The kidney may be even more challenging than the liver, as gene expression varies along the renal tubule, a structure that is compartmentally demarcated with respect to function (Brenner, 2000). Compounds that affect the proximal renal tubule will show different gene changes than compounds affecting the glomeruli, distal tubule, or collecting ducts. The anatomy of the kidney suggests that great care needs to be taken to evaluate gene expression for specific cell populations. The inner and outer cortex, medulla, and renal papilla may be difficult to collect consistently because the demarcations are not clear. Homogenization of the whole kidney may mask region-specific gene expression changes, but microdissection is certainly worth a try (Figs. 8-4, 8-5, and 8-6). Laser dissection approaches permit the detection of gene expression changes in specific tissue locations or cell types within a tissue (Sirivatanauksorn et al., 1999), but they are time-consuming and technically challenging.

There needs to be interactive dialogue between pathologists, biochemists, toxicologists, molecular biologists, and other members of the toxicogenomics team. Gene expression alterations can provide clues for pathologists to go back and reexamine tissues. For example, gene expression alterations suggestive of mitochondrial injury at a dose where pathology and clinical chemistry appeared normal may prompt a fruitful ultrastructural examination of the liver. Alternatively, morphological evidence of inflammatory infiltrates may lead to multiple analyses of differential gene expression to include and exclude genes thought to be associated with inflammatory cells. Dysregulation of genes associated with specific aspects of intermediary metabolism may indicate pertinent biochemical studies. For instance, concomitant dysregulation of malic enzyme and phosphoenolpyruvate carboxykinase would lead you to consider changes in energy metabolism and carbohydrate or lipid storage (Murray et al., 2000).

Gene expression data are strengthened when linked to morphological or biochemical data. We have moved from studying one gene at a time to thousands of genes at a time. A recent review suggests that our attention needs to move toward pathways, networks, and eventually the entire cell (Duyk, 2002). It is clearly time to consider pathways and networks in the context of whole animals. For example, liver disease causes dysfunction in the kidney (Epstein, 1994), and damage to the brain resulting from liver damage—hepatic encephalopathy—is a well-known clinical entity (Butterworth, 1994). Conversely, renal disease or cardiac dysfunction will affect the liver. Many toxins primarily affect one organ but also cause toxicity in other organs. For example, acetaminophen has been used primarily to study hepatotoxicity, but this chemical also causes renal damage (Lucas et al., 2000; Rocha et al., 2001). Furthermore, normal physiology requires constant communication between many organs (Murray et al., 2000) to maintain homeostasis in hostile environments, such as the surface of planet Earth. Understanding the roles normally played by individual organ transcriptomes will take us a long way toward understanding their roles in disease states.

ANALYSIS AND INTERPRETATION OF TRANSCRIPTOME DATA

It is important to distinguish between data analysis and data interpretation. Once you are in possession of a toxicogenomics data set, you will begin a process of data filtering or triage, in order to reduce large gene lists to manageable proportions, while minimizing loss of useful information. We recommend that you carefully define the selected approach to this process, an example of which is shown in Figure 8-7. This process involves experimental design, data quality assessment, statistical analysis, data visualization (Figs. 8-4 and 8-5) and interpretation, signal confirmation (e.g., RT-PCR; Fig. 8-6), and functional follow-up studies. The latter studies are probably the most important feature of this process, as they test interpretations and provide a measure of the value of these techniques. For instance, when gene expression patterns induced in cell cultures by the inhibitor of carnitine palmitoyl-transferase-1 indicated the presence of oxidative stress, a range of other analyses, including measurements of glutathione status and superoxide generation, confirmed this observation (Merrill et al., 2002). The latter study thus demonstrates the effectiveness of transcriptional studies for the detection of this mechanism of toxicity under the conditions used.

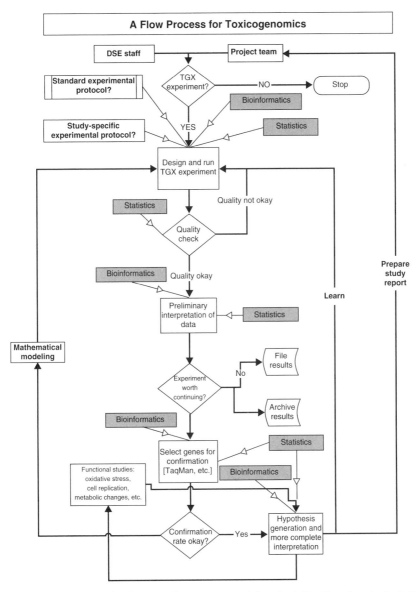

Figure 8-7 *A suggested toxicogenomics program work flowchart. Creation of such charts provides structure to the endeavor, reveals resource needs, and may provide a basis for workflow analysis. DSE = drug safety evaluation; TGX = toxicogenomics. (Source: From unpublished work by Kevin Morgan, Nanxiang Ge, and Zaid Jayyosi [Aventis Pharmaceuticals].)*

There is a growing number of tools for the analysis and visualization of toxicogenomics data (Baldi and Søren, 1998; Baxevanis and Ouellette, 2001). The use of toxicogenomics databases can also provide a valuable aid at this stage, especially for in vivo studies. Such databases consist of a library of large-scale gene expression responses to defined toxins and drugs, combined with powerful statistical tools, against which you can compare a test compound to look for similarities of mode of action or mechanism of toxicity. When undertaking toxicogenomics studies, we strongly recommend (from experience) that you do not underestimate the challenge of selecting an appropriate data normalization procedure, as "there is no fixed point in transcript space" (Morgan et al., 2002a). Furthermore, do not rely too heavily on ratios (Fig. 8-6), as they can be misleading (Crosby et al., 2000; Morgan et al., 2003). Check the validity of gene annotation lists, and carefully cross-check your work for errors that can lead to much wasted time and effort. Once you have exhausted the power of bioinformatics tools, you will be faced with an important decision: whether to attempt a detailed functional interpretation of the data. This takes weeks to months to accomplish for the average toxicogenomics data set.

ONE-GENE-AT-A-TIME INTERPRETATION OF TRANSCRIPTOME DATA

Interpretation of transcriptome data is both fascinating and educational, and rather like piecing together a jigsaw puzzle. This area of biology has not been tackled by many people as of yet, due, we suspect, to the daunting size and complexity of the data sets involved and to the current preference for pattern recognition approaches that can be undertaken largely by computers employing sophisticated mathematical algorithms (Burczynski et al., 2000; Eisen et al., 1998; Tamayo et al., 1999). This section is considered the logical progression of toxicogenomics, after computer-assisted pattern recognition approaches have been completed. With experience and time, this process is less daunting, interrelationships between gene expression changes become apparent, and with some research the underlying patterns of treatment-related effects become apparent. You will learn a great deal of biology in the process. You must, however, assign a structure to the various components of this exercise in order to tackle the problem efficiently. Here are some suggestions to guide you through the exercise.

Begin with a Data Set That Has Been Properly "Triaged"

The first step of interpretation is to be sure your data are sound. A simple test to ensure that treated and control have not been intermixed is to briefly scan the data tables and select a small subset of transcripts that show a characteristic expression pattern in one set but not the other, and use these as a reference point as you work through the data; do not simply trust the bioinformatics tools, as mistakes can be made at many stages in the data generation process. If one array stands out as not fitting the trend, return to the original source data for that array and check that it is a bona fide member of the treatment set. The procedure is a precautionary gross scan. Such outliers may also be detected during the application of PCA, heat maps, or statistical analyses referred to above. Once the arrays are confidently allocated to their respective treatment groups, recheck the documentary chain that each treatment is correctly specified, as you are about to commit a great deal of time to this work.

Getting to know the behavior of a small subset of transcripts in a tissue with which you are familiar can provide a valuable point of reference in these large data sets. It is also helpful to start with a chemical for which there are some existing data. Confirmation of expected expression patterns for known genes provides a degree of confidence that changes found for unknown genes are also valid. It is important to screen for platform-specific artifacts, such as "bleed" between spots on a nylon cDNA array (Crosby et al., 2000). Be aware of which array you are using; take nothing for granted, including the manufacturer's annotations.

You must also select a statistical cutoff, which is the level below which it is assumed that noise overwhelmingly obscures signal. Apply cutoffs uniformly within an experiment, and, if possible, from experiment to experiment, but learn from experience. Minimum fold-change cutoffs are to be discouraged! In these early times, the biologic variance, which varies widely (Bobashev et al., 2002), and the dynamics range of most genes are unknown. Some transcripts exhibit a normal distribution while others are bimodal (Bobashev et al., 2002). Therefore, in a gene (e.g., mdm2) that shows a fold change of no more than 1.2-fold, the control level may still be considered important, especially in light of other, related changes. Do not discard small fold changes.

Fold change can be altogether misleading as a measure of expression change (Fig. 8-6). Statistical analysis is a more reliable data filter and must be performed before you attempt to make sense of the responses. The ultimate test of such statistical procedures is a high correlation between array data and other measures of transcript abundance, such as RT-PCR (Fig. 8-6). We also recommend working with two complementary tables: one containing the statistically triaged data by group and the other comprising normalized individual culture or animal data in a form permitting rapid generation of bar graphs to check whether the statistics make sense.

Use Familiar Areas of Biology as a Guide to the Unfamiliar

Interpretation of experimental results can be colored by a priori expectations. Fortunately, in gene expression interpretation, you will be faced with overwhelming amounts of new information. Existing knowledge, rather than misleading your interpretation, can act as a valuable starting point within the data. Start with what you do know and work toward understanding unfamiliar areas of biology. In fact, you need not know anything about the system under study before you begin, because you will most certainly learn a great deal about it by the time you are finished.

As an example, consider a data set analyzed by one of the coauthors (LMC) in rat primary hepatocytes cultured in vitro for only 2 days before treatment.

Primary rat hepatocytes were treated for 24 hours with media made up in water variously chlorinated, or chlorinated *and* ozonated. Control water was from the tap. All water was centrifuged and filtered, then made into culture media by the addition of nutrients/growth factors and a bicarbonate buffer. Interestingly, the results indicated an unexpected alteration in β-oxidation of fatty acids (Table 8-2 and Fig. 8-9). These results matched the reported gene expression pattern seen at the cellular level after exposure of tissues to plasticizing agents such as DEHP (diethylhexylphthalate), which are found in the environment. It appears that these molecules may act as catecholamine mimics in the cell (so-called endocrine disruptors), and thus the observation might be worthy of further study (compare Table 8-2 with Fig. 8-8).

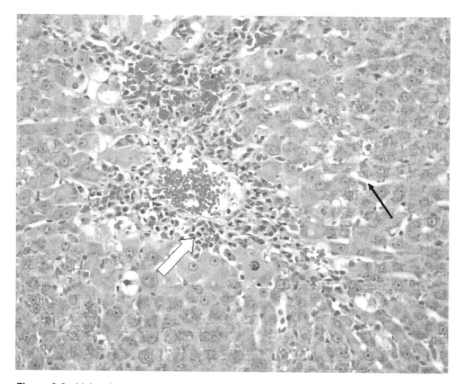

Figure 8-8 *Light micrograph of a portal triad from a rat exposed to acetaminophen, showing infiltration by inflammatory cells (open arrow), which can clearly contribute mRNA to your sample. The large cells that dominate this picture are hepatocytes; however, sinusoids throughout the liver (black arrow) are lined by endothelial cells and Kupffer cells, contain blood, and other cell types hidden from view but present throughout this section include a significant number of stellate cells (see Arias et al., 2001, p. 456)—all of which also contribute significant amounts of mRNA to your sample.*

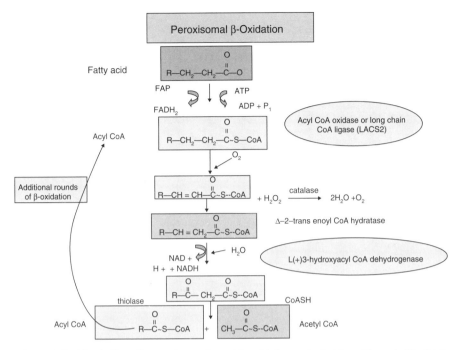

Figure 8-9 *Metabolic pathway of peroxisomal β-oxidation, adapted from* Harper's Biochemistry *(Murray et al., 2000), showing some genes of interest (see Table 8-1) in relation to the effects of PPARα activation.*

Search the Literature

In our opinion, extensive reading is the key to successful gene expression interpretation. This is where the real learning occurs, as you prepare to integrate new information with an established understanding derived from many years of effort by researchers in the relevant field. Bioinformatics tools provide rapid access to information, often as a brief annotation; however, information is not knowledge. It is in textbooks and detailed review articles that you will find insights derived from years of study and synthesis by individuals immersed in the field. It is in textbooks and review articles that you will find knowledge that is so essential for the interpretation of the "info-bullets" of bioinformatics. A great deal of gene expression information is of recent origin, however, and sometimes the significance or function of a particular gene is as of yet unknown. Clearly, a balanced approach, which effectively combines new and old approaches to information management and knowledge acquisition, is ideal.

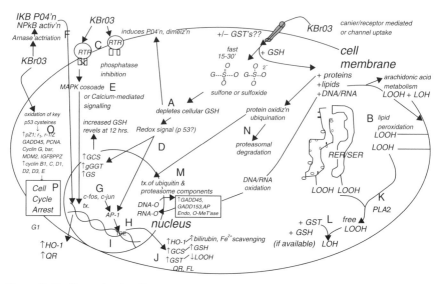

Figure 8-10 *Example of a diagram presenting a synthesis of interpretation for rat mesothelial cells exposed to potassium bromate.* (Source: *From Crosby et al., 2000, with permission.*)

Construct Diagrams as a Synthesis of Your Interpretation

Finally, the development of diagrams of your interpretation, such as those shown in Figures 8-10, 8-11, is a valuable aid to both understanding and explaining these data sets. Diagrams encapsulating data interpretations are difficult to generate but extremely rewarding. If you cannot construct a meaningful diagram of your interpretations and hypotheses, you will have difficulty convincing others of your ideas. Pathway diagrams often illustrate areas that will benefit from confirmatory functional studies. Many such pathways are available on the Internet and are being effectively exploited by bioinformaticians and toxicogenomics database providers.

CONCLUSIONS

A new generation of biologists will have the interesting task of integrating these massive new data sets with knowledge derived from existing approaches to toxicology. This issue is nicely stated in a recent discussion of the medical applications of gene expression [micro]array technology, as follows: "The development of a cooperative framework among regulators, product sponsors, and technology experts will be essential for realizing the revolutionary promise that microarrays hold for drug development, regulatory science, medical practice and public health" (Petricoin III et al., 2002).

Toxicogenomics is a truly interdisciplinary exercise that requires numerous experts to handle all facets of this work. Optimal experimental design requires

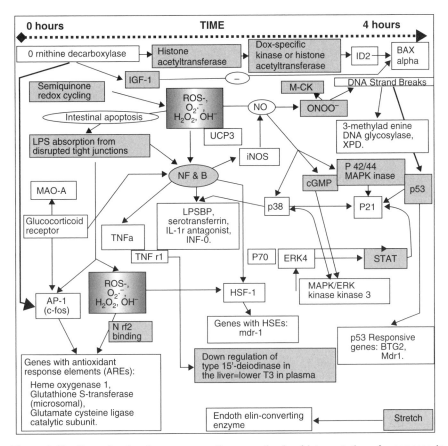

Figure 8-11 *Example of a diagram presenting a synthesis of interpretation of responses in the heart of rats treated with doxorubicin. (*Source*: From Brown et al., 2002, with permission.)*

the input of toxicologists and statisticians to ask the right questions in the right way. Data analysis and validation require the skills of molecular biologists and bioinformaticians (Baxevanis and Ouellette, 2001; Goodman, 2002). Large-scale transcript data sets are necessarily interpreted in the context of many other end points, including in-life data such as body and organ weights, hematology, clinical chemistry, pharmacology, and histopathology. The latter data require the input of clinical chemists, pharmacologists, medical and veterinary clinicians, toxicologists, and pathologists.

Hypothesis generation based on these interpretations can lead to studies that include the application of chemistry, biochemistry (Kuile and Westerhoff, 2001), physics, mathematics (Coffey, 1998; Holter et al., 2000; Kauffman, 1969), engineering, and many other scientific disciplines. Finally, the conclusions based on such studies will be used by risk assessors (Aardema and MacGregor, 2002; Morgan et al., 2002a; Waring and Ulrich, 2000) and risk managers (Petricoin III et al., 2002), leading ultimately to societal decisions concerning the use of drugs and other chemicals, with respect to animal and human health concerns. If this work is done intelligently and with care, these powerful new technologies will markedly improve our ability to make effective decisions with respect to health risk assessment.

ACKNOWLEDGMENTS

We are grateful to the more than 100 people in multiple institutions, including Glaxo-SmithKline, Aventis Pharmaceuticals, the U.S. Environmental Protection Agency, the National Institutes for Environmental Health Sciences, North Carolina State University, and the University of North Carolina at Chapel Hill who made our studies of the transcriptome both possible and enjoyable. Figures 8-4, 8-5, and 8-6 were kindly contributed by Wenyue Hu and Margaret Wojke of Aventis Pharmaceuticals. We also wish to thank Byron Butterworth for helpful editorial advice.

REFERENCES

Aardema MJ, MacGregor JT (2002). Toxicology and genetic toxicology in the new era of "toxicogenomics": impact of "-omics" technologies. *Mutation Res* **499**:13–25.

Abbas AK, Lichtman AH, Pober JS (2000). *Cellular and Molecular Immunology*. Philadelphia: W. B. Saunders Company.

Akhtar RA, Reddy AB, Maywood ES, Clayton JD, King VM, Smith AG, Gant TW, Hastings MH, Kyriacou CP (2002). Circadian cycling of the mouse liver transcriptome, as revealed by cDNA microarray, is driven by the suprachiasmatic nucleus. *Curr Biol* **12**:540–550.

Alberts B, Johnson A, Lewis J, Raff M, Roberts K, Walter P (2001). *Molecular Biology of the Cell*. New York: Garland Science.

Anisimov SV, Tarasov KV, Stern MD, Lakatta EG, Boheler KR (2002). A quantitative and validated SAGE transcriptome reference for adult mouse heart. *Genomics* **80**: 213–222.

Arias IM, Boyer JL, Chisari FV, Fausto N, Schachter D, Shafritz DA (2001). *The Liver: Biology and Pathobiology*. Philadelphia: Lippincott Williams & Wilkins, p 1064.

Baldi P, Søren B (1998). *Bioinformatics, The Machine Learning Approach*. Cambridge, MA: The MIT Press.

Ballatori N, Villalobos AR (2002). Defining the molecular and cellular basis of toxicity using comparative models. *Toxicol Applied Pharmacol* **183**:207–220.

Baxevanis AD, Ouellette BFF (2001). *Bioinformatics. A Practical Guide to the Analysis of Genes and Proteins*. New York: John Wiley & Sons, Inc., p 470.

Boorman GA, Anderson SP, Casey W, Brown HR, Crosby LM, Gottschalk K, Easton M, Ni H, Morgan KT (2002a). Toxicogenomics, drug discovery, and the pathologist. *Toxicol Pathol* **30**:15–27.

Boorman GA, Haseman JK, Waters MD, Hardisty JF, Sills RC (2002b). Quality review procedures necessary for rodent pathology databases and toxicogenomic studies: the National Toxicology Program experience. *Toxicol Pathol* **30**:88–92.

Brazma A, Hingamp P, Quackenbush J, Sherlock G, Spellman P, Stoechert C, Aach J, Ansorge W, Ball CA, Causton HC, Gaasterland T, Glenisson P, Holstege FCP, Kim IF, Markowitz V, Matese JC, Parkinson H, Robinson A, Sarkans U, Schulze-Kremer S, Stewart J, Taylor R, Vilo J, Vingron M (2001). Minimum information about a microarray experiment (MIAME)—toward standards for microarray data. *Nature Genet* **29**:365–371.

Brenner BM (2000). *The Kidney*, vol. 1. Philadelphia: W. B. Saunders Company.

Brown HR, Ni H, Benavides G, Yoon L, Hyder K, Giridhar J, Gardner G, Tyler R, Morgan KT (2002). Correlation of simultaneous differential gene expression in the blood and heart with known mechanisms of adriamycin-induced cardiomyopathy in the rat. *Toxicol Pathol* **30**:452–469.

Burczynski ME, McMillian M, Ciervo J, Li L, Parker JB, Dunn RTI, Hicken S, Farr S, Johnson MD (2000). Toxicogenomics-based discrimination of toxic mechanism in HepG2 human hepatoma cells. *Toxicol Sci* **58**:399–415.

Butterworth RF (1994). Hepatic encephalopathy. In: IM Arias, JL Boyer, N Fausto, WB Jacoby, D Schachter, DA Shafritz (eds), *The Liver: Biology and Pathobiology*. New York: Raven Press, Ltd., pp 1193–1208.

Cassiman D, Libbrecht L, Desmet V, Denef C, Roskams T (2002). Hepatic stellate cell/myofibroblast subpopulations in fibrotic human and rat livers. *J Hepatol* **36**: 200–209.

Castle AL, Caiver MP, Mendrick D (2002). Toxicogenomics: a new revolution in drug safety. *Drug Discovery Today* **7**:728–736.

Chevalier S, Macdonald N, Tonge R, Rayner S, Rowlinson R, Shaw J, Young J, Davison M, Roberts RA (2000). Proteomic analysis of differential protein expression in primary hepatocytes induced by EGF, tumour necrosis factor α or the peroxisome proliferator nafenopin. *Eur J Biochem* **1487**:17–56.

Churchill GA (2002). Fundamentals of experimental design for cDNA microarrays. *Nature Genet* **32**:490–495.

Coffey DS (1998). Self-organization, complexity and chaos: the new biology for medicine. *Nature Med* **4**:882–885.

Cornish-Bowden A, Cardenas ML (2000). From genome to cellular phenotype—a role for metabolic flux analysis? *Nature Biotechnol* **18**:267–268.

Corton CJ, Stauber AJ (2000). Toxicological highlight: toward construction of a transcript profile database predictive of chemical toxicity. *Toxicol Sci* **58**:217–219.

Crosby LM, Benavides G, Yoon L, Hyder K, DeAngelo AB, Morgan KT (2000). Morphologic analysis correlates with gene expression changes in cultured F344 rat mesothelial cells. *Toxicol Appl Pharmacol* **169**:205–221.

Dam K, Seidler FJ, Slotkin TA (2003). Transcriptional biomarkers distinguish between vulnerable periods for developmental neurotoxicty of chlorpyrifos: implications for toxicogenomics. *Brain Res Bull* **59**:261–265.

Drawid A, Jansen R, Gerstein M (2000). Genome-wide analysis relating expression level with protein subcellular localization. *Genome Anal* **16**:426–430.

Duyk GM (2002). Sharper tools and simpler methods. *Nature Genet* **32**:465–468.

Eisen MB, Spellman PT, Brown PO, Botstein D (1998). Cluster analysis and display of genome-wide expression patterns. *Proc Natl Acad Sci (USA)* **95**:14863–14868.

Epstein M (1994). The kidney in liver disease. In: IM Arias, JL Boyer, N Fausto, WB Jacoby, D Schachter, DA Shafritz (eds), *The Liver: Biology and Pathobiology*. New York: Raven Press, Ltd., pp 1235–1256.

Evans R, Kamdar SJ (1990). Stability of RNA isolated from macrophages depends on the removal of an RNA-degrading activity early in the extraction procedure. *Biotechniques* **8**:357–360.

Farr S, Dunn RTI (1999). Concise review: gene expression applied to toxicology. *Toxicol Sci* **50**:1–9.

Fountoulakis M, de Vera M, Crameri F, Boess F, Gasser R, Albertini S, Suter L (2002). Modulation of gene and protein expression by carbon tetrachloride in the rat liver. *Toxicol Appl Pharmacol* **183**:71–80.

Geertz A (2001). History, heterogeneity, developmental biology and functions of quiescent hepatic stellate cells. *Semin Liver Dis* **21**:311–335.

Geertz A, de Bleser P, Hautekeete ML, Niki T, Wisse E (1994). Fat-storing (Ito) cell biology. In: IM Arias, JL Boyer, N Fausto, WB Jacoby, D Schachter, DA Shafritz (eds), *The Liver: Biology and Pathobiology*. New York: Raven Press, Inc., pp 819–838.

Golub TR, Slonim DK, Tamayo P, Huard C, et al. (1999). Molecular classification of cancer: class discovery and class prediction by gene expression monitoring. *Science* **286**:531.

Goodman N (2002). Biological data becomes computer literate: new advances in bioinformatics. *Curr Opin Biotechnol* **13**:68–71.

Gumucio JJ, Bilier BM, Moseley RH, Berkowitz CM (1994). The biology of the liver cell plate. In: IM Arias, JL Boyer, N Fausto, WB Jacoby, DA Schachter, DA Shafritz (eds), *The Liver: Biology and Pathobiology*. New York: Raven Press, Ltd., pp 1142–1163.

Heller RA, Allard J, Zuo F, Lock C, Wilson S, Klonowski P, Gmuender H, van Wart H, Booth R (2000). Gene chips and microarrays: applications in disease profiles, drug target discovery, and drug action and toxicity. In: M Schena (ed), *DNA Microarrays*. Leeds, UK: Oxford University Press, pp 187–202.

Ho MW (1993). *The Rainbow and the Worm: The Physics of Organisms*. Singapore: World Scientific Publishing Co. Pte. Ltd.

Holter NS, Mitra M, Maritan A, Cieplak M, Banavar JR, Fedoroff NV (2000). Fundamental patterns underlying gene expression profiles: simplicity from complexity. *Proc Natl Acad Sci (USA)* **97**:8409–8414.

Idekar T, Thorsson V, Ranish JA, Christmas R, Buhler J, Eng JK, Bumgarner R, Goodlett DR, Aebersold R, Hood L (2001). Integrated genomic and proteomic analyses of a systematically perturbed metabolic network. *Science* **292**:929–934.

Iyer VR, Eisen MB, Ross DT, Schuler G, Moore T, Lee JCF, Trent JM, Staudt LM, Hudson JJ, Boguski MS, Lashkari D, Shalon D, Botstein D, Brown PO (1999). The transcriptional program in the response of human fibroblasts to serum. *Science* **283**:83.

Jacobsson H, Johansson L, Kimiaei S, Larsson SA (1999). Concentration of 123I-metaiodobenzylguanidine in left and right liver lobes. Findings indicate regional differences in function in the normal liver. *Acta Radiol* **40**:224–228.

Jelinsky SA, Harris HA, Brown EL, Flanagan K, Zhang X, Tunkey C, Lai K, Lane MV, Simcoe DK, Evans MJ (2003). Global transcription profiling of estrogen activity: estrogen receptor alpha regulates gene expression in the kidney. *Endocrinology* **144**:701–710.

Kauffman SA (1969). Metabolic stability and epigenesis in randomly constructed genetic nets. *Theoret Biol* **22**:437–467.

Kepler TB, Crosby LM, Morgan KT (2001). Normalization and analysis of DNA microarray data by self-consistency and local regression. *Genome Biol* **3**:1–12.

Kettman JR, Frey JR, Lefkovits I (2001). Proteome, transcriptome and genome: top down or bottom up analysis? *Biomolec Eng* **18**:207–212.

Kita Y, Shiozawa M, Jin W, Majewski RR, Besharse JC, Green AS, Jacob HJ (2002). Implications of circadian gene expression in kidney, liver and the effects of fasting on pharmacogenomic studies. *Pharmacogenetics* **12**:55–65.

Kuile BHT, Westerhoff HV (2001). Transcriptome meets metabolome: hierarchical and metabolic regulation of the glycolytic pathway. *FEBS Lett* **500**:169–171.

Kurachi M, Hashimoto S, Obata A, Nagai S, Nagahata T, Inadera H, Sone H, Tohyama C, Kaneko S, Kobayashi K, Matsushima K (2002). Identification of 2,3,7,8-tetrachlorodibenzo-*p*-dioxin-responsive genes in mouse liver by serial analysis of gene expression. *Biochem Biophys Res Commun* **292**:368–377.

Lencesova L, Ondrias K, Micutkova L, Filipenko M, Kvetnansky R, Krizanova O (2002). Immobilization stress elevates IP(3) receptor mRNA in adult rat hearts in a glucocorticoid-dependent manner. *FEBS Lett* **531**:432–436.

Lucas AM, Hennig G, Dominick PK, Whiteley HE, Roberts C, Cohen SD (2000). Ribose cysteine protects against acetaminophen-induced hepatic and renal toxicity. *Toxicol Pathol* **28**:697–704.

Marchant GE (2002). Toxicogenomics and toxic torts. *Trends Biotechnol* **20**:329–332.

Mathews MB, Sonenberg N, Hershey JWB (2000). Origins and principles of translational control. In: N Sonenberg, JWB Hershey, MB Mathews (eds), *Translational Control of Gene Expression*. Cold Spring Harbor, NY: Cold Spring Harbor Laboratory Press, pp 1–31.

Matsuzaki S, Onada M, Tajiri T, Kim DY (1997). Hepatic lobar differences in progression of chronic liver disease: correlation of asialoglycoprotein scintigraphy and hepatic functional reserve. *Hepatology* **25**:828–832.

Meador-Woodruff JH, Grandy DK, van Tol HH, Damask SP, Little KY, Civelli O, Watson SJJ (1994). Dopamine receptor gene expression in the human medial temporal lobe. *Neuropsychopharmacology* **10**:239–248.

Merrill C, Ni H, Yoon L, Tirmenstein MA, Narayanan P, Benavides G, Easton M, Creech D, Hu CX, Thomas HC, Morgan KT (2002). Etomoxir-induced oxidative stress detected by differential gene expression is confirmed biochemically. *Toxicol Sci* **68**:93–101.

Morgan KT, Brown HR, Benavides G, Crosby LM, Sprenger D, Yoon L, Ni H, Easton M, Morgan D, Laskowitz D, Tyler R (2002a). Toxicogenomics and human disease risk assessment. *Hum Ecol Risk Assess* **8**:1339–1353.

Morgan KT, Ni H, Brown HR, Yoon L, Qualls CWJ, Crosby LM, Renolds R, Gaskill B, Anderson SP, Kepler TB, Brainard T, Liv N, Easton M, Merrill C, Creech D, Sprenger D, Conner G, Johnson PR, Fox T, Sartor M, Richard E, Kuruvilla S, Casey W, Benavides G (2002b). Application of cDNA microarray technology to *in vitro* toxicology and the selection of genes for a real time RT-PCR-based screen for oxidative stress in Hep-G2 cells. *Toxicol Pathol* **30**:435–451.

Morgan KT, Casey W, Easton M, Creech D, Ni H, Yoon L, Anderson SP, Qualls CWJ, Crosby LM, MacPherson A, Bloomfield P, Elston TC (2003). Frequent sampling reveals dynamic responses by the transcriptome to routine media replacement in HepG2 cells. *Toxicol Pathol* **31**:1–14.

Murray RK, Granner DK, Mayes PA, Rodwell VW (2000). *Harper's Biochemistry*. Stamford CT: Appleton & Lange.

Nadal C (2000). Nonregenerative stimulation of hepatocyte proliferation in the rat: variable effects in relation to spontaneous liver growth; a possible link with metabolic induction. *Cell* **33**:287–300.

Nottermam DA, Uri A, Sierk AJ, Levine AJ (2001). Transcriptional gene expression profiles of colorectal adenoma, adenocarcinoma, and normal tissues examined by oligonucleotide arrays. *Cancer Res* **61**:3124–3130.

Olden K, Guthrie J (2001). Genomics: implications for toxicology. *Mutat Res* **473**:3–10.

Oliver DJ, Nikolau B, Wurtele ES (2002). Functional genomics: high-throughput mRNA, protein, and metabolite analyses. *Metabol Eng* **4**:98–106.

Petricoin III E, Hackett JL, Lesko LJ, Puri RK, Gutman SI, Chumakov K, Woodcock J, Feigal DW, Zoon KC, Sistare FD (2002). Medical applications of microarray technologies: a regulatory science perspective. *Nature Genet* **32**:474–479.

Quinn-Senger KE, Ramachandran R, Rininger JA, Kelly KM, Lewin DA (2002). Staking out novelty on the genomic frontier. *Curr Opin Chem Biol* **6**:418–426.

Rastan S (2001). Genomics: saviour or millstone? *Trends Genet* **17**:247–248.

Reynolds LJ, Richards RJ (2001). Can toxicogenomics provide information on the bioreactivity of diesal exhaust particles? *Toxicology* **165**:145–152.

Roberts GC, Smith CWJ (2002). Alternative splicing: combinatorial output from the genome. *Curr Opin Chem Biol* **6**:375–383.

Rocha GM, Michea LF, Peters EM, Kirby M, Xu Y, Ferguson DR, Burg MB (2001). Direct toxicity of nonsteroidal antiinflammatory drugs for renal medullary cells. *Proc Natl Acad Sci* (*USA*) **98**:5317–5322.

Ruepp SU, Tonge RP, Shaw J, Wallis N, Pognan F (2002). Genomics and proteomics analysis of acetaminophen toxicity in mouse liver. *Toxicol Sci* **65**:135–150.

Russell AC, Bekkedal MY-V, Mann TT, Ritchie GD, Rossi III J, Stenger DA, Pancrazio JJ, Andreadis JD (2002). Gene modulation in total brain induced by exposure to the bicyclic phosphorus ester trimethylolpropane phosphate (TMPP). *Neurotoxicology* **23**:215–221.

Seta K, Kim HW, Ferguson T, Kim R, Pathrose P, Yuan Y, Lu G, Spicer Z, Milhorn DE (2002). Genomic and physiological analysis of oxygen sensitivity and hypoxia tolerance in PC12 cells. *Annals NY Acad Sci* **971**:379–388.

Sirivatanauksorn Y, Drury R, Crnogorac-Jurcevic T, Sirivatanauksorn V, Lemoine NR (1999). Laser-assisted microdissection: applications in molecular pathology. *J Pathol* **189**:150–154.

Smith LL (2001). Key challenges for toxicologists in the 21st century. *Trends Pharmacol Sci* **22**:281–285.

Storch KF, Lipan O, Leykin I, Viswanathan N, Davis FC, Wong WH, Weitz CJ (2002). Extensive and divergent circadian gene expression in liver and heart. *Nature* **417**: 78–83.

Tamayo P, Slonin D, Mesirov J, Zhu Q, Kitareewan S, Dmitrovsky E, Lander ES, Gobub TR (1999). Interpreting patterns of gene expression with self-organizing maps: methods and application to hematopoietic differentiation. *Proc Natl Acad Sci (USA)* **96**:2907–2912.

Tsangaris GT, Botsonis A, Politis I, Tzortzatou-Stathopoulou F (2002). Evaluation of cadmium-induced transcriptome alterations by three color cDNA labeling microarray analysis on a T-cell line. *Toxicology* **178**:135–160.

Vidal-Vanaclocha F (1997). *Functional Heterogeneity of Liver Tissue: From Cell Lineage Diversity to Sublobular Compartment-Specific Pathogenesis.* Austin, TX: R. G. Landes Company.

Walhout AJM, Reboul J, Shtanko O, Bertin N, Vaglio P, Ge H, Lee H, Doucette-Stamm L, Gunsalus KC, Schetter AJ, Morton DG, Kemphues KJ, Reinke V, Kim SK, Piano F, Vidal M (2002). Integrating interactome, phenome, and transcriptome mapping data for the *C. elegans* germline. *Curr Biol* **12**:1952–1958.

Waring JF, Ulrich RG (2000). The impact of genomics-based technologies on drug safety evaluation. *Ann Rev Pharmacol Toxicol* **40**:335–352.

Watford M, Chellaraj V, Ismat A, Brown P, Raman P (2002). Hepatic glutamine metabolism. *Nutrition* **18**:301–303.

9

Introduction to Relationships Between Toxicology and Gene Expression

Hisham K. Hamadeh and Robert T. Dunn II

INTRODUCTION

Before understanding or pursuing potential relationships between toxicology and gene expression, each has to be understood individually. Chapter 1 provides a very good overview of toxicology principles and basics that should help the reader understand concepts in this and other chapters. In this chapter, we provide a very basic overview of gene expression, then introduce potential benefits that could be gained by marrying toxicology and high-density gene expression into one discipline. In addition, we highlight organ-specific approaches that aid in the interpretation of gene expression data from a toxicology perspective.

Toxicogenomics: Principles and Applications. Edited by Hamadeh and Afshari
ISBN 0-471-43417-5   Copyright © 2004 John Wiley & Sons, Inc.

TOXICOLOGY AND GENE EXPRESSION

Transcription involves chromosomal DNA acting as a template for the synthesis of RNA (Fig. 9-1). In most mammalian cells, only a small fraction of the DNA sequence is copied into protein encoding RNA. Double-stranded DNA is transcribed to produce nuclear RNA, and only a minor part of the latter survives the RNA-processing steps that include subtraction of noncoding sequences of base pairs (introns) from coding sequences (exons). The edited sequence, called messenger RNA (mRNA), leaves the nucleus and travels to the cytoplasm, where it encounters cellular bodies called ribosomes. The mRNA, which carries the gene's instructions, dictates the production of proteins by the ribosomes.

In the field of toxicology, most agents (drugs, chemicals, radiation, etc.) are anticipated to exert their adverse effects directly or indirectly by perturbing "normal" signaling processes in the cell. Under the premise that agents elicit cascades of events that start with receptor binding/activation leading to gene and protein expression, studying gene expression perturbations might provide insight and increased understanding of their mechanism(s) of action by highlighting which enzymes or proteins are targeted. Because mRNAs are precursors to enzymes, measurement of their relative abundance in agent-exposed versus control tissue/cell samples often reveals potential

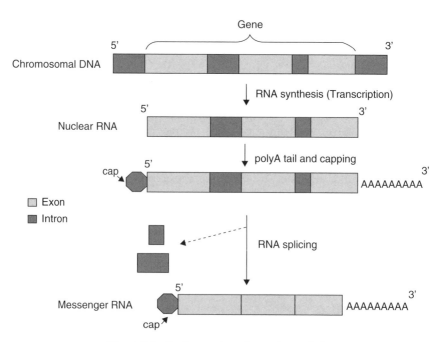

Figure 9-1 *Basic overview of gene transcription.*

involvement of genes in the etiology of the adverse effects associated with the agent of interest. This is not a novel idea, however; traditionally, mechanisms of stressor action were pursued by validating the involvement of individual candidate genes resulting from hypotheses that drew on historical reports.

A standard method used to monitor changes in mRNA levels of genes is a Northern blot. Although this traditional molecular technique definitively shows the expression level of all transcripts (including splice variants) for a particular gene, it is labor intensive and is only practical for examining the expression level of a limited number of genes. The process of gene expression measurements was undoubtedly accelerated in the early 1990s with the introduction of DNA microarrays capable of measuring the mRNA levels corresponding to thousands of genes simultaneously (Schena et al., 1995; DeRisi et al., 1996; Shalon et al., 1996). Thus, DNA microarrays aim at revealing gene–environment interactions on a much larger scale than previously possible.

The premise behind DNA microarrays is that DNA material (probe) placed on a certain substrate can be used to bind complementary DNA from reverse-transcribed RNA (target) corresponding to biological samples of choice. Target cDNA is in theory a quantitative copy of genes expressed at the time of sample collection. It can be radiolabeled or tagged with fluorescent dyes, thus enabling quantitation using scanning devices that can excite fluorescent material and monitor the emitted signals (Fig. 9-2). Scanners are described in more detail in Chapter 4. DNA microarrays come in various versions that differ in their substrates, target preparation, probe, and methods of signal detection. Chapter 3 provides a comprehensive overview of available platforms and summarizes their major properties. All of these platforms aim at

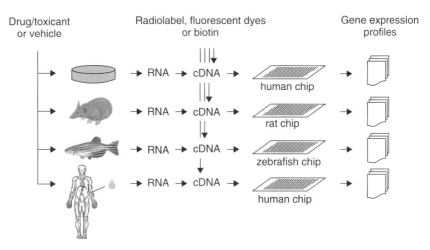

Figure 9-2 *Gene expression can be performed using samples derived from a wide array of tissues from different species.*

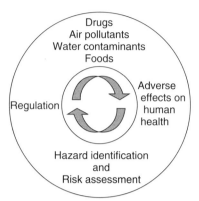

Figure 9-3 *Optimization of safety evaluation processes involve hazard identification, mechanistic inference, and risk assessment.*

measuring the relative abundance of mRNA levels corresponding to thousands of genes in a biological sample. In toxicology, gene expression profiles can be generated corresponding to drug- or toxicant-treated biological samples versus untreated or vehicle-treated controls to study global gene expression alteration in response to the exposure. Moreover, adverse end points resulting from toxicant exposure may be profiled and compared to normal tissue samples in order to discern differences in gene expression between the two states (Hamadeh et al., 2002a, b, c; Waring et al., 2001a, b, 2003).

Toxicological research evolves sequentially initiated with drugs, toxicants, and toxins, which directly or indirectly lead to adverse effects or diseases in biological systems (Fig. 9-3). These perturbations trigger efforts that aim at identifying the hazards by epidemiological or other means and their relationship to the disease(s), followed by experimental analysis designed to assist in assessing the risk involved with exposure to the agent. Risk assessment data are then used for regulatory purposes that limit or reduce exposure to the agent based on safety margins and data extrapolations. This process is not a perfect one and is a continuous exercise in learning that is often limited by our current state of knowledge. Several cycles are needed to achieve knowledge-based and "safe" levels of agents under study, depending on the level of understanding of the ensuing disease processes and the availability of adequate effect detection approaches. The continuous refinement of these processes inadvertently leads to more informed hazard identification and risk assessment, ultimately leading to better and more evolved regulatory decisions.

Organ Injury Detection

Detection of organ injury is a central part of risk assessment, and different approaches have been employed to test for organ damage. These approaches include organ imaging to study gross abnormalities, drug and dye clearance tests to investigate organ function, analysis of urinary components indicative of organ damage, and measurements of plasma markers such as alanine amino-transferase (ALT) and aspartate aminotransferase (AST) as indirect markers of organ damage. For example, the traditional laboratory approach for detection of renal disease involves determination of serum creatinine, blood urea nitrogen, creatinine clearance, urinary electrolytes, microscopic examination of urine sediment, and radiological studies. While most of the current traditional assays for detecting organ damage are useful for detecting different organ toxicities in rodents and humans, some of these indicators are insensitive and nonspecific, not allowing for early detection of adverse effects such as renal disease. The major deficiencies of the aforementioned assays include sensitivity, specificity, and predictability mandating the development of novel approaches for identification of organ injury.

On the issue of sensitivity, tissue imaging (Wagner et al., 2002) is associated with considerable subjectivity and reveals abnormalities only when appreciable organ damage has occurred, and measurement of circulating or urinary levels of certain enzymes or biochemical entities is indicative of late-stage/high-incidence organ injury. While dye and drug clearance tests are more sensitive than other tests for some types of hepatic damage, they still lack the sensitivity needed to monitor early stages of injury, where patients are most likely to benefit from medical intervention.

As for specificity, historically there have been successful attributions of some serological markers to certain types of injury. Alkaline phosphatase (ALP) and gamma-glutamyl transpeptidase (GGT) are supposed to be indicators of hepatic cholestasis, whereas alanine aminotransferase (ALT) and aspartate aminotransferase (AST) are considered hallmarks of hepatocellular damage. While ALT is found primarily in liver cells and therefore has greater specificity for liver disease, AST is present in the heart, skeletal muscle, brain, and kidney as well as the liver, and levels are known to increase in myocardial infarction, heart failure, muscle injury, central nervous system disease, and other nonhepatic disorders. Despite some nonspecificity, high levels of AST are indicative of liver cell injury. Values of >500 IU/L (>400 u/mL) suggest acute viral or toxic hepatitis. Such high values also occur in marked heart failure (ischemic hepatitis) and even with common duct stones. The magnitude of the elevation has little prognostic value and does not usually correlate with the degree of liver damage. Despite their vast utility in adverse effect detection, these and other markers are not very specific and could be altered by an array of other toxicities originating from extrahepatic organs. Thus a good portion of similar marker disease associations remain largely unspecific.

Predictability of chemically induced disease progression before its occurrence can provide for a revolutionary process by which potential adverse effects of drugs are identified. From a clinical perspective, drug-associated hepatotoxicity is often grounds for special monitoring of patients (ALT, AST levels, inpatient facility treatment only) and use restrictions (dose, time, liver function impairment), or in the cases of severe toxicity, removal from the market. During drug development, toxicologists conduct animal tests that include the assessment of potential hepatotoxic effects, some of which may prevent the drug from advancing to clinical development. However, in some cases potential adverse effects are not realized until after a drug has reached clinical testing or has been approved for more general use. This is due to the fact that clinical trials are designed to pick up common problems affecting 1 in 500 or 1 in 1,000 people and are unlikely to pick up rare adverse effects occurring at a rate of 1 in 50,000, for example. This latter rate, in some cases, is enough grounds for severely limiting the use of the drug or even withdrawal from the market.

Thus the need arises for alternate methodologies that can address the three objectives of sensitivity, specificity, and predictability when testing for organ injury.

How DNA Microarrays Can Benefit Toxicology

Sensitivity, Specificity, and Predictability. Measuring gene expression in a large-scale fashion addresses the aforementioned deficiencies—namely, sensitivity, specificity, and predictability—because very subtle alterations in gene expression can be measured. This sensitivity allows for the derivation of information from low-dose challenges, where most of the traditional assays are often inconclusive. With this increased sensitivity comes the need to draw the line where excessive sensitivity is undesirable (i.e., the genes observed are not related to adverse outcomes). It is thus the responsibility of the involved scientist to evaluate gene expression in conjunction with other ancillary data to derive the most balanced message from analyzed studies.

An expression pattern, which is a collective profile of multiple genes, is potentially more unique in its representation of a specific type of injury than a single enzyme or metabolite measurement. Because metabolites and enzyme levels can be influenced by many factors and have a role in multiple pathways leading to disease end points, a combination of genes is often necessary to define complex disease states. This has been experienced firsthand in the field of oncology, where different cancers have been classified based on the expression pattern of groups of genes (Perou et al., 2000). This paradigm is also applicable to the field of toxicogenomics, where disease states are often a result of interaction of different cell types expressing an array of markers that need to be monitored to correctly identify the respective disease with an increased degree of specificity.

The importance of prediction of adverse effects has been discussed previously in this chapter. The most frequently used serologic indicators for the diagnosis of renal disease are serum creatinine, serum urea nitrogen, and creatinine clearance, all of which are insensitive and nonspecific for detection of renal injury, due in part to the organ's superb compensatory abilities. By the time these markers are detected, damage to the kidney is usually extensive.

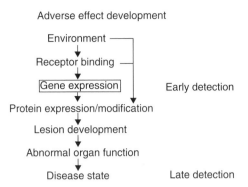

Adverse effect development

Environment
↓
Receptor binding
↓
Gene expression   Early detection
↓
Protein expression/modification
↓
Lesion development
↓
Abnormal organ function
↓
Disease state   Late detection

Figure 9-4 *Gene expression as an early step in the path to many disease end points.*

Gene expression has the potential to precede early signs of injury and be predictive of adverse events (Fig. 9-4). This has been nicely elucidated by a group at Harvard Medical School (Han et al., 2002) studying kidney injury molecule 1 gene (*KIM-1*) expression as a surrogate for acute renal failure (ARF), which is an important cause of morbidity and mortality in hospitalized patients.

Identification of reliable biomarkers for tubule injury would be useful to facilitate early intervention, evaluate the effectiveness of therapeutic interventions, and guide pharmaceutical development by providing an indicator for nephrotoxicity. KIM-1, a novel molecule, was cloned from rats, mice, and humans and shown to be markedly upregulated in postischemic rat kidney (Ichimura et al., 1998). Researchers monitored the KIM-1 ectodomain normally shed into the extracellular milieu of human 769-P (human kidney adenocarcinoma cells) and HK-2 (human kidney proximal tubular cells) cell lines, which express *KIM-1* under normal culture conditions.

The utility of KIM-1 as a marker of acute tubular damage in humans was demonstrated by examining kidney biopsy sections and urine samples for abundance of the KIM-1 protein in patients with various acute and chronic renal diseases. The study demonstrated that urinary KIM-1 protein concentration was significantly higher in urine samples from patients with ischemic acute tubular necrosis (ATN) compared to urine samples from healthy volunteers and patients with other forms of acute and chronic renal failure. Furthermore, the authors compared KIM-1 with other traditional markers of kidney injury (e.g., BUN) and concluded that KIM-1 was a more specific marker for ischemic tubule injury.

Mechanism(s) of Adverse Action. The mechanism(s) of toxicity of numerous drugs, industrial chemicals, and environmental contaminants remains largely unknown, hindering appropriate preventive and regulatory measures. Mechanisms in hepatocellular injury include metabolic activation of chemicals by cytochrome P450s, glutathione-S-transferases (GSTs), and alcohol dehydrogenase, leading to covalent binding of compounds to proteins, lipids, or

DNA and resulting in lipid peroxidation, protein thiol depletion, or calcium transport impairment. High-density gene expression measurements will offer insight into which mechanisms are associated with agent exposure. This has potential to facilitate more efficient biomarker development and more informed realization of drug-drug interactions resulting from the excessive activation of toxic pathways, or, alternatively, suppression of detoxification pathways.

Gene expression studies will further promote our understanding of the underlying mechanisms of drug-induced toxicity such as allowing the elucidation of activated signaling cascades or inhibition of protective mechanisms. An increased understanding of these mechanisms should aid in further understanding of the factors that contribute to risk. In the case of pharmaceuticals, development of early, sensitive markers from these studies will aid in the screening of compounds early in the development process and may ultimately provide sensitive surrogate markers for clinical studies. Similarly, in the case of industrial chemicals and environmental contaminants, mechanistic understanding will help accelerate processes such as the Environmental Protection Agency's (EPA) assessment of safety.

Dose Response and Time Course. As detailed in Chapter 1, dose responses and time courses are central components of toxicological studies. While dose-response analyses aim at studying whether an effect is correlated to increased exposure to a compound, a time course establishes causality between events and reveals transient, continuous, or delayed responses that may signify initial reactions or adaptive responses to exposure.

In order to define these responses and reactions in terms of gene expression alterations, multiple doses and time points should be included in the design of toxicogenomic investigations. However, these study designs quickly result in very large data sets that contain hundreds of thousands of data points and can be challenging to analyze efficiently. The use of various computational tools such as self-organizing maps (SOMs) and K-means clustering (discussed in more detail in Chapter 5) allows for the discovery of gene subsets that are altered in a dose response or a time course manner or both concomitantly (Fig. 9-5). Hence, genes that are altered only at postcompound exposure might be related to adaptation events or lesions that might have developed close to that time frame. Likewise, genes altered in a dose- or time-dependent manner could lead to increased understanding of the mechanisms of action of toxicants or drugs, especially if they precede lesions that develop in response to exposure to the studied agent(s).

Correlation of Gene Expression to Phenotype

Gene expression profiles derived from tissues are often telling of the state of the respective organ. As discussed earlier, this phenotype can be defined by several means including clinical observations (mottled kidney, pale liver), func-

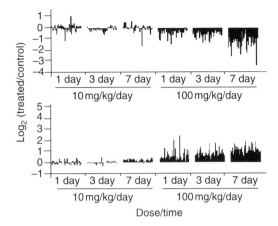

Figure 9-5 *Liver gene expression profiles elicited by daily oral administration of methapyrilene to Sprague-Dawley rats for 1, 3, or 7 days at doses of 10 or 100 mg/kg/day. The upper and lower panels show groups of genes decreased and increased respectively relative to time-matched controls in a time- and dose-dependent fashion (Hamadeh et al., 2002d).*

tional tests, measurement of serum components, or histopathological evaluation. Anchoring the gene expression to those alterations is key to finding biomarkers of adverse end points and furnishing a testable hypothesis regarding a mechanism of action.

Nephrotoxicity Case Studies. Several examples exist in the literature detailing studies where gene expression was utilized to understand the mechanism of kidney-specific lesion development (Huang et al., 2001; Luhe et al., 2003). We provide a brief description of examples dealing with nephrotoxicity. Before attempting to correlate gene expression alterations with phenotypic endpoints, the physiology of the organ in question must be understood. Chapter 1 offers a comprehensive overview of target organ toxicity, which should aid in understanding concepts in this chapter.

In the case of the kidney, it is a frequent target of both environmental toxins and certain classes of therapeutic compounds. The susceptibility of the kidney to toxic insult is due, in part, to the unique anatomical and physiological functionality of the organ. Each kidney receives a significant portion of the cardiac output ($\sim$125 mL/min in humans); thus toxins present in the blood have a high likelihood of coming into contact with the various cell types of the kidney. A second feature of the kidney that enhances its susceptibility is the numerous transport functions that are properties of the kidney cells. Cells of the proximal convoluted tubule reabsorb most of the water and associated ions (Na^+, Mg^{2+}, Ca^{2+}, K^+, Cl^-, HCO_3^-, PO_4^{3-}) in addition to smaller proteins, amino acids, and other metabolites. The reabsorptive function and capacity of the proximal tubule renders it uniquely susceptible to injury by toxins. A brief list of renal toxins is presented in Table 9-1. The heterogeneity of the toxins both within

TABLE 9-1 Abbreviated list of renal toxicants

Metals	*Analgesics*
Cadmium	Acetaminophen
Mercury	Nonsteroidal antiinflammatory drugs (NSAIDS)
Chromium	
Lead	*Antibiotics/Chemotherapeutics*
Uranium	
	Amphotericin B
Halogenated Hydrocarbons	Cephaloridine
	Gentamycin
Chloroform	Mitomycin
Tetrafluoroethylene	Puromycin
Bromobenzene	Cyclosporine
	Cisplatin
Mycotoxins	
Ochratoxin A	
Citrinin	
Fumonisins	

and between classes points to the fact that more than just structural features of the toxin molecules are operational in the manifestation of kidney toxicity.

Two recent toxicogenomic studies have been published that used rat proximal tubule toxicity as a model to assess gene expression changes that occur following administration of classical renal toxicants (Huang et al., 2001; Luhe et al., 2003). Both cisplatin (*cis*-dichlorodiammine platinum (II)) (CIS) and ochratoxin A (OTA) target the S3 segment of the proximal tubule and reliably induce toxicity including elevations in creatinine and blood urea nitrogen, in addition to histological changes. The cancer chemotherapeutic CIS is one of the most well-studied kidney toxins. CIS is a useful compound to investigate renal toxicity for several reasons. First, CIS is toxic across species. Preclinical models including mouse, rat, dog, and monkey are similarly affected by CIS (Guarino et al., 1979). In addition, renal toxicity is the dose-limiting effect in cisplatin-based chemotherapy regimens. Second, and perhaps related to its pan-species effects, is that CIS does not appear to require enzymatic activation but rather is nonenzymatically converted to the ultimate toxic form. Third, CIS has a second isomeric configuration called transplatin that is not toxic. The dichotomous toxicities of cis- and transplatin make the compounds very useful in performing a well-controlled study. Finally, CIS toxicity can be monitored by several techniques (Table 9-2) including measurement of plasma creatinine, blood urea nitrogen (BUN), and histological changes. Taken together, the combined attributes of cisplatin make the drug an attractive candidate for organ-based toxicogenomic studies.

OTA is a toxin derived from particular species of fungus including *Aspergillus ochraeus* and *Penicillium verrucosum* that are present in a variety of different food and drink products such as cereal grains, nuts, spices, coffee, and red wine (Petrik, 2003; Taniwaki et al., 2003; Cabanes et al., 2002). Similar to CIS, OTA induces toxicity in the kidney proximal tubule at the S3 segment (Luhe et al., 2003).

TABLE 9-2 Serum and urinary markers of kidney toxicity[a]

Alkaline phosphatase (u)
γ-glutamyl transpeptidase (u)
Blood urea nitrogen (s)
Proteinuria (u)
Glucosuria (u)
Ketonuria (u)
Creatinine (s)
Polyuria (u)

[a] s, serum; u, urine.

The toxicogenomics case studies presented here are a comparison between two independent studies on the gene expression changes in kidney that were induced by toxic doses of cisplatin and ochratoxin A. Perhaps the most useful observations between these studies are the nearly identical histopathological changes that included S3 proximal tubule cell necrosis and scattered apoptosis. The necrotic and apoptotic cells were located in the outer stripe of the medulla. It is also worthwhile to note that the CIS study used male Sprague-Dawley rats dosed orally with drug in aqueous vehicle while the OTA study used male Wistar rats dosed orally with drug in corn oil vehicle. Cisplatin shares no structural homology to OTA, yet both compounds appear to accumulate in a specific region of the proximal tubule to elicit a toxic response. Thus cisplatin and ochratoxin A, while chemically distinct and unrelated, converge at a common histological end point. The common histology becomes the take-off point to examine relatedness in the gene expression profiles between these studies.

Table 9-3 lists the genes expressed in common between rats treated with CIS and rats treated with OTA (Huang et al., 2001; Luhe et al., 2003). Although the list is relatively small, due in part to data generated from a low-density array (250 genes in the CIS study), several functional classes of genes are represented including cellular transport, DNA damage, stress response, and those related to metabolism. Various hypotheses can be proposed from these data to explain why certain genes are being modulated. For DNA damage genes, it can be speculated that due to the damage incurred by each agent on DNA, the cells have responded by attempting to control cell cycle in order to prevent fixation of DNA defects. Metabolism genes are altered because the cell has shut down certain functions in order to preserve critical processes and molecules such as cell cycle and DNA, respectively. Acute-phase response genes are induced following a number of different cellular stresses (Seshadri et al., 2002; Anderson et al., 1999) and in the present studies may represent a more generic response rather than a response directly linked to the toxins.

The paradigm of biomarker discovery through finding similarly regulated genes across studies that share similar pathological end points is plausible as illustrated in Figure 9-6, where multiple structurally and pharmacologically unrelated compounds sharing the same toxicological end point might eluci-

TABLE 9-3 Genes exhibiting a concordance in qualitative expression changes between two compounds—cisplatin and ochratoxins—that induce a similar type of kidney toxicity

Gene Accession Number	Gene Name	Direction of Change
M11670	Catalase	↓
L33869	Ceruloplasmin	↑
M64723	Clusterin	↑
M33821	Gamma-glutamyl transpeptidase	↓
D79981	Organic anion transporter-K1	↓
D83044	Organic cation transporter (OCT2)	↓
X69021	Senescence marker protein-30 (SMP-30)	↓
M21730	Annexin V	↑
J02640	NADPH quinone oxidoreductase (DT-diaphorase)	↑CIS ↓OTA
U36944	Gadd153	↑
L32591	Gadd45	↑

Source: Genes were compiled from published reports (Huang et al., 2001; Luhe et al., 2003).

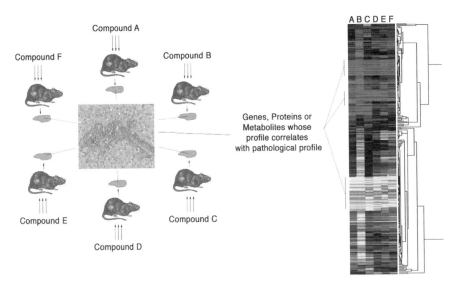

Figure 9-6 Schematic diagram outlining a strategy for the phenotypic anchorage of gene expression alteration data. A hypothetical analysis of tissue derived from animals treated with compounds A, B, C, D, E, or F reveals an overlap among the effects manifested in one common pathological end point. Commonality across animals revealed by cluster analysis of gene expression levels would indicate a potential association between the altered gene expression and the shared histopathological end point.

date similarity in expression of a subset of genes. These genes then traverse a lengthy path of validation to become acceptable biomarkers. Candidates have to satisfy many requirements including early detection, sensitivity, specificity, reversibility, and degree of severity. In addition, genes whose protein products are secreted into the bloodstream are much better candidates than others as adequate biomarkers for human applications because a noninvasive measure of these biomarkers would be ideal. Biomarker discovery is being accelerated with the participation of commercial companies such as Gene Logic Inc. (Gaithersburg, MD; www.genelogic.com) and Iconix Pharmaceuticals (Mountain View, CA; www.iconixpharm.com) that are building larger and larger data sets of related experiments to interrogate multiple toxicities affecting various organs following the strategy detailed earlier.

Tissue Distribution of Gene Expression. It is intriguing to note that in the studies of both Huang and Luhe (Huang et al., 2001; Luhe et al., 2003) decreased expression of *gamma-glutamyl transpeptidase*, a gene found mainly in the tubular S3 segment, occurs after treatment with either CIS or OTA. This sparks a novel way of localizing kidney toxicity within the mass of the tissue by utilizing knowledge of the relative abundance of genes in various sections of the functional unit of the kidney, the nephron. Table 9-4 shows the relative abundance of several genes across different segments of the proximal tubule of the nephron (Amin et al., 2004). A decrease in the expression of those genes

TABLE 9-4   Kidney proximal tubular tissue distribution of selected genes

Gene	Proximal Tubule Segment Distribution
Phosphoenolpyruvate carboxylkinase	S1 > S2 > S3
Fructose-1,6-bisphosphatase	S1 < S2 > S3
Glucose-6-phosphatase	S1 > S2 > S3
Fructokinase	S1 = S2 < S3
Fructose-1-phosphate aldolase	S1 = S2 > S3
Glycerokinase	S1 = S2 > S3
Glycerol-3-phosphate DH	S1 = S2 < S3
Glutamine synthetase	S3
Alanine aminotransferase	S1 = S2 < S3
Ornithine amintotransferase	S1 = S2 < S3
Gamma-glutamyl transpeptidase	S1 < S2 < S3
Gamma-glutamyl cysteine synthetase	S3
Gluthathione-S-transferase	S1 < S2 < S3
Cytochrome P450	S1 = S2 > S3
L-hydroxyacid oxidase	S1 = S2 < S3
Peroxisomes (D-amino acid oxidase and CAT)	S1 = S2 < S3
Aminopeptidase	S1 < S2 < S3
Alkaline phosphatase	S1 = S2 = S3
Fatty acyl CoA oxidase	S1 = S2 < S3
Choline oxidase	S1 < S2 = S3
25(OH)-D3-1 alpha-hydroxylase/DBP	S1 < S2 = S3

Source: Data abstracted from Amin et al., 2004.

might indicate true transcriptional downregulation or merely loss of cells expressing those genes in specific segments of the nephron. In either situation, the decrease suggests insult to the cells localized in the nephron segment under investigation (Amin et al., 2004). Although the example provided here relates to the kidney, this paradigm can be applied to many other organs with specialized sites and where information on region-specific gene expression is available. For example, in the liver, sharp gradients of expression exist for genes belonging to the P450 family or flavin-containing monooxygenases exist along the periportal-centrilobular chord length. Similarly, monitoring of decreased abundance of these genes relative to control might indicate site-specific loss of cells expressing the respective genes such as in the case of periportal or centrilobular necrosis.

Liver Case Study. Another demonstration of gene expression data interpretation in the context of phenotype outcome is evident in a study investigating methapyrilene-induced acute toxicity in rats (Hamadeh et al., 2002d). In that study, a numerical representation was devised to represent the severity of various histopathologic lesions observed in livers of exposed animals. This histopathologic scoring system (Fig. 9-7) was in agreement with the pattern of similarity generated by unsupervised hierarchical clustering of genes altered by chemical exposure. This agreement was true in both cases where all lesions or hepatocellular-specific lesions were used for scoring, suggesting that gene expression profiles were a culmination of all lesions even when alterations emanating from nonhepatocellular cell populations were diluted in the total mass of the liver.

Arguably, distinction between samples derived from the same compound exposure conditions with different grades of toxicity is more challenging than separating samples derived from tissues exposed to different compounds, because in the latter case the procedure is aided by unique metabolism and signaling profiles inherent to the pharmacology or modality of the various compounds. This issue has been visited in detail by Hamadeh and coworkers in a study that investigated the separation of gene expression profiles related to pharmacology from those related to the ensuing toxicity (Hamadeh et al., 2002e). In addition, gene expression derived from this study corroborated the main lesions observed in the livers of methapyrilene-treated rats. Gene expression observed to be modulated in a dose- and time-dependent fashion fell

$$\text{Sample score} = \sum_{\theta=1}^{x} \sum_{s=1}^{x} n_{s\theta}s$$

n is number of animals with lesion θ and severity s
x is number of replicate animals within dose-time group

Figure 9-7 *Formula used to derive numerical histopathological scores for toxicant-exposed replicate animals (Hamadeh et al., 2002d).*

loosely into four major categories: cell death, cell division or regeneration, impaired fatty acid metabolism/transport, and inflammatory pathways. These observations mirrored the hepatocellular necrosis, bile duct hyperplasia, lipid vacuolation, and inflammation observed in a dose- and time-dependent manner.

The combination of anchorage of pathologic observation with gene expression alterations and the demonstration of computational procedures for ranking toxicity incurred in chemically exposed animals hold promise to reveal toxicity-specific biomarkers that may aid in risk assessment processes.

CHALLENGES AND LIMITATIONS

Dilution Effects

When profiling tissues, several considerations must be kept in mind in order to gain the most information from the data set being analyzed. Tissues, unlike cells in culture, are a heterogeneous collection of cells that interact with each other and are involved in very complex signaling pathways that affect homeostasis. Liver tissue is used as an example to highlight the importance of dilution of gene expression events when analyzing gene expression sets, but this paradigm is applicable to most other organs due to their heterogeneous nature.

Liver tissue is made up of a variety of cell types including (1) hepatocytes, which comprise the majority of the liver and are responsible for many functions including metabolism and detoxification; (2) Kupffer cells, which constitute about 2% of the liver volume and are the macrophages that reside in the subepithelial spaces; (3) stellate cells, which constitute about 1.4% of the total liver volume and are fat-storing cells which when activated can be transformed into fibroblast-like cells that can deposit collagen; (4) endothelial cells, which constitute about 2.8% of liver tissue and are present in the sinusoidal spaces; and (5) infiltrating blood cells, including red and white blood cells, which can vary in numbers depending on how well the tissue is drained of blood at the time of harvest. Consequently, gene expression profiling of whole tissue will reflect the status of mRNA levels corresponding to genes in the multiple cell types that constitute that tissue. This challenge is faced in multiple situations, and if not considered, could lead to erroneous interpretation of high-density gene expression data. Below are general cases where certain parameters should be collected to help in the sound interpretation of data from heterogeneous tissue samples.

Localized Lesion. Profiling of different tumor types was made very popular by a series of manuscripts during the infancy of microarray technology. A parallel effort in toxicology would be the profiling of lesions or histopathological alterations incurred in livers, kidneys, or other organs from toxicant-exposed animal models. While tumor lesions are of sufficient volume and relatively easy to extract/carve from surrounding normal tissue, that is not the case with lesions such as bile duct hyperplasia, single-cell necrosis, or lipid vacuolization, which arguably are more challenging to isolate and affect a much smaller population of cells than a tumor. Laser capture microdissection (LCM) offers the luxury of isolating cells deemed relevant to the observed lesion; however, this practice remains somewhat expensive and labor intensive, and for some lesions sufficient quantities of RNA are very hard to obtain. Further refinement of the ensuing quality and quantity of RNA will undoubtedly evolve this tool into an indispensable aid in studies of this nature.

But in the absence of LCM, profiling tissue with sporadic lesions creates a situation where gene expression from the lesion and surrounding cells is diluted in the gene expression resulting from the whole tissue. In the hypothetical example depicted in Figure 9-8, gene x is induced 150-fold in cells within the developed lesion but is unchanged in the surrounding normal tissue. If mRNA were to be prepared from a portion of the liver of which the lesion comprises only 1% in volume/weight, then a quick mathematical analysis would indicate that theoretically the 150-fold increase would appear only as a 1.5-fold induction. This phenomenon is referred to as the dilution effect, where the alteration of gene expression by a relatively small population of cells is diluted by the mass of surrounding tissues that do not regulate those genes in the same fashion.

Hamadeh and coworkers capitulated this phenomenon in an in vitro experiment by measuring global gene expression alterations as a result of spiked concentrations of RNA from one cell type into another. This experiment facilitated the monitoring of the amplitude of original gene expression changes as a function of dilution in a simplified model (Hamadeh et al., 2002f). The authors found that gene expression differences between the two cell lines

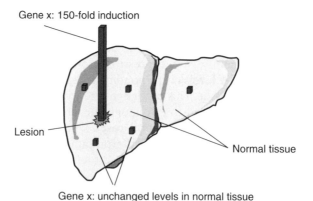

Figure 9-8 *Schematic diagram illustrating a theoretical yet plausible example where a gene might be elevated in a localized manner in liver tissue and how this elevation might be diluted if whole liver tissue was sampled for further gene expression studies.*

(HaCaT and MCF-7) were detected even after a 20-fold dilution factor was implemented. These results better our appreciation of the potential significance of biological alterations that comprise a relatively small percentage of an assayed organ and aid in the interpretation of the respective gene expression data. Pathologists involved in reporting the microscopic observations for toxicogenomics studies can play a major role in helping with this challenge by routinely recording the approximate percentages of tissue that display the reported lesion.

Peripheral Blood Cell Profiling. White blood cell gene expression profiling is gaining popularity due to its noninvasive nature, especially in human clinical studies where it is not possible to collect tissue samples. Gene expression profiles in those blood cells are used either as surrogates for toxicity in other organs from which it is a challenge to sample tissue or as surrogate markers of efficacy for drug exposure. The challenge of dilution becomes apparent again, since white blood cells are a heterogeneous mixture of numerous cell types such as neutrophils, basophils, eosinophils, and lymphocytes. Severe fluctuations in percentages of each cell type can significantly affect the interpretation of gene expression data, since these cell types might regulate different genes in various fashions. Thus alterations in gene expression profiles might be due to cell population shifts rather than induction or repression at the transcription level.

Separation of different subgroups of white blood cells is definitely possible using automated cell sorting equipment; however, the stress incurred by the cells might alter their original expression profile and compromise the quality of the harvested RNA. In the absence of separation of different cell types in the blood, the researcher is advised to collect data on the percentage of various cell types in the samples being studied, which will allow the determination of whether cell population fluctuations are the cause for the observed alterations in gene expression profiles. Alterations in blood cell populations, when they exist, are surely much cheaper and more efficient to monitor using cell-counting methods than implementing gene expression profiling endeavors.

Expressed Sequence Tags

Expressed sequence tags (ESTs) are small pieces of DNA sequence ($\sim$200–500 nucleotides) that are generated quite rapidly by sequencing the 3′, 5′, or both ends of an expressed portion of the genome. This is done under the premise that sequencing short pieces of DNA that represent genes expressed in certain tissues from different organisms will allow the discovery of novel genes in chromosomal DNA. EST analysis has already aided in the isolation of genes involved in Alzheimer's disease, colon cancer, and many other diseases (De Young et al., 2002; Brett et al., 2001; Vinals et al., 2001).

Similar opportunities are presenting themselves in the study of high-density gene expression and toxicology. Most DNA microarray chips contain an abundance of sequences corresponding to ESTs. Actually, ESTs often constitute the majority of clones present on a chip, and data generated surpass those corresponding to known genes. Thus in many instances the most altered clones in response to toxicant/drug exposure are nonannotated sequences, and they have also been shown to discriminate between different disease states of various toxicological end points resulting from drug/toxicant treatment of biological models. These phenomena are common in organisms such as mice, rats, and humans, but less prominent in others such as yeast, where a better, more complete understanding of the genome sequence exists.

CONCLUSIONS AND EXPECTATIONS

It has already been shown that high-density gene expression technologies are able to find signature profiles that are unique to specific toxicants (Waring et al., 2001a, b, 2003; Hamadeh et al., 2002a, b, c), and expectations remain high. It is anticipated that this technology will elucidate whether and how different cells in different tissues result in profoundly different response signatures for a given toxicant. It will also interrogate how different species show similar overlapping or distinct patterns of gene responses to a toxicant and the implications of that in terms of disease outcome and risk assessment. We will also learn about the dependency of the toxicant signature on the developmental stage of the biological system or the preexisting health status of the subject. Ultimately, the hope is that toxicogenomics will be able to highlight signatures as better indicators of responses to complex chemical mixtures. A very important outcome of this technology is the elucidation of unique responses to chronic low doses of toxicants or environmental pollutants that can be defined by gene expression profiling. An indirect outcome of this type of analysis is the ability to discover specific gene polymorphisms characteristic of an increased susceptibility to the pathology of environmental health diseases.

High-density gene expression profiling of toxicology studies aims at discovering gene targets that correlate with adverse effects. Those genes are usually part of complex physiological processes and are related to a multitude of toxicological end points. Thus interpretation of gene expression perturbation should be guided by other data such as clinical chemistry, microscopic observations, drug exposures, and other phenotype-defining parameters, some of which are described in more detail in chapter 8.

REFERENCES

Amin RP, Vickers AE, Sistare F, Thompson K, Roman RJ, Lawton M, Kramer J, Hamadeh HK, Collins J, Grissom S, Bennett L, Tucker CJ, Wilde S, Oreffo V, Davis J, Curtiss S, Naciff J, Cunningham M, Tennant R, Stevens J, Car B, Bertram TA, Afshari CA (2004). Determination of putative gene based markers of renal toxicity. *EHP Toxicogenomics*.

Anderson SP, Cattley RC, Corton JC (1999). Hepatic expression of acute-phase protein genes during carcinogenesis induced by peroxisome proliferators. *Mol Carcinogen* **26**:226–238.

Brett D, Kemmner W, Koch G, Roefzaad C, Gross S, Schlag PM (2001). A rapid bioinformatic method identifies novel genes with direct clinical relevance to colon cancer. *Oncogene* **20**(33):4581–4585.

Cabanes FJ, Accensi F, Bragulat MR, Castella G, Minguez S, Pons S (2002). What is the source of ochratoxin A in wine? *Int J Food Microbiol* **79**:213–215.

De Young MP, Damania H, Scheurle D, Zylberberg C, Narayanan R (2002). Bioinformatics-based discovery of a novel factor with apparent specificity to colon cancer. *In Vivo* **16**(4):239–248.

DeRisi J, Penland L, Brown PO, Bittner ML, Meltzer PS, Ray M, Chen Y, Su YA, Trent JM (1996). Use of a cDNA microarray to analyze gene expression patterns in human cancer. *Nat Genet* **14**(4):457–460.

Guarino AM, Miller DS, Arnold ST, Pritchard JB, Davis RD, Urbanek MA, Miller TJ, Litterst CL (1979). Platinate toxicity: past, present, and prospects. *Cancer Treat Rep* **63**:1475–1483.

Hamadeh HK, Knight BL, Haugen AC, Sieber S, Amin RP, Bushel P, Stoll R, Blanchard K, Jayadev S, Tennant R, Cunningham M, Afshari CA, Paules RP (2002a). Methapyrilene toxicity: anchorage of pathologic observations to gene expression alterations. *Toxicol Pathol* **30**(4):470–482.

Hamadeh HK, Bushel P, Martin K, Bennett L, Paules R, Afshari CA (2002b). Detection of diluted gene expression alterations using cDNA microarrays. *BioTechniques* **32**:322–329.

Hamadeh HK, Bushel PB, Jayadev S, Martin K, DiSorbo O, Sieber S, Bennett L, Tennant R, Stoll R, Barrett JC, Blanchard K, Paules RS, Afshari CA (2002c). Gene expression analysis reveals chemical-specific profiles. *Toxicol Sci* **67**:219–231.

Hamadeh HK, Bushel PB, Jayadev S, DiSorbo O, Bennett L, Li L, Tennant R, Stoll R, Barrett JC, Blanchard K, Paules RS, Afshari CA (2002d). Prediction of compound signature using high density gene expression profiling. *Toxicol Sci* **67**:232–240.

Hamadeh HK, Li L, Stoltz J, Bushel PR, Stoll R, Blanchard K, Jayadev S, Afshari CA (2002). Elucidation of signal versus effect in toxicogenomic studies. *J Appl Genomics Proteomics* **1**(2):109–121.

Hamadeh HK, Amin RP, Paules RS, Afshari CA (2002). An overview of toxicogenomics. *Curr Issues Mol Biol* **4**:45–56.

Han WK, Bailly V, Abichandani R, Thadhani R, Bonventre JV (2002). kidney injury molecule-1 (KIM-1): a novel biomarker for human renal proximal tubule injury. *Kidney Int* **62**(1):237–244.

Huang Q, Dunn RT, II, Jayadev S, DiSorbo O, Pack FD, Farr SB, Stoll RE, Blanchard KT (2001). *Toxicol Sci* **63**:196–207.

Ichimura T, Bonventre JV, Bailly V, Wei H, Hession CA, Cate RL, Sanicola M (1998). Kidney injury molecule-1 (KIM-1), a putative epithelial cell adhesion molecule containing a novel immunoglobulin domain, is up-regulated in renal cells after injury. *J Biol Chem* **273**(7):4135–4142.

Luhe A, Hildebrand H, Bach U, Dingermann T, Ahr H-J (2003). A new approach to studying ochratoxin A (OTA)-induced nephrotoxicity: expression profiling in vivo and in vitro employing cDNA microarrays. *Toxicol Sci* **73**:315–328.

Perou CM, Sorlie T, Eisen MB, van de Rijn M, Jeffrey SS, Rees CA, Pollack JR, Ross DT, Johnsen H, Akslen LA, Fluge O, Pergamenschikov A, Williams C, Zhu SX, Lonning PE, Borresen-Dale AL, Brown PO, Botstein D (2000). Molecular portraits of human breast tumours. *Nature* **406**(6797):747–752.

Petrik J, Zanic-Grubusic T, Barisic K, Pepeljnjak S, Radic B, Ferencic Z, Cepelak I (2003). Apoptosis and oxidative stress induced by ochratoxin A in rat kidney. *Arch Toxicol* **77**(12):685–693.

Schena M, Shalon D, Davis RW, Brown PO (1995). Quantitative monitoring of gene expression patterns with a complementary DNA microarray. *Science* **270** (5235):467–470.

Seshadri V, Fox PL, Mukhopadhyay CK (2002). Dual role of insulin in transcriptional regulation of the acute phase reactant ceruloplasmin. *J Biol Chem* **227** (31):27903–27911.

Shalon D, Smith SJ, Brown PO (1996). A DNA microarray system for analyzing complex DNA samples using two-color fluorescent probe hybridization. *Genome Res* **6**(7):639–645.

Taniwaki MH, Pitt JI, Teixeira AA, Iaminaka BT (2003). The source of ochratoxin A in Brazilian coffee and its formation in relation to processing methods. *Int J Food Microbiol* **82**:173–179.

Vinals C, Gaulis S, Coche T (2001). Using in silico transcriptomics to search for tumor-associated antigens for immunotherapy. *Vaccine* **19**(17–19):2607–2614.

Wagner RF, Beiden SV, Campbell G, Metz CE, Sacks WM (2002). Assessment of medical imaging and computer-assist systems: lessons from recent experience. *Acad Radiol* **9**(11):1264–1277

Waring JF, Ciurlionis R, Jolly RA, Heindel M, Ulrich RG (2001a). Microarray analysis of hepatotoxins in vitro reveals a correlation between gene expression profiles and mechanisms of toxicity. *Toxicol Lett* **120**(1–3):359–368.

Waring JF, Jolly RA, Ciurlionis R, Lum PY, Praestgaard JT, Morfitt DC, Buratto B, Roberts C, Schadt E, Ulrich RG (2001b). Clustering of hepatotoxins based on mechanism of toxicity using gene expression profiles. *Toxicol Appl Pharmacol* **175**(1):28–42.

Waring JF, Cavet G, Jolly RA, McDowell J, Dai H, Ciurlionis R, Zhang C, Stoughton R, Lum P, Ferguson A, Roberts CJ, Ulrich RG (2003). Development of a DNA microarray for toxicology based on hepatotoxin-regulated sequences. *EHP Toxicogenomics* **111**(1T):53–60.

10

The Use of a Compendium of Expression Profiles for Mechanism of Toxicity Predictions

Jeffrey F. Waring, Xudong Dai, Yudong He, Pek Yee Lum, Christopher J. Roberts, and Roger Ulrich

HISTORICAL PERSPECTIVE ON USE OF MICROARRAYS IN TOXICOLOGY

Humble Beginnings: Using Yeast as a Test System

The yeast *Saccharomyces cerevisiae* was the first organism for which the entire genomic sequence was available (Goffeau et al., 1996). In addition, the paucity of introns in the yeast genome made identifying the ~6,000 yeast genes fairly easy, and thus yeast was the first organism for which a whole-genome DNA microarray was available (DeRisi et al., 1997). The first study utilizing these microarrays involved profiling a time course of the yeast vegetative growth cycle from early lag phase to stationary phase (DeRisi et al., 1997).

Toxicogenomics: Principles and Applications. Edited by Hamadeh and Afshari
ISBN 0-471-43417-5   Copyright © 2004 John Wiley & Sons, Inc.

Yeast cells transition from glucose fermentation to ethanol oxidation upon the exhaustion of glucose from the culture. Whole-genome expression profiling showed that this transition involves a very large transcriptional program of hundreds of genes. Because even in the late 1990s the annotation of yeast genes was fairly extensive, a detailed biological pathway analysis was possible and showed that nearly every gene in the glycolytic pathway was downregulated, whereas genes involved in the TCA and glyoxylate cycles were upregulated. A similar key study in yeast was performed using Affymetrix microarrays, in which the growth of yeast cells in rich versus synthetic minimal medium was compared (Lockhart et al., 1997). Other important studies in yeast that demonstrate different important applications of whole-genome expression profiling included the comprehensive analysis of other developmental or cellular perturbations, such as sporulation (Chu et al., 1998) and mating and MAPK signaling pathways (Roberts et al., 2000), mapping known and unknown promotor binding sites through bioinformatics analysis of the promoter regions of coregulated genes (Tavazoie et al., 1999), as well as demonstrating the power of combining expression profiling and proteomic approaches (Ideker et al., 2001).

Gene Expression in Toxicology Research

The use of gene expression monitoring in toxicology research is not new. This is illustrated by at least two decades of research on the induction of genes by the environmental toxicant 2,3,7,8-tetrachlorodibenzo-p-dioxin (TCDD) using such techniques as Northern blots. Differential display techniques were later used to identify other genes regulated by TCDD; for example, four genes were identified to be regulated by TCDD in HepG2 cells using polymerase chain reaction (PCR) and differential display (Wang et al, 1996). Though labor intensive by today's standards, these approaches built the foundation for understanding at the molecular level how polycyclic aromatic hydrocarbons exert their effects on cells. More recently, using a cDNA array containing 9,182 probes, Puga et al. (2000) reported the transcriptional regulation of several hundred genes by TCDD in HepG2 cells that could be categorized according to known multiple effects of this toxin.

Currently gene expression profiling is being used to examine the effects of toxins on a variety of cell and tissue types. The focus has been on target organs for toxicity, primarily liver (Bartosiewicz et al., 2001; Bulera et al., 2001; Gerhold et al., 2001; Waring et al., 2001a) and kidney (Huang et al., 2001). Though most published studies on toxicogenomics are directed toward showing that gene expression results obtained using microarrays and standard compounds match those results previously published, investigators are increasingly applying microarray analysis to experimental drugs.

On-Target and Off-Target Drug Effects

Drugs have both desirable and undesirable effects (Fig. 10-1). In drug discovery, scientists search for potent inhibitors (poisons) for enzymes and other proteins that will interfere with or correct a disease pathway. Intended outcomes based on interaction with the intended target are referred to as on-target effects. Chemical leads, typically identified through high-throughput screening with isolated targets, are modified and refined to increase potency toward the selected protein target and away from a few closely related targets. However, potency does not imply selectivity, and screening closely related targets does not imply safety; though each drug may have its intended target, there are inevitably multiple untoward targets.

Untoward drug responses, including toxicity and drug interactions, are usually due to interaction with unintended molecular targets and are described as off-target effects. It is typically the case that any drug will have multiple targets, including the initial therapeutic target and additional untoward targets that may contribute both to the efficacy and toxicity profile of the drug. For example, the nonsteroidal antiinflammatory and analgesic drug ibuprofen is believed to work by inhibition of prostaglandin H synthases (also called cyclooxygenases or COX-1 and COX-2), thus inhibiting the conversion of arachidonic acid to proinflammatory prostaglandins. Increased selectivity

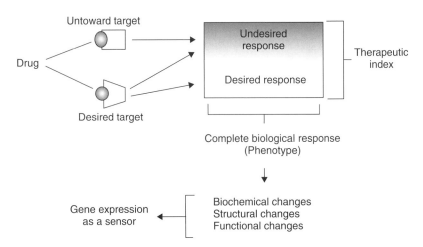

Figure 10-1 *Drugs are currently designed to interact with the desired molecular target, delivering the desired response (on-target effects). However, undesired responses or adverse effects can also be elicited through the intended target (mechanism-based toxicity). Most drugs also interact with several other unintentional targets, stimulating an undesired response or toxicity that could not be predicted based on the desired target (off-target effects). It is the combination of desired and undesired responses that make up the therapeutic index or ratio responsible for phenotype. Using the output of the genome as a sensor, it is possible to explore the molecular and cellular basis for the complete biological response to xenobiotic exposure.*

for COX-2 with new compounds has led to safer drugs (Hawkey, 1999). However, ibuprofen has also been shown to be an activator of the peroxisome proliferator-activated nuclear receptors PPARα and PPARγ, which may also contribute to its pharmacologic effects (Jaradat et al., 2001). Further, it has been shown to be a substrate for conjugation to coenzyme A (a step in its chiral inversion; Tracy et al., 1993) and an inhibitor of mitochondrial beta-oxidation (Browne et al., 1999). Hence, while the therapeutic activity of this drug is attributed to the cyclooxygenase target, the off-target effects may also contribute to efficacy and likely contribute to toxicity, which ranges from gastrointestinal ulcers to rare idiosyncratic toxicities in liver and muscle.

It has long been clear that DNA microarray expression profiling could provide an enormous amount of information regarding the effects of compounds on cells (Schena et al., 1995; Debouck and Goodfellow, 1999). An early study in yeast extended this notion significantly, bringing forth the decoder concept for determining the on-target and off-target effects of compounds. This approach uses cell lines harboring genetic deletions of the gene encoding the drug target to serve as a model for an ideal inhibitory compound, which gives 100% target inhibition with no off-target effects (Marton et al., 1998).

This study involved the immunosuppressive drugs FK506 and cyclosporin, which bind to FKBP or cyclophilins in yeast and mammalian cells, in turn inhibiting calcineurin (Hemenway and Heitman, 1999). The expression profiles of FK506 and cyclophilin treatments of yeast were compared to the calcineurin mutant expression profile. These profiles were found to be highly correlated, thus demonstrating that the FK506 target and calcineurin function along the same biochemical pathway.

This result validates the utility of using expression profiling to rapidly understand the effects of compounds on cells, as it took significant time and resources to determine through classical biochemistry that FK506 blocked calcineurin signaling (Liu et al., 1991). Furthermore, the calcineurin inhibition signature caused by the immunosuppressive drugs was eliminated in mutants lacking the FKBP or cyclophilin proteins. However, at high doses of FK506, a new, off-target signature for amino acid starvation was induced. This was consistent with the fact that high concentrations of FK506 inhibit amino acid import in yeast (Heitman et al., 1993). This off-target signature was eliminated in a mutant lacking the amino acid starvation response; however, this mutant showed a second off-target signature caused by high-concentration FK506 inhibition of a drug efflux pump (Egner et al., 1998). Thus, profiling the effects of compounds in mutant yeast strains was shown to be effective in parsing on-target versus off-target effects.

EXPERIMENTAL DESIGN FOR CONSTRUCTION OF A COMPENDIUM OF EXPRESSION PROFILES

Introduction

In constructing a compendium of expression profiles, a number of questions need to be addressed. The first and most obvious question is, What is a compendium of expression profiles? A compendium can be defined as a collection or database of gene expression profiles with multiple axes such as cell or tissue types, time points, and compound treatments. A compendium can have depth in all three axes or can be focused on one or two axes. In addition, a compendium can be extended with bridging or common compounds as controls.

Because a compendium is an agglomeration of many profiles, it is possible to obtain a high-level view of responses induced by a compound across many axes. Often, by examining the gene changes that occur from a number of similar tissue types or cellular conditions, it becomes possible to learn about entire biological pathways, to classify tissue types, and to identify the function of previously unknown genes (Golub et al., 1999; Hughes et al., 2000). This type of database is by no means limited to toxicogenomic analysis. Several groups have built compendiums of expression profiles from cardiovascular tissue (Hwang et al., 1997), multiple myeloma tissue (Claudio et al., 2002), and immune cells (Shaffer et al., 2001). In all of these cases, critical information was obtained from the compendium concerning gene expression signatures, cell differentiation stages, and signaling pathways.

How can a compendium assist in determining potential toxic liabilities of new drug candidates? The answer lies in the hypothesis that when cells are treated with a toxin, the resulting gene expression changes or gene expression profile will be reflective of the mechanism of toxicity of the compound. Thus, by comparing the gene expression profiles of an unknown compound against a reference database, it may be possible to infer potential toxic liabilities of a new compound based on guilt by association.

Of course, it is not always necessary to have a database of expression profiles in order to utilize gene expression in toxicology. If there is some indicator of what type of toxicity to expect, then a database becomes less important. If a compound is suspected to result in cell death, for instance, you could concentrate on gene changes involved in proliferation and cell cycle regulation. Likewise, if a compound is suspected to be a mitochondrial-damaging agent, you could concentrate on the regulation of genes involved in mitochondrial function.

However, much emphasis is placed on utilizing toxicogenomics at a very early stage of drug discovery before any knowledge of potential toxicity exists. At this point, a reference database becomes critical, since it allows for potential identification of a possible toxic liability without any previous knowledge of the compound. As already mentioned, if the expression profile from the test compound is statistically similar to a reference compound, the potential toxicity of the reference and test compound will very likely be related.

Before beginning construction of a compendium, a number of issues need to be addressed, including what controls should be used and how many replicates should be performed, what reference compounds should be used, how many reference compounds are necessary, which system should be used (cell lines, primary cells, rats, mice, etc.), what time points should be used, and what end points should be monitored for the in-life studies. There are no simple answers to these issues, and different groups have had success using different methods. This chapter reviews some of the work that has been done in this area and presents possible methods for construction of an expression database.

Controls and Replication

One of the most important requirements for performing a successful micro-array experiment is to include sufficient control hybridizations to ensure a complete understanding of the sources of variability inherent in the micro-array platform and the biological system being tested. Replication of the experiment needs to occur at both the level of the hybridization and at the level of the biological treatment. The number of replicates required for trustworthy results must be determined empirically. At Rosetta Inpharmatics, two-color hybridizations on cDNA or FlexJet microarrays have typically been run on pairs of microarrays, with the labeling of the samples reversed ("dye-flip") to eliminate color bias. Regarding biological variability, control hybridizations should include "same versus same" hybridizations, which control for the variability in gene expression observed in cells or tissues treated nominally identically (e.g., see fig. 2 in Hughes et al., 2001). Furthermore, the experimental perturbations being analyzed should be repeated a sufficient number of times so that the variability of the biological system is controlled for.

The number of replicates for a given biological system will depend on the question being addressed in the experiments, as well as the inherent variability in the system being tested. Controls should be dosed using the same vehicle and by the same route as those in experimental groups. When compared to a pooled control, each control animal produces a gene expression pattern that shows variation from the pool (Fig. 10-2). When several control samples are compared side-by-side (Fig. 10-3), it is seen that each sample differs not only from the pool but from the other samples. Since many of these same genes are also seen to fluctuate in experimental animals, it is essential to recognize that some gene changes, though significant, are not part of the compound signature and can be observed in controls.

System Selection

Another fundamental question to be addressed before construction of a compendium is, What species or cell line should be used? If performing the gene expression analysis in an animal such as rat or mouse, what organs should be

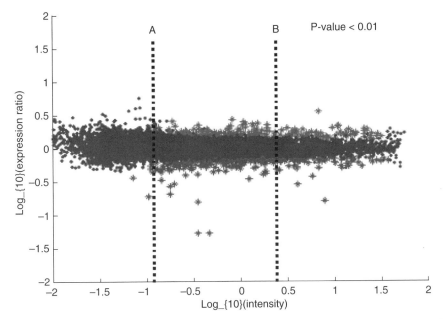

Figure 10-2 *A "same versus same" comparison of a control rat against a control mRNA pool illustrating normal biological fluctuations in a single animal compared to the average levels. Black reporters along the horizontal axis represent genes that do not differ from the pool in this control animal; top gray reporters indicate genes that are upregulated and bottom gray reporters shows genes that are downregulated. Line A represents about 1 mRNA copy per cell; line B represents about 10 mRNA copies per cell.*

assayed? There have not been any formal studies conducted on these questions. Certainly, the question of what species to use is an important one. Is a gene expression pattern observed in a rat predictive of what would occur in humans? Gene expression is in the same position as other more traditional methods of toxicity assessment. Compounds that may result in elevated AST and ALT measurements in rats, for instance, do not always do so in humans, and vice versa. Similarly, a gene expression pattern observed in a rat may not occur with the same compound in another species. If treatment with a compound results in similar toxicity across multiple species, the type of species used is not as critical. Many of the compendiums for toxicology have been constructed in rat, since the rat is the usual species evaluated for toxicity evaluation. However, if you wish to study a particular toxicity that occurs only in certain species, then the compendium should be constructed using expression profiles from that species.

The question regarding what organ to assay for gene changes again depends on what type of toxicity is being studied. Is it necessary to construct a compendium for every potential target organ, or will expression profiles translate across organs? Several studies have shown that treatment with a compound

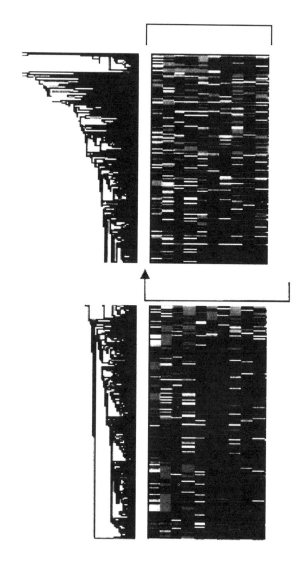

Figure 10-3 Heat map showing differences in gene expression between control rat liver samples (same versus same comparison). About 1,900 genes (horizontal axis) from a 25,000-gene microarray show differential regulation in 10 controls (vertical axis); these "jumpy" genes are thought to represent normal biological fluctuations in message levels.

results in very different gene changes from organ to organ. Huang et al. (2001) studied the gene expression changes resulting from treatment with the renal toxin cisplatin in Sprague-Dawley rats. Microarray results revealed the regulation of genes involved in apoptosis, calcium homeostasis, and tissue remodeling in the kidney. None of these changes were observed, however, in the liver. Similarly, a microarray study performed on ob/ob mice treated with an antisense to protein tyrosine phosphatase 1B revealed numerous gene changes in adipose tissue and liver involved in lipogenesis and gluconeogenesis. However, none of these changes were observed in muscle (Waring et al., 2003b). Thus, while there may be some exceptions, gene changes resulting from treatment with a xenobiotic do not generally translate across different organs. Therefore, a compendium dedicated to hepatotoxicity would likely not be predictive for nephrotoxicity and vice versa.

These findings strongly support the need for a compendium that surveys across many multiple tissues, especially when an unbiased or unfiltered view of the response of a certain drug is needed and when the target tissues are not clearly defined. A compendium deep in the tissue axis will be able to give information on target tissue(s) as well as unresponsive tissues. When these data are obtained, another compendium focusing on the target tissue, such as the liver for hepatotoxicity, can be built, this time expanding on another critical axis. This build-on approach is critical for the economics of building a database, especially when budget is a major concern.

There has been a strong desire to move toxicogenomics into in vitro settings. This would allow for the testing of new drug candidates at a very early stage of drug discovery, since the assay could be done using milligram amounts of compound. The approach here would be to first identify a cell line that would suitably reflect the expression in vivo. There is little disagreement that expression profiling in an in vitro system would not result in the exact same gene expression changes as observed in vivo. The key issue would be to determine whether the number of common changes between in vivo and in vitro is sufficient to allow the in vitro system to be used as a less material and less time-consuming screen.

Several groups have compared gene expression changes between in vitro and in vivo settings. Microarray analysis was used to identify gene expression changes resulting from treatment with 15 hepatotoxins in isolated rat hepatocytes (Waring et al., 2001b) or in livers from rats treated for 3 days (Waring et al., 2001a). In this case, allyl alcohol and carbon tetrachloride gave very similar expression profiles in vivo yet did not regulate many genes in common in vitro. You will have to judge if the number of commonly regulated signatures is sufficient to utilize the in vitro screen as a surrogate for screening in vivo.

Input Selection

The next question to be addressed when constructing a database of expression profiles is, How large of a database needs to be constructed? How many reference compounds should be used, and how many genes or sequences should be assayed for regulation? The common belief is that a database needs to be very large in order to be able to predict toxicity with any degree of accuracy. However, a number of groups have had success constructing relatively small databases.

A database composed of only 15 hepatotoxins using a microarray chip with only 1,000 genes was large enough to distinguish different mechanisms of toxicity (Waring et al., 2001a). This same database was used to determine the toxic mechanism behind a previously unexplored drug lead, an experimental thienopyridine inhibitor of NFkB-regulated adhesion proteins. The results showed that the expression profiles for the thienopyridine compound clustered with 3-methylcholanthrene and aroclor from both high- and low-dose treated animals. In addition, individual gene changes induced by the thienopyridine compound suggested that it resulted in DNA damage through an interaction with the aromatic hydrocarbon receptor (Waring et al., 2002). Thus, a viable hypothesis regarding the mechanisms of toxicity was devised, and importantly, this was done using a database composed of relatively few reference compounds and a small number of genes.

A similar study was performed by Bulera et al. (2001). A database was composed of expression profiles from rats treated with six hepatotoxins: microcystin, phenobarbital, lipopolysaccharide, carbon tetrachloride, thioacetamide, and cyproterone acetate. These compounds represented a range of different toxicities, including enzyme induction, neutrophil inflammation, and reactive metabolite induction. The studies were performed using a custom-designed microarray chip containing 1,600 rat genes. Despite the small number of compounds, expression profiling distinguished these groups of toxicants and correctly identified a blind sample by correlating its expression profile to the expression profiles in the database.

The success of these studies demonstrates that a large database may not always be necessary in order to utilize gene expression to predict toxicity. In fact, several studies have shown that the number of compounds and the size of the microarray chip are not the critical parameters; rather the critical parameter is that the database must be large enough to generate predictive gene sets. Several studies have demonstrated that focusing on a smaller number of genes can increase the predictive power of expression databases. This was shown in a study using expression profiling to predict the clinical outcome of breast cancer (van't Veer et al., 2002). The authors found that by using unsupervised clustering of 25,000 genes on a microarray chip, they were able to separate tumors from patients who developed distant metastases within 5 years from those who did not with approximately 70% accuracy. However, when the authors focused on 70 genes from the 25,000 that were highly correlative for tumor progression, the accuracy of the expression profiling increased to approximately 90%.

A similar result relating to chemical toxicity was obtained by Burczynski et al. (2000). In this study, HepG2 cells were treated with 100 different toxic compounds representing cytotoxic antiinflammatory drugs and DNA-damaging agents and were analyzed for gene expression using a 250-gene chip. The study showed that using all 250 genes on the chip, in this case the authors were unable to distinguish the two different toxicant classes. This was attributed to a high degree of variability in gene expression from one study to the next. However, when the authors conducted multiple repeat studies using one compound from each class, they were able to identify a much smaller number of genes that proved highly accurate in distinguishing the two classes of hepatotoxins.

Predictive gene sets for toxicity were also identified in a recent study by Hamadeh et al. (2002). In this study, the authors used a database composed of four different hepatotoxins representing two classes of hepatotoxicity: enzyme induction and peroxisomal proliferation. Despite the fact that only four reference compounds were used, it was possible to identify genes unique for each class; these genes were then used to interrogate expression profiles from blinded samples composed of phenytoin, diethylhexylpthalate, and hexobarbital. The unknown samples were correctly classified as enzyme inducers or peroxisome proliferators.

Another example of how predictive gene sets were identified from a small reference database was demonstrated by Thomas et al. (2001). The authors treated mice with 24 model compounds that fell into five different categories of hepatotoxicity: peroxisome proliferation, aryl hydrocarbon receptor agonists, noncoplanar polycholorinated biphenyls, inflammatory agents, and hypoxia-inducing agents. None of these categories had more than three compounds representing it, and most had only two. Nonetheless, the authors were able to identify a set of 12 predictor transcripts that theoretically could classify new compounds that fall within these classes. More information on the statistics used to identify the 12 predictor transcripts can be found under "Modeling and Prediction Based on Gene Expression."

Thus in most situations the number of reference compounds and interrogated genes should be large enough to allow for the identification of predictive gene sets. The question then is, How many compounds and genes are necessary for this? In terms of the number of genes, it is preferable to start with as large a number as possible. This increases the likelihood that a gene, which may be very significantly regulated by a class of toxins, is represented. We constructed a microarray chip composed of 25,000 oligonucleotides representing genes and ESTs. Analysis of different hepatotoxins revealed that the compounds regulated a number of sequences that have never been associated with toxicity. Had we attempted to construct a microarray chip composed entirely of genes that we expected to be important for toxicity, many of these sequences would not have been included (Waring et al., 2003a).

The question regarding the necessary number of reference compounds for construction of a compendium depends largely on the class of toxins. We have constructed a reference database for hepatotoxicity composed of over 70 compounds, both hepatotoxic and nonhepatotoxic. We have attempted to isolate predictive gene sets for different classes of hepatotoxicity. For some of these classes, very few compounds were necessary. For example, identifying a set of genes highly predictive for compounds that are ligands for the Ah receptor was possible using 2–3 reference compounds. Identifying a predictive gene set for phospholipidosis, however, was not possible using 3 reference compounds. Thus at the present time the only way to identify the necessary number of reference compounds for a given class of hepatotoxicity is by performing background experiments and statistical analysis for each class.

Treatment Duration and Dose Levels

In the making of a compendium, there are many axes that need to be considered. Needless to say, two important axes are treatment duration and dose levels. The optimal treatment and dose level really depends on the question being asked. For instance, the optimal treatment duration for an acute toxin may differ from that of a chronic toxin. A key issue when determining length and dose of treatment period is to ensure that a measurable outcome is achieved. This is necessary in order to link the gene expression data with another measurement. Thus if you are profiling compounds that result in DNA damage, it is essential that the dose level of the compound will result in this outcome.

In many cases, the RNA should be harvested prior to the observation of DNA damage, since once DNA damage has occurred, expression changes unique to this mechanism of toxicity may be masked by other changes related to such things as necrosis, fibrosis, and inflammation. Thus it is often helpful to perform smaller pilot experiments in order to determine the dose and duration range for larger studies that will comprise the compendium. In addition, it is also possible to determine optimal dose and duration parameters from previous experiments reported in the literature. More information regarding linking gene expression to other end points is discussed in the next section.

In summary, the overall goal of constructing a compendium for toxicology is to be able to identify gene sets that are predictive for a given toxicity. The type of system, number of compounds and probe sets, and dose levels and treatment duration can all vary depending on the questions being asked. Constructing a smaller compendium, which previous studies have shown can be highly effective in identifying and predicting toxicity, may be an optimal starting point. You can then build onto the compendium, adding in additional compounds, dose levels, and treatment periods.

MODELING AND PREDICTION BASED ON GENE EXPRESSION

Unsupervised Approaches

Algorithms used for gene expression analysis can be divided into two major categories: unsupervised and supervised approaches. Unsupervised approaches include various types of clustering algorithms such as hierarchical clustering (bottom-up agglomerative and top-down divisive), partitioning clustering (k-means and k-medians), and self-organizing maps (Eisen et al., 1998, Alon et al., 1999, Tamayo et al., 1999). Genes are clustered together based on their similarity calculated from gene expression changes over a set of experiment conditions. At the same time, experiment conditions are grouped based on their similarity calculated according to gene expression. From the groups of gene sets or experiment condition, it is possible to study the function of genes and relevant biological pathways. Independent of clustering algorithms, the similarity metric plays an important role in defining clustering patterns (Roberts et al., 2000). Among many choices of similarity metrics are distance-based metrics such as Euclidean, city block, and Minkowski; correlation-based metrics such as Pearson correlation coefficient and rank correlation; and density-based metrics such as mutual information.

Classification Methods

Supervised approaches take the advantage of end-point data in selecting genes whose expression changes correlate with the end-point data of interest. Classification methods for category end-point data include the naive Bayesian classifier, support vector machine-based classifier, nearest neighbor classifier, linear discriminator classifier, and neural net–based classifier. Most of these algorithms have been used traditionally in other fields (Alizadeh et al., 2000; Golub et al., 1999; Khan et al., 2001; van't Veer et al., 2002).

The use of these classification methods based on expression data has been recently demonstrated for diagnostics and prognosis in clinics, for chemical compound classification in drug discovery, for biomarker discovery in assay screening, and for target discovery for certain therapeutic areas. A trend of using biomarkers is to report the end-point data of interest based on gene expression of a set of biomarkers. Various types of linear regression models have been used. Often the end-point data are not one single variable. Instead, the interest is in multiple variables simultaneously associated with gene expression, so it is crucial to identify a set of biomarkers that report for a group of variables of interest. Partial least square methods have been used to report multiple variables as end-point data of interest based on expression data.

Expression Profiles and Toxicity End Points

Methods to associate expression profile with traditional toxicity end points are provided to demonstrate the utilities of these analytical methods in the field of toxicogenomics. Although clinical chemistry indicators are the only end points in the following specific examples, the same analytic method can be directly applied to other clinical end points such as degree of mutagenicity. In the liver minicompendium study of Waring and Ulrich, the correlation coefficients were calculated between the clinical chemistry and gene expression regulation (Waring, 2001a). Specifically, eight clinical chemistry measurements—ALT, AST, ALP, cholesterol, bilirubin, glucose, GGT, and BUN—were used. The logarithmic ratio of the treated and control clinical chemistry measurements was regressed with the error-weighted logarithmic ratio values for each gene across 15 treatments. The correlation coefficient was obtained for each pair of clinical chemistry and individual genes by least squares fitting. Based on the correlation coefficient, genes whose expression regulation was highly correlated with clinical chemistry measurements or other toxicity end points could be selected as candidate predictors for the degree of toxicity.

Signature Gene for Better Classification of Unknown Compounds

Classification of compound toxicity is another focus of toxicogenomics. To predict the class of toxicity associated with an unknown compound, a set of signature genes can be identified by an anchor compound-based approach. A variety of studies have demonstrated the feasibility of such a strategy (Thomas et al., 2001; Hamadeh et al., 2002).

Hamadeh et al. (2002) reported a discriminative gene set between the peroxisome proliferator and enzyme inducer phenobarbital. There were three compounds within the peroxisome proliferator group and one compound in the enzyme inducer. Genes significantly different between the peroxisome proliferator–treated group and the enzyme inducer were identified by an ANOVA test. Then a linear discriminator function was established based on the training set consisting of a peroxisome proliferator anchor compound and phenobarbital. This study illustrates the feasibility of the anchor compound approach.

Work by Thomas et al. (2001) identified gene sets that differentiated more comprehensive classes of compounds. Five classes of compounds were included in their data set. Two or three anchor compounds were selected for each class. To screen for genes that could discriminate between groups, a Bayesian statistics–based method was applied. In particular, it was assumed that predictor variables $X1, \ldots Xk$ were independent of each other when conditioned on the class variable C. The classification model was constructed by the joint probability distribution for an expression compound class vector $(x,c) = (X1 = x1, \ldots Xk = xk, C = c)$, where X is the gene or transcription-based predictor as written in the following equation:

$$P(x,c) = P(C = c)\prod_{i=1}^{k} P(X_i = x_i | C = c)$$

where X is the gene or transcription based predictors , C is defined as the class of the compound that causes the corresponding regulation in expression.

The conditional predictive distribution for the class c given x and data set D, therefore, is:

$$P(c|x, D) = \frac{P(c, x|D)}{P(x|D)}$$

where the numerator is defined as

$$P(c,x|D) = \frac{t_c + 1}{N_t + NC} \prod_{i=1}^{k} \frac{f_{cxi} + 1}{F_c + V_{xi}}$$

where tc is the number of treatment in class c, Nt is the total number of treatments, NC is the total number of classes, fcxi is the number of cases in which predictor variables x equal to xi in each class c,Fc is the number of treatments in class c and Vxi is the number of possible values of xi.

The denominator is defined as

$$P(x|D) = \sum_{c'} (c', x|D)$$

The resulting conditional distribution is the classification function. A forward selection method was employed to optimize classification with leave-one-out validation as illustrated in Figure 4. All the individual genes were first ranked according to their discriminative power based on the Bayes model. Then the one with the best internal classification rate and most stable classification result was selected as a seed. The second best gene was chosen to join with

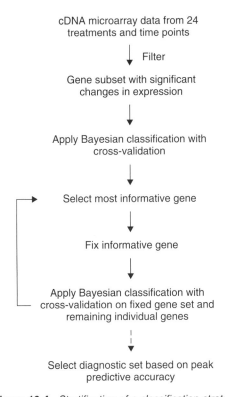

Figure 10-4 *Stratification of a classification strategy.*

the seed in the model. The classification power of the combination was examined by the leave-one-out validation method in which one treatment is left out when a model for the new set of predictor variables is built. The final gene set is established based on the peak prediction power and confidence. With the defining final gene set, the classifier was built on all the expression profiles in their study.

Utilizing this approach, a set of 12 genes/transcripts was identified as a signature gene set to distinguish peroxisome proliferators, aryl hydrocarbon receptor agonists, noncoplanar polychlorinated biphenyls, inflammatory agents, and hypoxia-inducing agents.

CONCLUSION

There is little doubt that toxicology is undergoing a considerable change. Toxicogenomics provides both the opportunity and mechanism for converting a largely descriptive science into one that is quantitative and predictive. This change is not gradual, and like other revolutions in science, it has its share of skeptics. Skepticism is beneficial, however, since it challenges the toxicologist to apply the same basic principles of scientific inquiry to this new field as to any other field of study. Ultimately the goal is to increase our ability to more accurately identify hazards and predict risk.

REFERENCES

Alizadeh AA, et al. (2000). Distinct types of diffuse large B-cell lymphoma identified by gene expression profiling. *Nature* **403**(6769):503–511.

Alon U, et al. (1999). Broad patterns of gene expression revealed by clustering analysis of tumor and normal colon tissues probed by oligonucleotide arrays. *Proc Natl Acad Sci USA* **96**(12):6745–6750.

Bartosiewicz M, Penn S, Buckpitt A (2001). Applications of gene arrays in environmental toxicology: fingerprints of gene regulation associated with cadmium chloride, benzo(a)pyrene, and trichloroethylene. *Environ Health Perspect* **109**:71–74.

Browne GS, Nelson C, Nguyen T, Ellis BA, Day RO, Williams KM (1999). Stereoselective and substrate-dependent inhibition of hepatic mitochondria beta-oxidation and oxidative phosphorylation by the non-steroidal anti-inflammatory drugs ibuprofen, flurbiprofen, and ketorolac. *Biochem Pharmacol* **57**:837–844.

Bulera SJ, Eddy SM, Ferguson E, Jatkoe TA, Reindel JF, Bleavins MR, De La Iglesia FA (2001). RNA expression in the early characterization of hepatotoxicants in Wistar rats by high-density DNA microarrays. *Hepatology* **33**:1239–1258.

Burczynski ME, McMillian M, Ciervo J, Li L, Parker JB, Dunn RT II, Hicken S, Farr S, Johnson MD (2000). Toxicogenomics-based discrimination of toxic mechanism in HepG2 human hepatoma cells. *Toxicol Sci* **58**:399–415.

Chu S, DeRisi J, Eisen M, Mulholland J, Botstein D, Brown PO, Herskowitz I (1998). The transcriptional program of sporulation in budding yeast. *Science* **282**:699–705.

Claudio JO, Masih-Khan E, Tang H, Goncalves J, Voralia M, Li ZH, Nadeem V, Cukerman E, Francisco-Pabalan O, Liew CC, Woodgett JR, Stewart AK (2002). A molecular compendium of genes expressed in multiple myeloma. *Blood* **100**: 2175–2186.

Debouck C, Goodfellow PN (1999). DNA microarrays in drug discovery and development. *Nat Genet* (suppl 1):48–50.

DeRisi JL, Iyer VR, Brown PO (1997). Exploring the metabolic and genetic control of gene expression on a genomic scale. *Science* **278**:680–686.

Egner R, Rosenthal FE, Kralli A, Sanglard D, Kuchler K (1998). Genetic separation of FK506 susceptibility and drug transport in the yeast Pdr5 ATP-binding cassette multidrug resistance transporter. *Mol Biol Cell* **9**:523–543.

Eisen MB, Spellman PT, Brown PO, Botstein D (1998). Cluster analysis and display of genome-wide expression patterns. *Proc Natl Acad Sci USA* **95**:14863–14868.

Gerhold D, Lu M, Xu J, Austin C, Caskey CT, Rushmore T (2001). Monitoring expression of genes involved in drug metabolism and toxicology using DNA microarrays. *Physiol Genomics* **5**:161–170.

Goffeau A, Barrell BG, Bussey H, Davis RW, Dujon B, Feldmann H, Galibert F, Hoheisel JD, Jacq C, Johnston M, Louis EJ, Mewes HW, Murakami Y, Philippsen P, Tettelin H, Oliver SG (1996). Life with 6000 genes. *Science* **274**:563–567.

Golub TR, Slonim DK, Tamayo P, Huard C, Gaasenbeek M, Mesirov JP, Coller H, Loh ML, Downing JR, Caligiuri MA, Bloomfield CD, Lander ES (1999). Molecular classification of cancer: class discovery and class prediction by gene expression monitoring. *Science* **286**:531–537.

Hamadeh HK, Bushel PR, Jayadev S, DiSorbo O, Bennett L, Li L, Tennant R, Stoll R, Barrett JC, Paules RS, Blanchard K, Afshari CA (2002). Prediction of compound signature using high density gene expression profiling. *Toxicol Sci* **67**:232–240.

Hawkey CJ (1999). COX-2 inhibitors. *Lancet* **353**:307–314.

Hemenway CS, Heitman J (1999). Calcineurin. Structure, function, and inhibition. *Cell Biochem Biophys* **30**:115–151.

Huang Q, Dunn RT II, Jayadev S, DiSorbo O, Pack FD, Farr SB, Stoll RE, Blanchard KT (2001). Assessment of cisplatin-induced nephrotoxicity by microarray technology. *Toxicol Sci* **63**:196–207.

Hughes TR, Marton MJ, Jones AR, Roberts CJ, Stoughton R, Armour CD, Bennett HA, Coffey E, Dai H, He YD, Kidd MJ, King AM, Meyer MR, Slade D, Lum PY, Stepaniants SB, Shoemaker DD, Gachotte D, Chakraburtty K, Simon J, Bard M, Friend SH (2000). Functional discovery via a compendium of expression profiles. *Cell* **102**:109–126.

Hughes TR, Mao M, Jones AR, Burchard J, Marton MJ, Shannon KW, Lefkowitz SM, Ziman M, Schelter JM, Meyer MR, Kobayashi S, Davis C, Dai H, He YD, Stephaniants SB, Cavet G, Walker WL, West A, Coffey E, Shoemaker DD, Stoughton R, Blanchard AP, Friend SH, Linsley PS (2001). Expression profiling using microarrays fabricated by an ink-jet oligonucleotide synthesizer. *Nat Biotechnol* **19**:342–347.

Hwang DM, Dempsey AA, Wang RX, Rezvani M, Barrans JD, Dai KS, Wang HY, Ma H, Cukerman E, Liu YQ, Gu JR, Zhang JH, Tsui SK, Waye MM, Fung KP, Lee CY,

Liew CC (1997). A genome-based resource for molecular cardiovascular medicine: toward a compendium of cardiovascular genes. *Circulation* **96**:4146–4203.

Ideker T, Thorsson V, Ranish JA, Christmas R, Buhler J, Eng JK, Bumgarner R, Goodlett DR, Aebersold R, Hood L (2001). Integrated genomic and proteomic analyses of a systematically perturbed metabolic network. *Science* **292**:929–934.

Jaradat MS, Wongsud B, Phornchirasilp S, Rangwala SM, Shams G, Sutton M, Romstedt KJ, Noonan DJ, Feller DR (2001). Activation of peroxisome proliferator-activated receptor isoforms and inhibition of prostaglandin H(2) synthases by ibuprofen, naproxen, and indomethacin. *Biochem Pharmacol* **62**:1587–1595.

Khan J, Wei JS, Ringner M, Saal LH, Ladanyi M, Westermann F, Berthold F, Schwab M, Antonescu CR, Peterson C, Meltzer PS (2001). Classification and diagnostic prediction of cancers using gene expression profiling and artificial neural networks. *Nat Med* **7**:673–679.

Liu J, Farmer JD Jr, Lane WS, Friedman J, Weissman I, Schreiber SL (1991). Calcineurin is a common target of cyclophilin-cyclosporin A and FKBP-FK506 complexes. *Cell* **66**:807–815.

Marton MJ, DeRisi JL, Bennett HA, Iyer VR, Meyer MR, Roberts CJ, Stoughton R, Burchard J, Slade D, Dai H, Bassett DE Jr, Hartwell LH, Brown PO, Friend SH (1998). Drug target validation and identification of secondary drug target effects using DNA microarrays. *Nat Med* **4**:1293–1301.

Roberts CJ, Nelson B, Marton MJ, Stoughton R, Meyer MR, Bennett HA, He YD, Dai H, Walker WL, Hughes TR, Tyers M, Boone C, Friend SH (2000). Signaling and circuitry of multiple MAPK pathways revealed by a matrix of global gene expression profiles. *Science* **287**:873–880.

Schena M, Shalon D, Davis RW, Brown PO (1995). Quantitative monitoring of gene expression patterns with a complementary DNA microarray. *Science* **270**:467–470.

Shaffer AL, Rosenwald A, Hurt EM, Giltnane JM, Lam LT, Pickeral OK, Staudt LM (2001). Signatures of the immune response. *Immunity* **15**:375–385.

Tamayo P, Slonim D, Mesirov J, Zhu Q, Kitareewan S, Dmitrovsky E, Lander ES, Golub TR (1999). Interpreting patterns of gene expression with self-organizing maps: methods and application to hematopoietic differentiation. *Proc Natl Acad Sci USA* **96**:2907–2912.

Tavazoie S, Hughes JD, Campbell MJ, Cho RJ, Church GM (1999). Systematic determination of genetic network architecture. *Nat Genet* **22**:281–285.

Thomas RS, Rank DR, Penn SG, Zastrow GM, Hayes KR, Pande K, Glover E, Silander T, Craven MW, Reddy JK, Jovanovich SB, Bradfield CA (2001). Identification of toxicologically predictive gene sets using cDNA microarrays. *Mol Pharmacol* **60**:1189–1194.

van't Veer LJ, Dai H, van de Vijver MJ, He YD, Hart AA, Mao M, Peterse HL, van der Kooy K, Marton MJ, Witteveen AT, Schreiber GJ, Kerkhoven RM, Roberts C, Linsley PS, Bernards R, Friend SH (2002). Gene expression profiling predicts clinical outcome of breast cancer. *Nature* **415**:530–536.

Wang X, Harris PK, Ulrich RG, Voorman RL (1996). Identification of dioxin-responsive genes in Hep G2 cells using differential mRNA display RT-PCR. *Biochem Biophys Res Commun* **220**:784–788.

Waring JF, Jolly RA, Ciurlionis R, Lum PY, Praestgaard JT, Morfitt DC, Buratto B, Roberts C, Schadt E, Ulrich RG (2001a). Clustering of hepatotoxins based on mechanism of toxicity using gene expression profiles. *Toxicol Appl Pharmacol* **175**: 28–42.

Waring JF, Ciurlionis R, Jolly RA, Heindel M, Ulrich RG (2001b). Microarray analysis of hepatotoxins in vitro reveals a correlation between gene expression profiles and mechanisms of toxicity. *Toxicol Lett* **120**:359–368.

Waring JF, Gum R, Morfitt DC, Jolly RA, Ciurlionis R, Heindel M, Gallenberg L, Buratto B, Ulrich RG (2002). Identifying toxic mechanisms using DNA microarrays: evidence that and experimental inhibitor of cell adhesion molecule expression signals through the aryl hydrocarbon nuclear receptor. *Toxicology* **181–182**:537–550.

Waring JF, Cavet G, Jolly RA, McDowell J, Dai H, Ciurlionis R, Zhang C, Stoughton R, Lum P, Ferguson A, Roberts CJ, Ulrich RG (2003a). Development of a DNA microarray for toxicology based on hepatotoxin-regulated sequences. *Environ Health Perspect* **111**(6):863–870.

Waring JF, Ciurlionis R, Clampit JE, Morgan S, Gum RJ, Jolly RA, Kroeger P, Frost L, Trevillyan J, Zinker BA, Jirousek M, Ulrich RG, Rondinone CM (2003b). PTP1B antisense-treated mice show regulation of genes involved in lipogenesis in liver and fat. *Mol Cell Endo* **203**(1–2):155–168.

Wodicka L, Dong H, Mittmann M, Ho MH, Lockhart DJ (1997). Genome-wide expression monitoring in *Saccharomyces cerevisiae*. *Nat Biotechnol* **15**:1359–1367.

11

Using Genetically Altered Mice in Toxicogenomic Analysis of Chemical Exposure

J. Christopher Corton and Steven P. Anderson

INTRODUCTION

We now have in hand the entire sequence of the human genome, and inventories of other mammalian organisms are nearing completion. This surfeit of information has fueled an exponential increase in functional genomic analyses in all subdisciplines of biology and medicine, including toxicology. These technologies, when paired with sophisticated computational and statistical analyses, hold great promise as catalysts for a phase transition in toxicology—a shift from toxicology as a primarily descriptive discipline to a more mechanistically based science.

As described elsewhere in this volume, evolving techniques are allowing for the simultaneous analysis of messenger RNA (mRNA) levels of thousands of genes, covering in many cases all or most of the queried genome (Chapters 7–11). The global analyses of proteins or metabolites, referred to as proteomics (Chapters 12 and 13) and metabonomics (Chapter 14), respectively, are being

Toxicogenomics: Principles and Applications. Edited by Hamadeh and Afshari
ISBN 0-471-43417-5   Copyright © 2004 John Wiley & Sons, Inc.

increasingly used by toxicologists. The data generated from these functional genomics approaches and interpreted with the help of bioinformatics tools will be used to integrate the changes observed at the tissue, cellular, and molecular levels of organization to create comprehensive biologically based models of chemical action.

Functional genomics approaches are allowing us an unprecedented view of the global changes that occur after chemical exposure. As many chemicals can induce both adverse and therapeutic or adaptive effects depending on the exposure conditions, the challenge for the toxicogenomicist is to correctly interpret the large number of changes anticipated after chemical exposure.

This review specifically focuses on how genetically altered mice can be used to help the toxicogenomicist interpret the mechanistic basis and biological significance of global changes in cellular components that occur after chemical exposure. We describe the types of genetically altered mice and end with a discussion of the prospects for a coordinated and systematic effort to increase the speed, accuracy, and detail of predicting molecular mechanisms of chemical action.

MICE AS EXPERIMENTAL MODELS TO CHARACTERIZE CHEMICAL TOXICITY

The universality of the genetic code and the high degree of conservation of key biochemical reactions and developmental processes emphasize the close relationship of humans to species that are morphologically distinct and evolutionarily divergent. In theory, primates should provide the best animal models of human disease because they are so closely related to us. However, because primates are comparatively long lived and are less fecund than rodents, breeding experiments are more difficult to organize, and thus primates are not well suited to experimentation. Instead, mice have been the most widely used animal models of human disease because of their short life span (2–3 years), short generation time (90 days), and high fecundity.

Genetic knowledge of the laboratory mouse is secondary only to knowledge of our own genetics, and the phenotypes of many mouse mutants have been characterized (Denny and Justice, 2000; http://www.informatics.jax.org/). A general comparison of the mouse and human genomes reveals similar gene numbers and genome size (Waterston et al., 2002; Okazaki et al., 2002). Gene family organization is very similar for individual families, although recent duplications have led to differences in gene numbers within these families.

At the DNA level, coding sequences are generally 70–90% homologous; polypeptides are even more similar (75–95%). Likewise, gene expression is very similar between the two species, although for orthologous genes there are sometimes differences in the choice of alternative promoters and patterns of splicing, or in imprinting. Almost all genes in humans have functional homologs in the mouse. In many cases, these homologs are located in syntenic regions of the corresponding mouse chromosome. It is remarkable that there are large numbers of genes that are highly homologous between humans and mice but have no known function, opening up the prospects of identification of new models of human disease.

The power of mouse genetics for functional studies lies in the multitude of sophisticated tools available to the geneticist. From replacing one nucleotide in a specific gene to replacing whole chromosomes, current technologies make it possible to engineer almost any genetic change in mice. Specific genes are routinely inactivated by homologous recombination in embryonic stem cells. At the opposite end of the spectrum, transgenic mice are generated by inserting functional mouse or human genes into the mouse genome. These technologies for generating novel phenotypes complement rather than supersede the more classical genetic approaches such as generating mutants by exposing mice to chemical mutagens. With the first draft of the mouse genome completed, these approaches will keep the mouse at the forefront of experimental genetics and functional genomics, providing powerful tools for understanding mechanisms of toxicity.

TYPES OF GENETICALLY ALTERED MICE

Targeted Mouse Mutants

Any foreign DNA can be transferred into embryonic stem (ES) cells. ES cells are derived from 3.5–4.5 day postcoitum embryos and are isolated from the inner cell mass of the blastocyst. A major advantage of using ES cells is that they can be maintained as undifferentiated pluripotent cells in culture indefinitely. A number of strategies are available for creating genomic alterations in these cells and are discussed below. When genetically altered ES cells are injected into a host blastocyst, they have the capacity to contribute to all tissues, including the germ line, of the developing mouse, where they can transmit the foreign gene to their progeny. The total number of gene knockouts that have been made to date is approaching 5,000, with the number of laboratories applying this technology steadily rising. This large number of mouse mutants has provided valuable information about the role of the corresponding genes in development and adult homeostasis.

The most powerful genetic use of ES cells is in gene targeting (Capecchi, 1996; Woychik et al., 1998). In this approach, targeting vectors are constructed that can be specifically integrated at a desired genomic location by homologous recombination in ES cells (Fig. 11-1). These cells are subsequently injected into blastocysts, where they contribute to the developing embryo in various tissues including the germ cells.

Chimeric mice are backcrossed to wild-type mice to obtain heterozygotes that are mated to each other to obtain homozygote knockouts. The gene target is usually disrupted by introduction of a neomycin resistance marker within

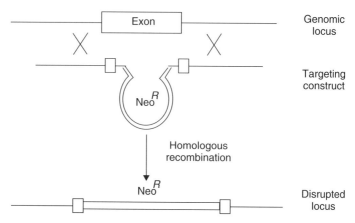

Figure 11-1 *Functional inactivation of genes by homologous recombination. Targeting constructs are made in which some or all of the gene's exons are deleted and replaced with the neomycin resistance cassette (NeoR). The construct is transfected into embryonic stem cells integrated into the genome through homologous recombination, and cells resistant to neomycin are selected. Cells that carry a disrupted targeted locus are identified and used to make chimeric mice.*

the coding region of the gene, resulting in deletion of part of the gene. Mouse mutations may be studied in the heterozygous state (to assess potential gene dosage effects) or the homozygous state (for analysis of the null phenotype). The phenotypic changes observed in these knockout mice often provide clues about the function of the genes in wild-type animals.

Targeted gene inactivation in mice has significantly accelerated our ability to determine the role of the gene products in normal physiology. At the same time, these types of manipulations have demonstrated the extreme complexity of genetic regulation and gene function. However, targeted inactivation of many genes results in no discernible phenotype. On the simplest level, there are three possible explanations for this: (1) functional redundancy with a different gene, (2) variation in susceptibility alleles, and (3) subtlety of phenotypic alteration. In addition, several other complications may arise, including an early embryonically lethal phenotype preventing determination of the role of the gene product in the adult. Also, the phenotype can affect multiple cellular lineages, making it difficult to determine functions in different cell types or tissues. Fortunately, we now have at our disposal a battery of well-characterized tools to circumvent these problems by allowing spatial and temporal gene inactivation.

Conditional Knockouts

The most popular conditional gene knockout strategy combines homologous and site-specific recombination (Sauer, 1993; Dymecki, 1996; Lobe and Nagy,

1998). Enzymes called site-specific recombinases recognize defined DNA sites and either delete or invert intervening sequences depending on the orientation of the sites. The Cre recombinase of the P1 bacteriophage and the FLP recombinase of yeast have been the most widely used for experiments in mice because they require only short (34 bp) consensus sequences to catalyze recombination.

The general strategy for conditional knockouts is to clone two recognition sites for a site-specific recombinase on each side of essential exons in a manner that does not alter the normal function of the gene (Fig. 11-2). Mice carrying this altered allele in homozygous form (floxed mice; *flanked by lox* sites, the *lox* site being the target of the Cre recombinase) can then be established without complications of an altered phenotype. To create the conditional knockout, mice possessing these recombinase sites are mated with mice that express the recombinase in a tissue-specific manner. Recombination will occur only in those tissues that express the recombinase. A drawback here is that the gene deletion could still occur in the developing embryo and, if essential for development, will prevent establishing lines of tissue-specific null mice. Ways to circumvent this problem are discussed below. There are now a growing number of commercially available mouse strains that express the Cre recombinase in a tissue-specific manner. These commercially available mice can be used in conjunction with a mouse line that possesses a floxed gene of interest to make a number of tissue-specific knockouts.

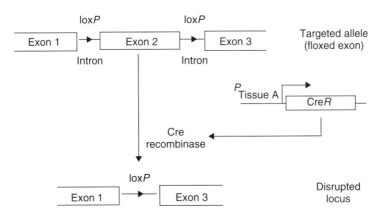

Figure 11-2 *Tissue-specific functional inactivation of genes using the Cre-lox system. Targeted homologous recombination is used to create a targeted allele containing sites recognized by the Cre-recombinase called loxP sites. The loxP sites are usually engineered into introns, where they leave the normal function of the gene intact and flank the exon to be deleted (said to be floxed, flanked by loxP sites). Mice containing these targeted genes are mated with mice that express the Cre recombinase expressed from a tissue-specific promoter (e.g., tissue A). Only in those cells expressing the Cre recombinase will recombination between the loxP sites occur, resulting in deletion of the exon.*

In addition to cell- and tissue-specific knockouts, inducible knockouts can be engineered that allow gene ablation at will. This strategy combines the Cre technology and the reversed tetracycline-controlled transactivator system (Kistner et al., 1996). The reversed tetracycline-controlled transactivator (rtTA) is a fusion protein composed of a mutant version of the Tn10 tetracycline repressor from *Escherichia coli* fused to a C-terminal portion of protein 16 from the herpes simplex virus that functions as a strong transcriptional activator. Under control of a tissue-specific promoter, the rtTA protein is active only in the presence of the antibiotic doxycycline, which leads to binding to the tetracycline operator (tetO) controlling expression of the Cre recombinase (Fig. 11-3). After doxycycline treatment, the targeted gene will be inactivated after recombination only in those cells in which rtTA and Cre are expressed.

The genetic techniques discussed here have been used to create mice in which the mouse gene is replaced by a mutant form of the same mouse gene by a homologous gene or by the corresponding human gene (Lovik, 1997). These knock-in mice can be constructed by homologous recombination as discussed using as a vector a cDNA of the gene to be expressed flanked by sequences of the gene to be replaced. After homologous recombination, the ES cells possess a copy of the target gene replaced by another gene. The ES cells can then be handled as already described.

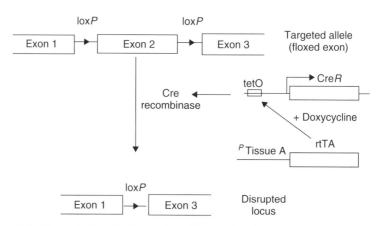

Figure 11-3 *The inducible Cre-lox system. Transgenic mice are created that express the reverse tetracycline-controlled transactivator (rtTA) protein from a tissue-specific promoter and a Cre recombinase gene under control of the tetracycline operator. The rtTA system consists of a fusion protein composed of a mutant version of the Tn10 tetracycline repressor fused to the transcriptional activation domain of protein 16 from herpes simplex virus. These mice are mated to mice that contain an exon(s) flanked by loxP sites in a targeted gene. When mice are exposed to doxycycline, the rtTA protein binds to the tetO, resulting in expression of the Cre recombinase. The recombinase then removes sequences between the loxP sites of the targeted locus. Recombination only occurs in cells that express the rtTA and Cre recombinase proteins and are assessable to doxycycline.*

Systematic Inactivation of Gene Function in the Mouse

Possessing a knockout for every gene together with a unique sequence identifier will be ideal for many biologists. On a large scale, gene trap mutagenesis aspires to this goal (Cecconi and Barbara, 2000). The gene to be trapped inactivates endogenous genes and consists of a drug-resistance marker gene (e.g., puromycin N-acetyl transferase, PURO) under control of a phosphoglycerase kinase (PGK) promoter. Instead of carrying a polyA addition signal, the transferase gene contains a consensus exon donor site at the end of its coding region. After transfection of ES cells, the DNA construct integrates into a gene at a site that contains at least one downstream exon and polyA addition site, producing a functional mRNA encoding the transferase.

Growth of ES cells in the presence of puromycin selects for vector insertion into genes. The random amplification of cDNA ends (RACE) strategy allows for isolation of sequences from the gene into which the DNA was inserted. Lexicon (Woodlands, TX) has a library of these sequence tag sites referred to as Omnibank Sequence Tags (OST) maintained in a proprietary database. Using this database, biologists can purchase the embryonic stem cell clones carrying the desired gene disruptions. The clone can then be injected into blastocytes, significantly reducing the time to obtain mutant nullizygous mice. The German Human Genome Project is funding a similar ES cell insertional mutagenesis program at GSF (Neuherberg, Germany). Web sites at each of these locations have searchable databases to find genes that have been disrupted.

Transgenic Mice

Transgenic mice express a foreign gene or transgene that is microinjected into fertilized mouse eggs. The transgene usually consists of a reporter gene or a cDNA under control of a general or tissue-specific promoter (Hofker and Breuer, 1998). The DNA is integrated randomly into chromosomal DNA, usually at a single site, although rarely two sites of integration are found in a single animal. The level of transgene expression depends on the site of insertion and determines the tissue distribution and level of expression. Thus, transgenic founder mice with the same transgene integrated at different sites may exhibit different phenotypes dependent on the site of integration. Mice generated using this approach most frequently harbor gain-of-function mutations associated with constitutive or tissue-specific overexpression.

In other cases, transgenic mice have been created that express dominant negative forms of proteins that abolish function of the wild-type protein. Other transgene constructs express ribozymes or cDNA in the antisense direction that abolish expression of the endogenous wild-type protein. Of particular interest to most toxicologists, it is also possible to generate transgenic rats (Heideman, 1991). Finally, mice that carry a transgene such as *lacZ* under control of a cell type–specific promoter have been used to isolate rare cell types by fluorescence-activated cell sorters that are subsequently used for transcript profiling (e.g., Chen et al., 2002a).

Inconveniently, constitutive expression of some transgenes in neonatal mice kills the developing embryo prior to parturition. A way to circumvent this is to induce expression of the gene postnatally, in a reversible fashion if need be, allowing the mouse to develop to term without expressing the fatal transcript. An advantage to this protocol is that each individual mouse serves as its own control animal, because the animal can be studied both before and after induction of the gene of interest. Mice that express a gene in a tissue-specific and inducible manner have been created using the binary tetracycline regulatory system as already described. When the *rtTA* gene under control of a tissue-specific promoter is activated in the presence of the antibiotic doxycycline, the protein product binds to the tetracycline operator (tetO) controlling expression of the inducible transgene and activates transcription from a minimal human cytomegalovirus promoter. Thus, when doxycycline is added to the drinking water of the mice, the gene of interest will be expressed only in those cells in which *rtTA* is expressed by means of the tissue-specific promoter (Fig. 11-4).

Conversely, it is also possible to shut down the otherwise constitutive expression of a transgene using the tetracycline repressor system (Kistner et al., 1996). With this technique the transgene and the tetracycline repressor are under control of a tissue-specific promoter on two separate constructs

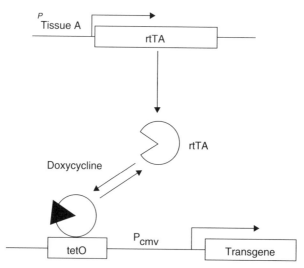

Figure 11-4 *Inducible expression of a transgene using the reverse tetracycline-controlled transactivator (rtTA) system. The rtTA gene is placed under control of a tissue-specific promoter ($P_{tissue\ A}$). When the rtTA gene is expressed in the presence of the antibiotic doxycycline, the rtTA protein binds to the tetracycline operator (tetO), controlling expression of the transgene from a minimal human cytomegalovirus (P_{cmv}).*

expressed in the same cell types. The transgene promoter contains a tetO engineered in a location such that binding by the tetracycline repressor blocks expression of the transgene. In the absence of doxycycline, the tetracycline repressor is expressed but does not bind to the tetO, allowing expression of the transgene. In the presence of doxycycline, the tetracycline repressor binds to the tetO, shutting down transgene expression.

As toxicologists are ultimately interested in determining the relevance of the mechanism of toxicity in mammalian models to human exposures, "humanized" mice have been created. Humanized mice are created by either homologous recombination, replacing the mouse gene with the human gene as already discussed, or by construction of transgenic mice expressing the human gene. In the latter example, the human and the mouse genes are coexpressed and may result in difficulties in interpreting the contribution of each gene to a phenotypic effect. Thus, transgenic mice possessing the human gene can be mated to mice in which the corresponding mouse gene has been inactivated, creating a mouse strain where only the human gene is expressed (Gonzalez and Kimura, 2003). In many cases these humanized mice more accurately mimic the interaction of drug-metabolizing systems and exposure to xenobiotic chemicals. These mice can also be used to determine if similarities or differences exist in the spectrum of genes regulated by the mouse and human genes after chemical exposure.

PHENOTYPE-DRIVEN MUTAGENESIS STRATEGIES IN THE MOUSE

The reverse genetics (i.e., moving from genotype to phenotype) approaches will allow the systematic production of mouse mutants for any gene that has been cloned. Complementary to this are more traditional forward genetics approaches, where the investigator first generates a phenotype (mutant mouse), then identifies the genotypic alteration that underlies that phenotype. The study of mouse mutants that develop diseases similar to those in humans but have mutations in different genes is essential for our understanding of the molecular mechanisms involved in the pathogenesis of diseases whose underlying basis relies on the interaction of multiple genes. The full power of genetic analysis of gene function requires the availability of multiple alleles of the same gene, including alleles of different strength and gain of functional alleles, since many human diseases result from only partial loss of gene function.

Large-scale systematic screening efforts to identify mouse mutants by chemical mutagenesis are now being carried out in a number of laboratories (de Angelis and Balling, 1998). The alkylating agent ethyl nitrosourea (ENU) is currently the most powerful mutagen used to produce germ-line mutations in mice. The mutations recovered after ENU mutagenesis are mainly point mutations or small intragenic lesions, resulting in hypomorphic (partial loss of function), gain-of-function, and complete loss-of-function mutations. Current protocols are available that allow very efficient germ cell mutagenesis in mice.

The frequency of mutant recovery is about 1/1,000 for a specific locus that can be scored phenotypically.

ENU mutagenizes premeiotic granule stem cells, allowing for the production of a large number of F1 founder animals from a single treated male, minimizing the number of animals required and the handling of ENU. A drawback of this approach is that mutations are not tagged. Although this is a disadvantage with respect to the cloning of the responsible genes, the availability of point mutations will be important for a more detailed functional analysis of many genes. It is anticipated that the advances currently made in the field of genomics, particularly the production of high-resolution genetic, physical, and transcript maps, will reduce the difficulties inherent in the identification and cloning of genes mutated after ENU treatment.

One approach for an ENU phenotype-based mutagenesis experiment is to screen the entire genome for a new dominant or recessive mutation (Justice, 2000). Mutagenized males are crossed with wild-type partners, and the offspring from this cross are evaluated for new dominant phenotypes. To detect new recessive phenotypes, it is necessary to make the offspring homozygous for the mutations by either intercrossing the F1 offspring or backcrossing F1 females with their mutagenized male parent. In addition, each recovered mutation can be mapped using standard mitotic procedures. Mating and the screening of prodigy for mutant phenotypes are required for each step and involve a considerable effort and a substantial commitment to the maintenance of large breeding colonies of mice.

A genomic scan for new ENU-induced mutations is currently the most efficient approach for screening for a single phenotype throughout the genome. These ENU mutagenesis protocols will likely lead to a comprehensive database of phenotypic abnormalities resulting from single- or multiple-point mutations. This information has the potential of being extremely valuable to the toxicologist in allowing a systematic search for phenotypes in the ENU mutagenized mice that mimic those caused by chemical toxicity. Subsequent mapping experiments could allow identification of key genes whose alteration in function results in the chemical-induced phenotype.

Unless a toxic insult is so egregious that it leads to acute death, different strains of mice almost always exhibit qualitative and quantitative differences to it. Researchers can take advantage of this genetic variability by surveying susceptibility in various strains of mice and identifying discretely measurable (i.e., quantitative or digital traits) rather than qualitative (analog, or continuously varying) phenotypes through selective breeding.

The next step after identifying susceptible and resistant strains is to map the genetic loci that segregate with the phenotype. In mice there are several recombinant inbred (RI) strains available that have been developed by crossing two standard inbred strains, then brother × sister mating for 20+ generations, to produce new strains in which the parental genes have recombined (Taylor, 1996). After testing susceptibility (i.e., phenotyping) of a subset containing many RI strains to a given toxin, it may be possible to identify genetic markers associated with this trait, thereby mapping some of the quantitative trait loci (QTLs) for a given trait.

A significant disadvantage of this approach is the large number of animals that must be crossed and phenotyped in order to narrow a susceptibility locus down to a reasonably small region of a chromosome. Furthermore, suitable RI strains are not always available. Nonetheless, these techniques have proven themselves extremely powerful, even if they are cumbersome and expensive. For example, between 10 and 12 loci that appear to be associated with lung tumor susceptibility in the mouse have been mapped (Festing et al., 1998). It is likely that the time to identify candidate disease genes will be significantly shortened by using a combination of expression profiling and linkage of the trait to a rapidly growing database of single-nucleotide polymorphisms (SNPs; Grupe et al., 2001).

In a relatively new approach combining ENU mutagenesis and ES cell technology, a number of groups are creating mouse strains with an allelic series of mutant genes (Chen et al., 2000; Munroe et al., 2000). ES cells are mutagenized with ENU, then plated out for single colonies. Colonies are picked and transferred to 96-well dishes for a semi-high-throughput method for mutation identification. This can be through selection for a genotypic change if a selection strategy can be devised for the gene of interest. If no selection strategy is available, then to screen colonies for mutations in the gene of interest using denaturing high-performance liquid chromatography–based heteroduplex analysis. Once the ES cell colony has been identified with the mutation of interest, the ES cells are injected into blastocysts, and chimeric mice are bred as already described to create heterozygous or homozygous mice for that gene. This strategy has been used to create an allelic series of mutations in the Smad2 and Smad4 genes (Vivian et al., 2002).

STRATEGIES FOR INTERPRETATION OF GLOBAL CHANGES IN COMPONENTS OF THE GENOME

There is a growing appreciation that genetically engineered mice can be used to identify and interpret consistent patterns of gene expression associated with specific types of toxic insult. Perhaps in the not-so-distant future, we will become sufficiently proficient in integrating these types of data to allow us to read the transcriptome in a manner analogous to a pathologist's interpreting lesions on a microscopic slide (Boorman et al., 2002). Our ability to do so will largely depend on our ability to separate adaptive (those not directly related to the mechanism of toxicity) from maladaptive (toxic) responses—for example, recognizing that changes in the protein products of the genes mediating the hypolipidemic effects of peroxisome proliferator drugs are not mechanistically responsible for the hepatocellular mitogenesis driving the hepatic carcinogenicity of these chemicals. At present, our ability to separate cause from effect is woefully inadequate, but using genetically defined mouse strains in which to link gene expression fingerprints to toxic mechanisms promises to greatly facilitate this process.

Using genetically altered mice that exhibit increased resistance or sensitivity to chemical exposure can help link the gene expression changes observed by genome-wide expression scanning to toxic effects of the chemical. The relationship between dose or time of chemical exposure and two phenotypic responses can be viewed as a two-dimensional landscape for a chemical (Corton et al., 2002). Gene expression changes that occur over a range of doses or times of chemical exposure could be used to make correlations between phenotypic changes and altered expression of batteries of genes.

Genetically altered mice may differ compared to the wild-type condition, ranging from complete insensitivity to hypersensitivity to chemical exposure. Adaptive responses may parallel or may occur independently of toxic responses. Confirmation that those genes are clearly associated with chemically induced toxicity can be made in genetically altered mice that exhibit differences in responsiveness to chemical exposure. If genes that were identified in the wild-type experiments associated with toxicity are mechanistically linked to toxicity, expression of those genes should track with the altered toxicity exhibited in the genetically altered mice. This type of approach of following up observations in genetically altered mice allows identification of genes associated with toxicity as well as those genes more closely associated with adaptive responses.

One of the most promising uses for genetically altered mice is to test hypotheses of the role of specific genes or functional pathways in mediating chemical toxicity from patterns of chemically induced gene expression changes in wild-type mice (Fig. 11-5). Bioinformatics tools can be used to group genes

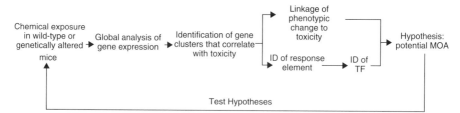

Figure 11-5 *Coupling genetically altered mice with global analysis of gene expression to increase mechanistic understanding of chemical action. Using bioinformatics approaches, global analysis of gene expression in chemically exposed wild-type mice will lead to the identification of clusters of genes that correlate with toxicity. The known or putative function of those gene products can lead to a prediction of phenotypic changes associated with chemical toxicity. Comparison of the sequence of the promoter regions of the clustered genes can be used to identify common response elements found within the promoter elements that determine coordinate expression. Candidate response elements can be tested for functional significance by assessing interaction with transcription factors that coordinate the expression of those clustered genes. Genetically altered mice can be used to test hypotheses of the functional significance of the transcription factors in regulating the clustered genes or in gene products that determine phenotypic responses linked to toxicity.*

into clusters based on their coordinated regulation over time or dose after exposure (Gilbert et al., 2000; Sherlock, 2000). The clustered genes most closely linked to toxicity can be used to define the mechanism of chemical action in two ways.

Clusters of genes can be used to make predictions of phenotypic changes that may be linked to toxicity. Work in a number of models has shown that coordinately regulated genes often have similar functional roles in the cell—for example, alteration in components of a metabolic pathway. This is especially true with yeast in which transcription is tightly regulated in a coordinated fashion (e.g., Kuhn et al., 2001). The known function of genes that cluster with toxicity could be used to predict phenotypic changes occurring during chemical exposure linked to or a manifestation of toxicity. In this context, clustering approaches have been used to predict the function of uncharacterized genes that fall into the same functional cluster (Hughes et al., 2000).

Cluster analysis can also be used to identify early events that precede altered gene expression. The nucleotide sequence of the promoters of the clustered genes can be compared to identify common sequences that may determine coordinate regulation (Zhang, 1999; Wolfsberg et al., 1999). A shared sequence can be compared to previously identified response elements recognized by one or more transcription factors. If the sequence is not characterized, in vitro approaches such as electrophoretic mobility shift assays or DNA affinity purification can be used to identify transcription factors that recognize the sequence. Once a transcription factor controlling response gene expression has been identified, events leading to activation or repression of that transcription factor can be characterized. This approach has been very effective in identifying response elements and the transcription factors that bind to them in batteries of yeast genes (Geraghty et al., 1999). It will be increasingly used to identify response elements and associated signal transduction pathways that mediate toxicity now that the mouse genome sequence is completed.

Hypotheses generated from clustering approaches can be tested by further iterations of expression scanning in appropriate genetically altered mice. For example, repeating the chemical exposure in mice that lack a key transcription factor involved in chemical responses may reveal either increased or decreased toxicity, depending on the role of that transcription factor in either damage or repair, respectively. As the number of genetically altered mouse strains available to the researcher increases, this type of analysis in which appropriate mutant mice are used in successive rounds of refining chemical mechanisms will be increasingly used.

TOXICOGENOMICS: FUTURE PROSPECTS USING GENETICALLY ALTERED MICE

In the near future, toxicologists may use two types of information derived from traditional toxicity tests to identify candidate target genes responsible for chemical-induced phenotypes (Fig. 11-6). First, gene expression profiles altered by an uncharacterized chemical could be compared to a database derived from appropriate tissues of chemically exposed wild-type mice. Profiles from untreated genetically altered mice would also add great value to this database. This approach will result in the identification of the chemical class into which the unknown chemical falls as well as the specific pathway(s) the chemical alters. Second, a set of phenotypic changes induced by the chemical could be used to find similarities with phenotypes from characterized mouse mutants. These two types of complementary strategies are described in greater detail below.

One of the more intriguing prospects for the use of genomic information in toxicology is the creation of databases useful for categorizing chemicals according to mechanism of action. Comparison of gene expression profiles of individual chemicals from many mechanism-of-action classes (e.g., peroxisome proliferators, estrogenic chemicals, cytotoxic chemicals) in wild-type mice will allow for the identification of common sets of genes whose expression is consistently linked to particular disease outcomes associated with classes of chemicals. When the database becomes sufficiently comprehensive, transcript

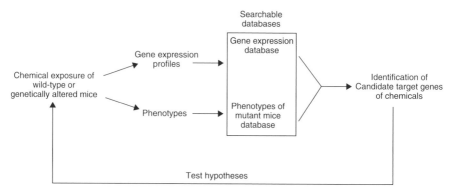

Figure 11-6 *Accelerating the identification of targets of toxicants using genetically altered mice. In the near future, two types of searchable databases will aid the toxicologist in understanding the mechanism of chemical-induced effects. One is a gene expression database that will be maximally effective when it contains two types of information: gene expression changes that occur after exposure to diverse chemicals from different mechanism-of-action classes (e.g., peroxisome proliferators, estrogenic chemicals, cytotoxic chemicals, etc.) and gene expression differences between wild-type and genetically altered mice. Gene expression in genetically altered mice will reveal the genetic networks that are perturbed when the level of a targeted gene is altered. The second type of database useful for understanding chemically induced toxicity is one in which various altered phenotypes in mutant mice have been annotated in a database that is directly searchable. This type of database is currently being constructed at a number of different institutions. Starting with conventional toxicity experiments, gene expression profiles and altered phenotypes in target tissues can be compared to databases of gene expression and phenotypes, leading to identification of candidate target genes that may mediate the effects of a chemical. The target genes hypothesized to be involved in chemical action can then be tested in appropriate genetically altered mice.*

profiles of new chemicals with uncharacterized toxicity could be compared to transcript profile databases, allowing the new chemicals to be provisionally placed into one or more mechanism-of-action classes. More directed studies could then be undertaken to confirm or refute the predicted mechanism of action of the new chemical. This type of analysis will potentially simplify the tests needed to characterize toxicity and could substantially reduce the time and resources needed to determine the potential toxicity of each new chemical.

A number of studies have recently described initial efforts to categorize chemicals based on transcript profiles (discussed in Chapters 6 and 14). A substantial matrix of data on many chemicals with known exposure-disease outcomes is needed to maximize the likelihood of detecting true positives and minimizing false negatives. This will require the evaluation of gene expression profiles of structurally related chemicals not causing disease, as well as those known to cause disease with varying potency. This approach to predicting toxicity has been thoroughly reviewed in a number of recent articles (Nuwaysir et al., 1999; Corton and Stauber, 2000; Henry et al., 2002; Boorman et al., 2002; Hamadeh et al., 2002).

In addition to information on chemical exposure of wild-type animals, the transcript profiles of control and chemically treated genetically altered mice will be useful to predict the signaling or metabolic pathways through which chemicals may be mediating their effects. A large number of studies have been published defining the gene expression differences between wild-type and genetically altered mice; some of these studies have included exposure to chemicals or other physical or biological stressors likely to be relevant for pathway evaluation (Table 11-1). A number of these studies have identified particular genes mediating chemical responses.

In our laboratories, we compared transcript profiles generated from the livers of wild-type and *PPARα*-null (peroxisome proliferator-activated receptor α) mice. PPARα mediates most if not all of the biologic effects of peroxisome proliferators (PP). PPARα heterodimerizes with another nuclear receptor, RXR, a target for a number of experimental drugs. Because RXR can heterodimerize with many nuclear receptors in the liver, we sought to determine the importance of PPARα in mediating the effects of exposure to a RXR agonist. Transcript profiles were generated from the livers of mice treated with either a PP or a RXR agonist. We were particularly interested in a group of genes that determine either activation or detoxification of endogenous and exogenous chemicals called phase I metabolism genes, which include the cytochrome P450 family of genes.

Phase I genes that exhibited a significant difference in expression between two or more groups were separately clustered, allowing a comparison of their expression behavior after chemical exposure in the two different strains (Fig. 11-7). A number of the genes, including *Cyp2f2* and *Cyp4a* members known to be regulated by PP (Ye et al., 1997; Okita and Okita, 2001), were shown to be altered by not only the PP but also by the RXR agonist in the wild-type

TABLE 11-1 Transcript profiling studies in nullizygous mice

Toxicological Category	Gene	Treatment	Tissue Profiled	References
Cancer	Car	Phenobarbital	Liver	Ueda et al., 2002
	Hif-1	Nickel	Fibroblasts	Salnikov et al., 2002
	Igf-1 (liver-specific)		Liver, bone	Sjogren et al., 2002a
	Jnk1; Jnk2	TPA	Fibroblasts	Chen et al., 2002a
	Nrf2	Sulforaphone	Small intestine	Thimmulappa et al., 2002
	Nrf2; Keap1	3H-1,2-dithiole-3-thione	Liver	Kwak et al., 2002
	Parp		Fibroblasts	Simbulan-Rosenthal et al., 2000
	PPARα	DEHP	Liver	Hasmall et al., 2002
	PPARα	WY-14,643; AGN194201	Liver	This chapter
Endocrine/Reproductive	Estrogen receptor α; Estrogen receptor β	Estrogen	Bone	Lindberg et al., 2002
Inflammation	Cox-1; Cox-2	Acetaminophen	Liver	Reilly et al., 2001
	Glucocorticoid receptor		Lung	Kaplan et al., 2003
	Il-10/4; Il-10/12	Parasitic trematode infection	Liver	Hoffmann et al., 2001
	Il-10	Il-10; LPS	Macrophage	Lang et al., 2002
	iNos; Phox	IFNg; tuberculosis	Macrophage	Ehrt et al., 2001
	iNos	Adenovirus-expressed hiNOS	Primary hepatocytes	Zamora et al., 2002
	Prostanoid receptor EP4		Intestine	Kabashima et al., 2002
Neurotoxicology	Cystatin-B		Brain	Lieuallen et al., 2001
	FMR1		Brain	D'Agata et al., 2002; Brown et al., 2001
	Superoxide dismutase 1		Spinal cord	Yoshihara et al., 2002
Steatosis	ApoA1		Liver	Callow et al., 2000
	Leptin		Liver	Ferrante et al., 2001; Liang and Tall, 2001
	LXRα; LXRβ	T0901317	Liver	Stulnig et al., 2002
	PPARα	Fasting	Liver	Kersten et al., 2001
Miscellaneous	Anx7	Fasting	Adrenal	Srivastava et al., 2002
	Crx		Eye	Livesey et al., 2000
	Erythropoietin		Brain	Maurer et al., 2002
	Frk; rak		Multiple tissues	Chandrasekharan et al., 2002
	Megalin		Kidney	Hilpert et al., 2002
	Ncor		Liver	Feng et al., 2001
	Rhodopsin		Retina	Kennan et al., 2002

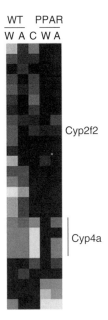

Figure 11-7 *Altered expression of phase I genes after exposure to PPARα or RXR agonists. Wild-type and PPARα-null mice were treated with 3 consecutive daily doses of either the PPARα agonist WY-14,643 (W) or the RXR agonist AGN194201 (A). Mice were sacrificed on the fourth day, and liver gene expression was determined using the U74v2 mouse chips (Affymetrix), normalized by scaling, then statistically filtered using the Resolver software (Rosetta). Phase I genes that exhibited a significant difference between two or more groups were separately clustered and visualized using CLUSTER and TreeView (Eisen et al., 1999). Also shown are the differences observed between the control PPARα-null mice and wild-type mice (C). The Cyp2f2 and Cyp4a family members known to be regulated by peroxisome proliferators are shown.*

mice. Additional phase I genes were identified as being regulated by these chemicals that play roles in the metabolism of fatty acids, steroids, and xenobiotics. This information and additional data (Anderson et al., submitted) indicate that there is a significant overlap in the effects of these two chemicals. The fact that most of the gene expression changes after chemical exposure were abolished in the PPARα-null mice indicates that PPARα is required for mediating the effects of the PP as well as the RXR agonist.

This strategy of comparing the gene expression patterns in wild-type and nullizygous mouse strains helps to identify the heterodimer complexes responsible for chemically induced gene expression. Thus the genes in which chemical-induced alterations in exposure is dependent on PPARα are likely mediated through PPARα-RXR heterodimers, regardless of whether activation is through PPARα or through RXR. When expression of the gene is maintained in the *PPARα*-null strain after exposure to the RXR agonist, we can hypothesize that expression is mediated by a heterodimer consisting of RXR and another to-be-identified nuclear receptor or with RXR itself. This approach can be used to define the mechanism of action through which the chemical induces toxicity and the relevance of that mechanism to humans.

Given the fact that many genes determine chemical toxicity, it is also possible to use the transcript profiles to make predictions of differences in chemical sensitivity in chemically treated wild-type mice or to control genetically altered mouse strains. For example, we predict that wild-type mice treated with a PP would be resistant to exposure to naphthalene, given that there is decreased expression in the livers of PP-treated wild-type mice of *Cyp2f2*, the enzyme responsible for activating naphthalene and a number of other toxicants (Shultz et al., 2001).

Beyond their use to predict chemical class, databases of genomic information might someday be used to identify specific targets of chemicals, allowing for precise determination of mechanism of action. Indications that this is a feasible goal came from recent work carried out by Friend and colleagues, who constructed a compendium of expression profiles from approximately 300 mutants and chemically treated strains of yeast (Hughes et al., 2000). An important finding from these studies is that the expression profiles obtained from chemical or drug-treated cells are similar to profiles from strains in which the genes encoding targets of the chemicals are mutated. This indicates that the cell responds to the two disparate perturbations (i.e., inactivation of the gene or the protein product of the gene) by regulating the same set of genes. This type of strategy could theoretically be used to create a database of gene expression profiles in toxicologically important tissues from genetically altered mice.

Once databases are sufficiently robust, it may be possible for toxicologists to compare an expression profile of a chemical under scrutiny to hundreds or thousands of these profiles in Web-based databases to identify likely gene or pathway targets that could be confirmed experimentally. Although many years off, this type of analysis may revolutionize toxicology by shifting research from a chemical category-based orientation to one in which mechanism of action can be more precisely defined.

Until we have the ability to predict chemical targets, toxicologists need to have information that complements the chemical profiles. This information would allow the identification of pathways whose genes are coordinately regulated by chemical exposure. At the present time our understanding of the genes that define metabolic or signal transduction pathways and the mechanisms by which these pathways are interconnected to regulate normal physiology is inadequate to predict mechanism of action of chemical toxicity. However, coordinated strategies to identify the genes regulated by defined mechanisms are growing. For example, batteries of gene targets have been identified by transcript profiling in a number of systems in which transcription factors or other proteins are overexpressed using transgenic technologies (Table 11-2). This information is potentially useful in defining the pathways through which chemicals work.

TABLE 11-2 Transcript profile studies in transgenic mice

Transgene	Treatment	Tissue Profiled	References
CETP	Cholic acid	Liver	Luo et al., 2001
hCOX-2		Brain	Mirjany et al., 2002
FoxM1B	Partial hepatectomy	Liver	Wang et al., 2001
Hormone-sensitive lipase		Heart	Suzuki et al., 2001
Calcineurin; calsequestrin; Galpha (q); protein kinase C epsilon activation peptide		Heart	Aronow et al., 2001
Insulin receptor substrate-2		Liver	Tobe et al., 2001
LMP2A		B cells	Portis and Longnecker, 2003
Mad1		Thymocytes	Iritani et al., 2002
Pax6		Lens	Chauhan et al., 2002
PPARγ1		Liver	Yu et al., 2003
SR-B1		Liver	Callow et al., 2000

It is unclear at this point what information is needed to accurately predict the mechanism of chemical action because we lack specific examples of comparisons between chemical profiles and information from genetically altered mice or other experimental systems. However, systems biology approaches in which these interactions are systematically uncovered through a comprehensive assessment of changes in genomic components after perturbation of wild-type and mutant strains will be instrumental in defining the pathways important in regulating the functions of the cell and in identifying all of the components with which chemicals could potentially interact.

There is also hope that the spectrum of adverse or adaptive phenotypic responses observed in chemically treated mice could be screened against databases of mouse genetic and phenotypic data (Fig. 11-7). An important mouse database is the Mouse Genome Database housed at the Jackson Laboratory. This database includes a mouse locus catalog listing existing mouse mutants and extensive map information describing their locations. To simplify the search for mutant mice, the International Mouse Strain Resource has recently been developed, which is mirrored at two Web sites: the Jackson Laboratory and the Medical Research Council, Harwell, UK. A great variety of phenotypic data is likely to accumulate on large numbers of mutant mouse strains, creating the need for phenotype databases that can be linked with gene map, mutagenesis, and gene expression databases.

To provide a genetic resource to the toxicology community, mutant mice must be readily available. Although many mutant stocks can be purchased from commercial vendors or from the Jackson Laboratory, additional distribution centers are required. To meet this demand, the National Institute of Environmental Health Sciences (NIEHS) has funded Mutant Mouse Resource Centers at a number of locations in the United States to serve as stock archives and regional distribution centers. In addition, the European Mutant Mouse Archive (EMMA) consists of a number of mutant mouse repositories.

SUMMARY

Transgenic mice are proving to be powerful tools in mechanistically based toxicology research. Through the inactivation of key genes by specific targeting, the role of a specific gene in regulating cellular responses to a wide variety of stressors can be assessed after toxicologically relevant exposures. Transgenic animals have been used to define genes that have a central role in the metabolic activation/detoxication of xenobiotics, regulate cellular responses to chemical carcinogens, and mediate the spectrum of biological responses induced by ligand-receptor interactions (Gonzalez, 2000). The incorporation of recoverable target genes in the genome of mice has been used to examine tissue-specific mechanisms of mutagenicity that may be involved in tumor development (Dean et al., 1999). Transgenic mice with altered oncogenes or tumor suppressor genes are being evaluated as appropriate short-term cancer models (Gulezian et al., 2000).

Transgenic mice are also being used to identify genes and specific genetic mutations that confer genetic susceptibility of humans to the toxic effects of xenobiotics. The enzyme systems involved in the metabolic bioactivation and detoxication of xenobiotics are significant determinants of interindividual variability and risk in response to the toxic effects of xenobiotic exposures. Polymorphisms in DNA repair systems are also being linked to cancer susceptibility. Animal models with specific single-nucleotide polymorphisms linked to human genetic susceptibility can be constructed in mice and directly tested for their sensitivity to various agents, thereby defining the significance of specific polymorphisms as determinants of human risk. These models can mimic aspects of human responsiveness and susceptibility, providing the toxicologist with humanized mouse models to directly test hypotheses regarding human genetic susceptibilities (Gonzalez, 2000).

In this chapter we have outlined approaches for the construction of a new generation of mouse models that exhibit cell type– or tissue type–specific gene expression. Models expressing altered gene products in a time- and tissue-specific manner will be valuable in focusing attention on relevant target tissues. Genes that can be induced to express normal or mutant proteins during specific periods of embryonic development will aid in the identification of critical target proteins as well as windows of sensitivity linked to adverse outcomes. Integrating genomic studies with transgenic mice to identify genetic pathways altered by toxic agents will likely significantly increase the depth and pace of our understanding of how chemicals interact with biological systems.

REFERENCES

Aronow BJ, Toyokawa T, Canning A, Haghighi K, Delling U, Kranias E, Molkentin JD, Dorn GW II (2001). Divergent transcriptional responses to independent genetic causes of cardiac hypertrophy. *Physiol Genomics* **6**:19–28.

Boorman GA, Anderson SP, Casey WM, Brown RH, Crosby LM, Gottschalk K, Easton M, Ni H, Morgan KT (2002). Toxicogenomics, drug discovery, and the pathologist. *Toxicol Pathol* **30**:15–27.

Brown V, Jin P, Ceman S, Darnell JC, O'Donnell WT, Tenenbaum SA, Jin X, Feng Y, Wilkinson KD, Keene JD, Darnell RB, Warren ST (2001). Microarray identification of FMRP-associated brain mRNAs and altered mRNA translational profiles in fragile X syndrome. *Cell* **107**:477–487.

Callow MJ, Dudoit S, Gong EL, Speed TP, Rubin EM (2000). Microarray expression profiling identifies genes with altered expression in HDL-deficient mice. *Genome Res* **10**:2022–2029.

Capecchi MR (1996). Altering the genome by homologous recombination. *Science* **244**: 1288–1292.

Cecconi F, Barbara IM (2000). Gene trap: a way to identify novel genes and unravel their biological function. *FEBS Lett* **480**:63–71.

Chandrasekharan S, Qiu TH, Alkharouf N, Brantley K, Mitchell JB, Liu ET (2002). Characterization of mice deficient in the Src family nonreceptor tyrosine kinase Frk/rak. *Mol Cell Biol* **22**:5235–5247.

Chauhan BK, Reed NA, Zhang W, Duncan MK, Kilimann MW, Cvekl A (2002). Identification of genes downstream of Pax6 in the mouse lens using cDNA microarrays. *J Biol Chem* **277**:11539–11548.

Chen CZ, Li M, de Graaf D, Monti S, Gottgens B, Sanchez MJ, Lander ES, Golub TR, Green AR, Lodish HF (2002a). Identification of endoglin as a functional marker that defines long-term repopulating hematopoietic stem cells. *Proc Natl Acad Sci USA* **99**:15468–15473.

Chen N, She QB, Bode AM, Dong Z (2002b). Differential gene expression profiles of Jnk1- and Jnk2-deficient murine fibroblast cells. *Cancer Res* **62**:1300–1304.

Chen Y, Yee D, Dains K, Chatterjee A, Cavalcoli J, Schneider E, Om J, Woychik RP, Magnuson T (2000). Genotype-based screen for ENU-induced mutations in mouse embryonic stem cells. *Nat Genet* **24**:314–317.

Corton JC, Stauber AJ (2000). Toward construction of a transcript profile database predictive of chemical toxicity. *Toxicol Sci* **58**:217–219.

Corton JC, Anderson SP, Recio L (2002). Interpretation of toxicogenomic data using genetically-altered mice. In: Vanden Heuvel, Greenlee, Perdew, Mattes (eds), *Comprehensive Toxicology*, vol XIV. Amsterdam: Elsevier Science, pp 515–526.

D'Agata V, Warren ST, Zhao W, Torre ER, Alkon DL, Cavallaro S (2002). Gene expression profiles in a transgenic animal model of fragile X syndrome. *Neurobiol Dis* **10**: 211–218.

Dean SW, Brooks TM, Burlinson B, Mirsalis J, Myhr B, Recio L, Thybaud V (1999). Transgenic mouse mutation assay systems can play an important role in regulatory mutagenicity testing in vivo for the detection of site-of-contact mutagens. *Mutagenesis* **14**:141–151.

de Angelis M, Balling R (1998). Large scale ENU screens in the mouse: genetics meets genomics. *Mutat Res* **400**:25–36.

Denny P, Justice MJ (2000). Mouse as the measure of man? *Trends Genet* **16**:283–287.

Dymecki SM (1996). Flp recombinase promotes site-specific DNA recombination in embryonic stem cells and transgenic mice. *Proc Natl Acad Sci USA* **93**:6191–6196.

Ehrt S, Schnappinger D, Bekiranov S, Drenkow J, Shi S, Gingeras TR, Gaasterland T, Schoolnik G, Nathan C (2001). Reprogramming of the macrophage transcriptome in response to interferon-gamma and *Mycobacterium tuberculosis*: signaling roles of nitric oxide synthase-2 and phagocyte oxidase. *J Exp Med* **194**:1123–1140.

Feng X, Jiang Y, Meltzer P, Yen PM (2001). Transgenic targeting of a dominant negative corepressor to liver blocks basal repression by thyroid hormone receptor and increases cell proliferation. *J Biol Chem* **276**:15066–15072.

Ferrante AW Jr, Thearle M, Liao T, Leibel RL (2001). Effects of leptin deficiency and short-term repletion on hepatic gene expression in genetically obese mice. *Diabetes* **50**:2268–2278.

Festing MFW, Lin L, Devereux TR, Gao F, Uang A, Malkinson AM, You M (1998). At least four loci and genes are associated with susceptibility to the chemical induction of lung adenomas in A/J × BALB/c mice. *Genomics* **53**:129–136.

Geraghty MT, Bassett D, Morrell JC, Gatto GJ Jr, Bai J, Geisbrecht BV, Hieter P, Gould SJ (1999). Detecting patterns of protein distribution and gene expression in silico. *Proc Natl Acad Sci USA* **96**:2937–2942.

Gilbert DR, Schroeder M, van Helden J (2000). Interactive visualization and exploration of relationships between biological objects. *Trends Biotechnol* **18**:487–494.

Gonzalez FJ (2001). The use of gene knockout mice to unravel the mechanisms of toxicity and chemical carcinogenesis. *Toxicol Lett* **120**:199–208.

Gonzalez FJ, Kimura S (2003). Study of P450 function using gene knockout and transgenic mice. *Arch Biochem Biophys* **409**:153–158.

Grupe A, Germer S, Usuka J, Aud D, Belknap JK, Klein RF, Ahluwalia MK, Higuchi R, Peltz G (2001). In silico mapping of complex disease-related traits in mice. *Science* **292**:1915–1918.

Gulezian D, Jacobson-Ram D, McCullough CB, Olson H, Recio L, Robinson D, Storer R, Tennant R, Ward JM, Neumann DA (2000). Use of transgenic animals for carcinogenicity testing: considerations and implications for risk assessment. *Toxicol Pathol* **28**:482–499.

Hasmall S, Orphanides G, James N, Pennie W, Hedley K, Soames A, Kimber I, Roberts R (2002). Downregulation of lactoferrin by PPARalpha ligands: role in perturbation of hepatocyte proliferation and apoptosis. *Toxicol Sci* **68**:304–313.

Heideman J (1991). Transgenic rats: a discussion. *Biotechnology* **16**:325–332.

Henry CJ, Phillips R, Carpanini F, Corton JC, Craig K, Igarashi K, Leboeuf R, Marchant G, Osborn K, Pennie WD, Smith LL, Teta MJ, Vu V (2002). Use of genomics in toxicology and epidemiology: findings and recommendations of a workshop. *Environ Health Perspect* **110**:1047–1050.

Hilpert J, Wogensen L, Thykjaer T, Wellner M, Schlichting U, Orntoft TF, Bachmann S, Nykjaer A, Willnow TE (2002). Expression profiling confirms the role of endocytic receptor megalin in renal vitamin D3 metabolism. *Kidney Int* **62**:1672–1681.

Hoffmann KF, McCarty TC, Segal DH, Chiaramonte M, Hesse M, Davis EM, Cheever AW, Meltzer PS, Morse HC III, Wynn TA (2001). Disease fingerprinting with cDNA microarrays reveals distinct gene expression profiles in lethal type 1 and type 2 cytokine-mediated inflammatory reactions. *FASEB J* **15**:2545–2547.

Hofker MH, Breuer M (1998). Generation of transgenic mice. *Methods Mol Biol* **110**: 63–78.

Hughes TR, Marton MJ, Jones AR, Roberts CJ, Stoughton R, Armour CD, Bennett HA, Coffey E, Dai H, He YD, Kidd MJ, King AM, Meyer MR, Slade D, Lum PY, Stepaniants SB, Shoemaker DD, Gachotte D, Chakraburtty K, Simon J, Bard M, Friend SH (2000). Functional discovery via a compendium of expression profiles. *Cell* **102**:109–126.

Iritani BM, Delrow J, Grandori C, Gomez I, Klacking M, Carlos LS, Eisenman RN (2002). Modulation of T-lymphocyte development, growth and cell size by the Myc antagonist and transcriptional repressor Mad1. *EMBO J* **21**:4820–4830.

Justice MJ (2000). Capitalizing on large-scale mouse mutagenesis screens. *Nat Rev Genet* **1**:109–115.

Kabashima K, Saji T, Murata T, Nagamachi M, Matsuoka T, Segi E, Tsuboi K, Sugimoto Y, Kobayashi T, Miyachi Y, Ichikawa A, Narumiya S (2002). The prostaglandin receptor EP4 suppresses colitis, mucosal damage and CD4 cell activation in the gut. *J Clin Invest* **109**:883–893.

Kaplan F, Comber J, Sladek R, Hudson TJ, Muglia LJ, Macrae T, Gagnon S, Asada M, Brewer JA, Sweezey NB (2003). The growth factor midkine is modulated by both glucocorticoid and retinoid in fetal lung development. *Am J Respir Cell Mol Biol* **28**:33–41.

Kennan A, Aherne A, Palfi A, Humphries M, McKee A, Stitt A, Simpson DA, Demtroder K, Orntoft T, Ayuso C, Kenna PF, Farrar GJ, Humphries P (2002). Identification of an IMPDH1 mutation in autosomal dominant retinitis pigmentosa (RP10) revealed following comparative microarray analysis of transcripts derived from retinas of wild-type and Rho(−/−) mice. *Hum Mol Genet* **11**:547–557.

Kersten S, Mandard S, Escher P, Gonzalez FJ, Tafuri S, Desvergne B, Wahli W (2001). The peroxisome proliferator-activated receptor alpha regulates amino acid metabolism. *FASEB J* **15**:1971–1978.

Kistner A, Gossen M, Zimmermann F, Jerecic J, Ullmer C, Lubbert H, Bujard H (1996). Doxycycline-mediated quantitative and tissue-specific control of gene expression in transgenic mice. *Proc Natl Acad Sci USA* **93**:10933–10938.

Kuhn KM, DeRisi JL, Brown PO, Sarnow P (2001). Global and specific translational regulation in the genomic response of *Saccharomyces cerevisiae* to a rapid transfer from a fermentable to a nonfermentable carbon source. *Mol Cell Biol* **21**:916–927.

Kwak MK, Wakabayashi N, Itoh K, Motohashi H, Yamamoto M, Kensler TW (2002). Modulation of gene expression by cancer chemopreventive dithiolethiones through the Keap1-Nrf2 pathway: identification of novel gene clusters for cell survival. *J Biol Chem*.

Lang R, Patel D, Morris JJ, Rutschman RL, Murray PJ (2002). Shaping gene expression in activated and resting primary macrophages by IL-10. *J Immunol* **169**(5): 2253–2263.

Liang CP, Tall AR (2001). Transcriptional profiling reveals global defects in energy metabolism, lipoprotein, and bile acid synthesis and transport with reversal by leptin treatment in ob/ob mouse liver. *J Biol Chem* **276**:49066–49076.

Lieuallen K, Pennacchio LA, Park M, Myers RM, Lennon GG (2001). Cystatin B-deficient mice have increased expression of apoptosis and glial activation genes. *Hum Mol Genet* **10**:1867–1871.

Lindberg MK, Weihua Z, Andersson N, Moverare S, Gao H, Vidal O, Erlandsson M, Windahl S, Andersson G, Lubahn DB, Carlsten H, Dahlman-Wright K, Gustafsson JA, Ohlsson C (2002). Estrogen receptor specificity for the effects of estrogen in ovariectomized mice. *J Endocrinol* **174**:167–178.

Livesey FJ, Furukawa T, Steffen MA, Church GM, Cepko CL (2000). Microarray analysis of the transcriptional network controlled by the photoreceptor homeobox gene. *Crx Curr Biol* **10**:301–310.

Lobe CG, Nagy A (1998). Conditional genome alteration in mice. *Bioessays* **20**:200–208.

Lovik M (1997). Mutant and transgenic mice in immunotoxicology: an introduction. *Toxicology* **119**:65–76.

Luo Y, Liang CP, Tall AR (2001). The orphan nuclear receptor LRH-1 potentiates the sterol-mediated induction of the human CETP gene by liver X receptor. *J Biol Chem* **276**:24767–24773.

Maurer MH, Frietsch T, Waschke KF, Kuschinsky W, Gassmann M, Schneider A (2002). Cerebral transcriptome analysis of transgenic mice overexpressing erythropoietin. *Neurosci Lett* **327**:181–184.

Mirjany M, Ho L, Pasinetti GM (2002). Role of cyclooxygenase-2 in neuronal cell cycle activity and glutamate-mediated excitotoxicity. *J Pharmacol Exp Ther* **301**:494–500.

Munroe RJ, Bergstrom RA, Zheng QY, Libby B, Smith R, John SW, Schimenti KJ, Browning VL, Schimenti JC (2000). Mouse mutants from chemically mutagenized embryonic stem cells. *Nat Genet* **24**:318–321.

Nuwaysir EF, Bittner M, Trent J, Barrett JC, Afshari CA (1999). Microarrays and toxicology: the advent of toxicogenomics. *Mol Carcinogen* **24**:153–159.

Okazaki Y, Furuno M, Kasukawa T, Adachi J, Bono H, Kondo S, Nikaido I, Osato N, Saito R, Suzuki H, Yamanaka I, Kiyosawa H, Yagi K, Tomaru Y, Hasegawa Y, Nogami A, Schonbach C, Gojobori T, Baldarelli R, Hill DP, Bult C, Hume DA, Quackenbush J, Schriml LM, Kanapin A, Matsuda H, Batalov S, Beisel KW, Blake JA, Bradt D, Brusic V, Chothia C, Corbani LE, Cousins S, Dalla E, Dragani TA, Fletcher CF, Forrest A, Frazer KS, Gaasterland T, Gariboldi M, Gissi C, Godzik A, Gough J, Grimmond S, Gustincich S, Hirokawa N, Jackson IJ, Jarvis ED, Kanai A, Kawaji H, Kawasawa Y, Kedzierski RM, King BL, Konagaya A, Kurochkin IV, Lee Y, Lenhard B, Lyons PA, Maglott DR, Maltais L, Marchionni L, McKenzie L, Miki H, Nagashima T, Numata K, Okido T, Pavan WJ, Pertea G, Pesole G, Petrovsky N, Pillai R, Pontius JU, Qi D, Ramachandran S, Ravasi T, Reed JC, Reed DJ, Reid J, Ring BZ, Ringwald M, Sandelin A, Schneider C, Semple CA, Setou M, Shimada K, Sultana R, Takenaka Y, Taylor MS, Teasdale RD, Tomita M, Verardo R, Wagner L, Wahlestedt C, Wang Y, Watanabe Y, Wells C, Wilming LG, Wynshaw-Boris A, Yanagisawa M, Yang I, Yang L, Yuan Z, Zavolan M, Zhu Y, Zimmer A, Carninci P, Hayatsu N, Hirozane-Kishikawa T, Konno H, Nakamura M, Sakazume N, Sato K, Shiraki T, Waki K, Kawai J, Aizawa K, Arakawa T, Fukuda S, Hara A, Hashizume W, Imotani K, Ishii Y, Itoh M, Kagawa I, Miyazaki A, Sakai K, Sasaki D, Shibata K,

Shinagawa A, Yasunishi A, Yoshino M, Waterston R, Lander ES, Rogers J, Birney E, Hayashizaki Y; The FANTOM Consortium.; RIKEN Genome Exploration Research Group Phase I & II Team (2002). Analysis of the mouse transcriptome based on functional annotation of 60,770 full-length cDNAs. *Nature* **420**:563–573.

Okita RT, Okita JR (2001). Cytochrome P450 4A fatty acid omega hydroxylases. *Curr Drug Metab* **2**:265–281.

Portis T, Longnecker R (2003). Epstein-Barr virus LMP2A interferes with global transcription factor regulation when expressed during B-lymphocyte development. *J Virol* **77**:105–114.

Reilly TP, Brady JN, Marchick MR, Bourdi M, George JW, Radonovich MF, Pise-Masison CA, Pohl LR (2001). A protective role for cyclooxygenase-2 in drug-induced liver injury in mice. *Chem Res Toxicol* **14**:1620–1628.

Salnikow K, Davidson T, Costa M (2002). The role of hypoxia-inducible signaling pathway in nickel carcinogenesis. *Environ Health Perspect* **110**(suppl 5):831–834.

Sauer B (1993). Manipulation of transgenes by site-specific recombination: use of Cre recombinase. *Methods Enzymol* **225**:890–900.

Sherlock G (2000). Analysis of large-scale gene expression data. *Curr Opin Immunol* **12**:201–205.

Shultz MA, Morin D, Chang AM, Buckpitt A (2001). Metabolic capabilities of CYP2F2 with various pulmonary toxicants and its relative abundance in mouse lung subcompartments. *J Pharmacol Exp Ther* **296**:510–519.

Simbulan-Rosenthal CM, Ly DH, Rosenthal DS, Konopka G, Luo R, Wang ZQ, Schultz PG, Smulson ME (2000). Misregulation of gene expression in primary fibroblasts lacking poly(ADP-ribose) polymerase. *Proc Natl Acad Sci USA* **97**:11274–11279.

Sjogren K, Sheng M, Moverare S, Liu JL, Wallenius K, Tornell J, Isaksson O, Jansson JO, Mohan S, Ohlsson C (2002). Effects of liver-derived insulin-like growth factor I on bone metabolism in mice. *J Bone Miner Res* **17**:1977–1987.

Srivastava M, Kumar P, Leighton X, Glasman M, Goping G, Eidelman O, Pollard HB (2002). Influence of the Anx7 (+/−) knockout mutation and fasting stress on the genomics of the mouse adrenal gland. *Ann NY Acad Sci* **971**:53–60.

Stulnig TM, Steffensen KR, Gao H, Reimers M, Dahlman-Wright K, Schuster GU, Gustafsson JA (2002). Novel roles of liver X receptors exposed by gene expression profiling in liver and adipose tissue. *Mol Pharmacol* **62**:1299–1305.

Suzuki J, Shen WJ, Nelson BD, Patel S, Veerkamp JH, Selwood SP, Murphy GM Jr, Reaven E, Kraemer FB (2001). Absence of cardiac lipid accumulation in transgenic mice with heart-specific HSL overexpression. *Am J Physiol Endocrinol Metab* **281**:E857–E866.

Taylor BA (1996). Recombinant inbred strains. In: MF Lyon, S Rastan, SDM Brown (eds), *Genetic Variants and Strains of the Laboratory Mouse*, vol 2. Oxford: Oxford University Press, pp 1597–1659.

Thimmulappa RK, Mai KH, Srisuma S, Kensler TW, Yamamoto M, Biswal S (2002). Identification of Nrf2-regulated genes induced by the chemopreventive agent sulforaphane by oligonucleotide microarray. *Cancer Res* **62**:5196–5203.

Tobe K, Suzuki R, Aoyama M, Yamauchi T, Kamon J, Kubota N, Terauchi Y, Matsui J, Akanuma Y, Kimura S, Tanaka J, Abe M, Ohsumi J, Nagai R, Kadowaki T (2001).

Increased expression of the sterol regulatory element-binding protein-1 gene in insulin receptor substrate-2(−/−) mouse liver. *J Biol Chem* **276**:38337–38340.

Ueda A, Hamadeh HK, Webb HK, Yamamoto Y, Sueyoshi T, Afshari CA, Lehmann JM, Negishi M (2002). Diverse roles of the nuclear orphan receptor CAR in regulating hepatic genes in response to phenobarbital. *Mol Pharmacol* **61**:1–6.

Vivian JL, Chen Y, Yee D, Schneider E, Magnuson T (2002). An allelic series of mutations in Smad2 and Smad4 identified in a genotype-based screen of N-ethyl-N-nitrosourea-mutagenized mouse embryonic stem cells. *Proc Natl Acad Sci USA* **99**: 15542–15547.

Wang X, Quail E, Hung NJ, Tan Y, Ye H, Costa RH (2001). Increased levels of forkhead box M1B transcription factor in transgenic mouse hepatocytes prevent age-related proliferation defects in regenerating liver. *Proc Natl Acad Sci USA* **98**:11468–11473.

Waterston RH, Lindblad-Toh K, Birney E, Rogers J, Abril JF, Agarwal P, Agarwala R, Ainscough R, Alexandersson M, An P, Antonarakis SE, Attwood J, Baertsch R, Bailey J, Barlow K, Beck S, Berry E, Birren B, Bloom T, Bork P, Botcherby M, Bray N, Brent MR, Brown DG, Brown SD, Bult C, Burton J, Butler J, Campbell RD, Carninci P, Cawley S, Chiaromonte F, Chinwalla AT, Church DM, Clamp M, Clee C, Collins FS, Cook LL, Copley RR, Coulson A, Couronne O, Cuff J, Curwen V, Cutts T, Daly M, David R, Davies J, Delehaunty KD, Deri J, Dermitzakis ET, Dewey C, Dickens NJ, Diekhans M, Dodge S, Dubchak I, Dunn DM, Eddy SR, Elnitski L, Emes RD, Eswara P, Eyras E, Felsenfeld A, Fewell GA, Flicek P, Foley K, Frankel WN, Fulton LA, Fulton RS, Furey TS, Gage D, Gibbs RA, Glusman G, Gnerre S, Goldman N, Goodstadt L, Grafham D, Graves TA, Green ED, Gregory S, Guigo R, Guyer M, Hardison RC, Haussler D, Hayashizaki Y, Hillier LW, Hinrichs A, Hlavina W, Holzer T, Hsu F, Hua A, Hubbard T, Hunt A, Jackson I, Jaffe DB, Johnson LS, Jones M, Jones TA, Joy A, Kamal M, Karlsson EK, Karolchik D, Kasprzyk A, Kawai J, Keibler E, Kells C, Kent WJ, Kirby A, Kolbe DL, Korf I, Kucherlapati RS, Kulbokas EJ, Kulp D, Landers T, Leger JP, Leonard S, Letunic I, Levine R, Li J, Li M, Lloyd C, Lucas S, Ma B, Maglott DR, Mardis ER, Matthews L, Mauceli E, Mayer JH, McCarthy M, McCombie WR, McLaren S, McLay K, McPherson JD, Meldrim J, Meredith B, Mesirov JP, Miller W, Miner TL, Mongin E, Montgomery KT, Morgan M, Mott R, Mullikin JC, Muzny DM, Nash WE, Nelson JO, Nhan MN, Nicol R, Ning Z, Nusbaum C, O'Connor MJ, Okazaki Y, Oliver K, Overton-Larty E, Pachter L, Parra G, Pepin KH, Peterson J, Pevzner P, Plumb R, Pohl CS, Poliakov A, Ponce TC, Ponting CP, Potter S, Quail M, Reymond A, Roe BA, Roskin KM, Rubin EM, Rust AG, Santos R, Sapojnikov V, Schultz B, Schultz J, Schwartz MS, Schwartz S, Scott C, Seaman S, Searle S, Sharpe T, Sheridan A, Shownkeen R, Sims S, Singer JB, Slater G, Smit A, Smith DR, Spencer B, Stabenau A, Stange-Thomann N, Sugnet C, Suyama M, Tesler G, Thompson J, Torrents D, Trevaskis E, Tromp J, Ucla C, Ureta-Vidal A, Vinson JP, Von Niederhausern AC, Wade CM, Wall M, Weber RJ, Weiss RB, Wendl MC, West AP, Wetterstrand K, Wheeler R, Whelan S, Wierzbowski J, Willey D, Williams S, Wilson RK, Winter E, Worley KC, Wyman D, Yang S, Yang SP, Zdobnov EM, Zody MC, Lander ES; Mouse Genome Sequencing Consortium (2002). Initial sequencing and comparative analysis of the mouse genome. *Nature* **420**:520–562.

Wolfsberg TG, Gabrielian AE, Campbell MJ, Cho RJ, Spouge JL, Landsman D (1999). Candidate regulatory sequence elements for cell cycle–dependent transcription in Saccharomyces cerevisiae. *Genome Res* **9**:775–792.

Woychik RP, Klebig ML, Justice MJ, Magnuson TR, Avrer ED (1998). Functional genomics in the post-genome era. *Mut Res* **400**:3–14.

Yoshihara T, Ishigaki S, Yamamoto M, Liang Y, Niwa J, Takeuchi H, Doyu M, Sobue G (2002). Differential expression of inflammation- and apoptosis-related genes in spinal cords of a mutant SOD1 transgenic mouse model of familial amyotrophic lateral sclerosis. *J Neurochem* **80**:158–167.

Ye X, Lu L, Gill SS (1997). Suppression of cytochrome P450 Cyp2f2 mRNA levels in mice by the peroxisome proliferator diethylhexylphthalate. *Biochem Biophys Res Commun* **239**:660–665.

Yu S, Matsusue K, Kashireddy P, Cao WQ, Yeldandi V, Yeldandi AV, Rao MS, Gonzalez FJ, Reddy JK (2003). Adipocyte-specific gene expression and adipogenic steatosis in the mouse liver due to peroxisome proliferator-activated receptor gamma 1 (PPARgamma 1) overexpression. *J Biol Chem* **278**:498–505.

Zamora R, Vodovotz Y, Aulak KS, Kim PK, Kane JM III, Alarcon L, Stuehr DJ, Billiar TR (2002). A DNA microarray study of nitric oxide–induced genes in mouse hepatocytes: implications for hepatic heme oxygenase-1 expression in ischemia/reperfusion. *Nitric Oxide* **7**:165–186.

Zhang MQ (1999). Promoter analysis of co-regulated genes in the yeast genome. *Comput Chem* **23**:233–250.

12

Introduction to High-Throughput Protein Expression

B. Alex Merrick

INTRODUCTION

In cells and tissues of complex organisms, only a portion of the genome is expressed at any one time as mRNA (transcriptome) or polypeptides (proteome). Gene expression constantly changes during health, adaptation, disease, and aging. The complexity of describing protein expression in health and disease on a global scale has led to the development of high-throughtput technologies called proteomics that detail the structure and functions of all proteins in an organism over time. As a refinement, toxicoproteomics is the identification of critical proteins and pathways in biological systems that are affected by and respond to adverse chemical and environmental exposures using global protein expression technologies. Knowledge of protein primary sequence, quantities, posttranslational modifications, structures, interactions, cellular spatial relationships and functions are important attributes needed for comprehensive protein expression analysis. It is this multifold and complex nature of protein attributes that has spawned the development of so many different proteomic technologies (Table 12-1).

Toxicogenomics: Principles and Applications. Edited by Hamadeh and Afshari
ISBN 0-471-43417-5 Copyright © 2004 John Wiley & Sons, Inc.

TABLE 12-1   **Protein expression technologies**[a]

Discovery Technologies

2D gel-MS
MALDI-Tof MS
LC-ESI-MS-MS
ICAT-MS

High-Throughput Technologies

Retentate chromatography-MS
Antibody, peptide, aptamer, protein microarrays
ToF ToF-MS
SPR-BIA

---

[a] Protein expression technologies are divided into discovery-based technologies, which contain the potential for discovery of new genes and proteins, and high-throughput technologies, for which known protein attribute information is being sought on a large scale containing high information content and large sample screening capability. 2D, two-dimensional gel; MS, mass spectrometry; MALDI-ToF, matrix-assisted laser desorption ionization–time of flight; LC, liquid chromatography; ESI, electrospray ionization; MS/MS, tandem mass spectrometry; ICAT, isotope-coded affinity tag; RC-MS, rententate chromatography mass spectrometry such as SELDI-ToF MS; SPR-BIA, surface plasmon resonance–biomolecular interaction analysis.

High-throughput protein expression is a relative term intended to denote a dramatic departure from the historical approach of comparing samples on the basis of a single gene product measurement to a global analysis of protein attributes. The nature of proteomic analysis should provide a large amount of data or high information density per individual sample, an objective that is just recently being realized. Therefore, ongoing efforts in high-throughput proteomics strive to fulfill four objectives: (1) high information density per sample, (2) high number of samples for analysis, (3) fine discrimination of altered protein expression, and (4) analysis in a timely manner. The current throughput capacity for transcript analysis puts these objectives in perspective.

DNA microarrays can evaluate >10,000–20,000 genes per chip, allow for hybridization of several chips per analysis session, provide information on abundant and low copy transcripts, determine the presence of new genes with unknown function or expressed sequence tags (ESTs), and provide a readout within a few days. Admittedly, the bioinformatics and biological interpretation of transcript data do take considerably longer than the collection of raw data. Yet the data output of hundreds of thousands of genes that can be routinely evaluated by multiple DNA chip hybridizations in one experiment is an achievement that is not yet possible in global protein expression analysis. However, recent advances in existing approaches and development of new technologies (Table 12-1) suggest that this gap for timely, high-density proteomic data is rapidly narrowing. This chapter reviews the advances in high-throughput protein expression methods and highlights promising future trends and technologies.

MASS SPECTROMETRY AND SEPARATION TECHNOLOGIES

There are several means to identify proteins including immunoreactivity, enzymatic activity, ligand affinity, pharmacologic response, and many others. However, mass spectrometry can more precisely identify proteins (1) by producing primary amino acid sequences for portions of protein or sequence tags and (2) by mass analysis of proteolytic digest fragments of proteins that collectively form a diagnostic peptide fingerprint for protein identity. Primary protein sequence is the venue for searching into the more extensive nucleic acid databases, which have been more highly characterized across many species compared to protein databases.

In addition to being a sensitive method of identification, mass spectrometry can reveal many fine structural details including posttranslational modifications and protein-protein interactions. When preceded by some form of protein separation, mass spectrometry is a highly preferred method of protein identification in proteomic analysis. Although two-dimensional (2D) gel electrophoresis has been well established for 25 years, the coupling of 2D gels with mass spectrometry (MS) has only more recently been joined together to form a powerful paradigm for analysis of protein expression. Advances in automation and miniaturization have made 2D-MS a key high-throughput protein expression technology (Fig. 12-1; Witzmann and Li, 2002).

How does mass spectrometry (MS) identify proteins? Mass spectrometers measure the mass to charge ratio (m/z) of ions from analytes ranging in mass from drugs to large polypeptides. Typically, however, purified proteins are enzymatically digested (i.e., trypsin) into 5mer to 20mer peptides for greater mass accuracy in protein identification by MS. MS instruments contain three basic elements: an ionization source to convert proteins (or their peptides) into a gas phase, a mass analyzer, and an ion detector (Fig. 12-2). The most common ionization devices are matrix-assisted laser desorption ionization (MALDI) and electrospray ionization (ESI). Once a sample is ionized, it must be mass analyzed. The most common mass analyzers for protein analysis are time of flight (ToF), triple quadrupole, quadrupole-ToF (hybrid), and ion trap devices .

The two general types of MS instruments are (1) MALDI-ToF MS, or matrix-assisted laser desorption ionization–time of flight mass spectrometer, and (2) LC-ESI-MS/MS, or liquid chromatography and electrospray ionization tandem mass spectrometer (Fig. 12-2). In MALDI-ToF, a laser provides the energy, and a matrix such as α-cyanohydroxycinnamic acid (CHCA) provides the environment to effectively ionize and volatilize peptides that are dried on an inert target surface (i.e., gold target surface). Once in gas phase,

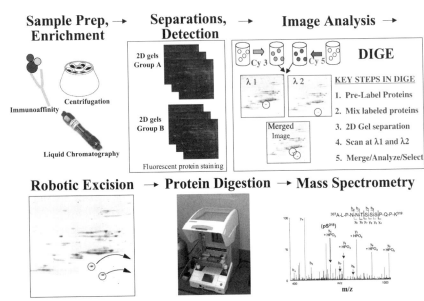

Figure 12-1   *Protein expression analysis by two-dimensional gel electrophoresis and mass spectrometry. The sequence of analysis is outlined for protein identification using two-dimensional (2D) gel electrophoresis and mass spectrometry. Sample preparation may involve various enrichment procedures to concentrate protein fractions or subsets of interest. All salts and ions must be removed prior to isoelectric focusing (IEF) and mass separation during 2D gel analysis. Two general approaches are used in separation and imaging: (1) The first is the standard method for comparison of individual 2D gel separations from group A (control) or group B (treatment) samples where images are electronically registered, then intensities of registered proteins are compared for relative difference from control. (2) An alternate second approach is the DIGE method, or difference in gel electrophoresis, whereby protein samples from groups A and B are prelabeled with Cy3 or Cy5 fluorescent dyes, mixed, and separated on the same gel to avoid inaccuracies of electronic registration and differing sample loads. Ratiometric differences in expression for each protein can be determined by scanning the same gel at different wavelengths (λ1 and λ2) to detect differences in fluorescent intensities. Up- or downregulated proteins are excised from gels, robotically digested with proteolytic enzymes, and analyzed by mass spectrometry to determine protein identity.*

the positively charged peptides are attracted to "fly" through a vacuum tube toward a negatively charged detector source so that the time of flight can be calibrated in proportional to peptide mass. The mass to charge ratio (m/z) values for each peptide are determined by measuring the time it takes ions to move from the ion source to the detector.

In electrospray ionization mass spectrometry (ESI-MS) a fine spray of solubilized peptides are vaporized in a high-voltage field; similarly peptides separated by liquid chromatography in LC-MS will elute from the column in a fine stream of effluent that is heated to vaporize and ionize peptides in a field of high voltage. Once ionized, the first mass analyzer isolates a particular ion

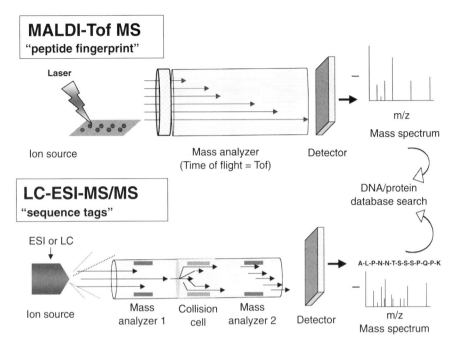

Figure 12-2 *Protein identification by mass spectrometry. Mass spectrometers consist of three components: an ionization source, mass analyzers, and a detector. Purified proteins are often enzymatically digested to obtain higher mass accuracy compared to analyzing the native protein. Two primary types of mass spectrometers are used for protein identifications: MALDI-ToF MS and tandem mass spectrometry, or MS/MS. MALDI-ToF utilizes a laser to ionize peptides into a time-of-flight mass analyzer. The mass spectrum from MALDI-ToF forms a peptide fingerprint of the native protein. The ionization source for tandem mass spectrometers is often an electrospray ionization device (ESI) or can be the vaporized eluant from a liquid chromatography (LC) separation (i.e., HPLC or capillary electrophoresis) directly connected to the instrument. Peptides are fragmented into smaller fragments in the collision chamber, then separated by a second mass analyzer to derive an amino acid sequence tag. See text for further details.*

for dissociation in a gas-phase collision cell whereupon the collision product ions are then analyzed in a second or tandem mass analyzer prior to striking the detector. Each of these ionization sources, a laser in MALDI-ToF or the high-voltage ionization of droplets in ESI-MS/MS, will produce a different spectrum of detectable masses and intensities because the effectiveness and nature peptide ionization are different for each source. In addition, the presence of multiple peptides that influence ionization potential make most peptide ion measurements only semiquantitative.

Protein identification is indirect using MALDI-ToF MS because the peptide fingerprint may be incomplete or the measured peptide ions may be almost the same value as peptides generated by other proteins (Table 12-2). The confidence in the match of measured ions obtained from the mass

TABLE 12-2 Capabilities and limitations of current protein expression technologies

2D-MS	MALDI-ToF MS	LC-ESI-MS/MS
High resolution	Rapid	Amino acid sequence tag information
Semiquantitative	Peptide fingerprinting	Posttranslational modifications
Differential expression	Sensitive	Multidimensional chromatography
Detect protein modifications	Automated	High information density possible
Semiglobal separation*	Uncertainty in protein ID*	Time-consuming analysis*
Low-abundance proteins*	Incomplete protein databases*	Lower sensitivity than MALDI*
Solubilization problems*	Variable ionization*	Interference from salts and ions*
Basic and hydrophobic proteins*		

ICAT	RC-MS	Antibody Microarrays
Sensitive	Sensitive, selective	Selective immunoaffinity recognition
Quantitative, wide dynamic range	Rapid analysis, medium throughout	Sensitive, high throughput
High information density per sample	Native proteins analyzed	High information density per sample
Differential protein expression	Patterns, not protein identification*	Validated antibodies are limited in supply*
Complex analysis*	Nonglobal*	Cross-reactivity of shared epitopes*
Variable ionization efficiency*	Limited mass range*	Semiquantitative*
Not all proteins cross-link*	Scale-up, other identification needed*	Protein-protein interaction interference*

The capabilities and drawbacks* of protein expression technologies are briefly highlighted and are more thoroughly discussed in the text. Abbreviations are in Table 12-1 legend. Asterisk indicates a limitation of the technology.

spectrum becomes greater with an increasing percentage of matches, with the theoretical masses obtained by a simulated digestion of a protein contained in a database. Of the two MS instruments, MALDI-ToF MS is the more sensitive for peptide fingerprinting, but tandem mass spectrometry provides more definitive information for identification (Table 12-2).

In tandem mass spectrometry, or MS/MS, peptides from protein enzymatic digests are ionized into a mass spectrum, and ions from this first spectrum are selected (manually or by computer algorithm) and further broken apart in a collision cell for analysis by a second (tandem) mass spectrometer (Fig. 12-2). In the collision chamber, fragmentation of the selected peptide occurs at or around the peptide bond to create a ladder of fragments that can be reconstructed into an amino acid sequence or sequence tag of the peptide. These peptide sequences derived from the digested protein can be searched in both protein and nucleic acid databases to assign a more definitive protein identification than peptide fingerprinting. Although more sample is required for tandem MS analysis, it has the major advantage of generating the amino acid sequence for a more definitive identification after a search in nucleic acid, protein, or expressed sequence tag (EST) nucleotide databases (Table 12-2).

Advances in the field of mass spectrometry have provided newer instruments such as the MALDI ToF-ToF tandem MS (Table 12-1; Huang et al., 2002), which can provide the sensitivity of MALDI and the sequence tag capability of earlier liquid-based MS/MS machines at a capacity of thousands of samples per day. The two challenges in protein profiling using separation-based MS have now become (1) providing sufficient numbers of purified proteins (feeding the instrument) and (2) the longstanding bottleneck of database searching and bioinformatics. Clearly, the ToF-ToF MS/MS instrument possesses extremely high-throughput capabilities that will have a large impact on protein expression analysis.

Areas of technical development are the introduction of microfluidics in multidimensional protein separations, protein processing, and introduction devices into mass spectrometers (Figeys, 2002). Biomolecular interaction analysis (BIA) can provide real-time kinetic information and rate constants of association using the optical method of surface plasmon resonance (SPR). Microfluidic connections between a SPR-BIA sensor and an ESI-MS/MS instrument can provide real-time interaction analysis and protein identification in the same run (Natsume et al., 2001).

TWO-DIMENSIONAL GEL SEPARATION

Two-dimensional (2D) gel electrophoresis is an established and rapid method to separate thousands of proteins from complex lysates derived from almost any source of organelle, cell, tissue, organ, or phyla (Garrels, 1979). 2D gel electrophoresis is a hybrid technique whereby proteins are resolved by charge using isoelectric focusing and by mass with SDS PAGE. The standard 2D gel (18 cm IPG strip; 20 × 20 cm SDS PAGE gel) can regularly separate 2,000–3,000 and up to 10,000 protein features depending on analytical conditions. Several 2D gels can be run in parallel at 12 to 20 gels per session. Four important advances in the 2D gel field are (1) development of immobilized pH gradient (IPG) strips for improved isoelectric focusing and larger sample loads (100–1,000 μg protein), (2) use of narrow-pH-range IPG gels, (3) introduction of MS-compatible, reversible, fluorescent dyes like Sypro ruby, and (4) a new protein display method called differential gel electrophoresis (DIGE) to more easily find protein expression difference in 2D gels.

The work flow for 2D-MS analysis is shown in Figure 12-1. Fluorescent staining of 2D gel separated proteins with Sypro ruby permits sensitive, quantitative detection, which is followed by comparative image analysis of many gels to determine which subset of proteins is differentially expressed among control and treated samples. A new technique called DIGE (Fig. 12-1) can improve the determination of up- or downregulated proteins by separately prelabeling each protein preparation from control and treatment groups with Cy2-, Cy3-, or C5-tagged cross-linking agents. The labeled protein lysates are mixed together, separated on a single 2D gel, scanned at different wavelengths for separate dye detection, and the images are electronically merged to determine common and differentially expressed proteins. For whichever staining method is used, proteins found to be differentially expressed can be excised from each gel, either manually or robotically, enzymatically digested into smaller fragments (i.e., trypsin) by an automated digestor, then robotically spotted onto MALDI targets or placed into automated sample loaders for MS/MS analysis.

Reasons for the popularity of the 2D gel proteomic approach are its adaptability to almost any protein sample from any organism, simultaneous separation of large numbers of intact proteins by pI and MW, differential display capabilities, and potential for indicating posttranslational modifications. There are some limitations (Table 12-2). While most steps downstream of protein separation can be automated, complete robotic handling of 2D gels has not been achieved. 2D gel separations are often operator dependent and can vary in each lab. Also, not all proteins will focus or solubilize equally well, since hydrophobic and basic proteins focus with difficulty and many low-copy regulatory proteins cannot be detected over the abundant structural and maintenance proteins.

In DNA microarray analysis, substantially up- or downregulated transcripts are of primary interest in toxicogenomics. Similarly, the primary object in toxicoproteomic studies is to determine the group of differentially expressed proteins as the basis of protein selection for excision and identification. Given the large dynamic range of protein expression and preponderance of structural and metabolic proteins, finding differentially expressed proteins in whole-cell or whole-organ homogenates can be quite challenging. Some form of protein enrichment, purification, and reduction of complexity of the proteome are common strategies to discriminate differentially expressed proteins from those with stable expression (Hunter et al., 2002; Nakai, 2001). The isolation of sub-proteomes enriched by subcellular fractionation or by immunoaffinity capture, ligand binding, and adsorption chromatography are very effective strategies for drilling down to lower copy proteins at a particular cellular spatial location or for those proteins containing special functional attributes (i.e., ATP-binding proteins).

CHROMATOGRAPHIC SEPARATIONS AND MASS SPECTROMETRY

Chromatographic devices connected online to mass spectrometers have become powerful tools for identifying large number of proteins from proteolytically digested samples or cell lysates. Reverse-phase liquid chromatography (LC) separation by itself cannot resolve complex protein digests, since many peptides coelute under one peak. LC can achieve high resolution of peptides from fractionated digests, the best-known system being termed MudPIT for multidimensional protein identification technique (Table 12-2; Hunter et al., 2002; Link et al., 1999). In the MudPIT approach, a reduced and denatured protein isolate is proteolytically digested into peptides that are loaded onto a microcapillary column containing a strong cation exchange (SCX) resin. Small fractions of peptides are eluted in a stepwise salt gradient from the SCX column onto a second reversed-phase LC column, whereupon salts are removed and the peptides are eluted by a slow organic solvent gradient directly connected to an MS/MS instrument. The sequence tags generated for each peptide are searched in protein, nucleic acid, and EST databases to identify the proteins from which they were originally derived.

This cycle is repeated 10–20 times to separate and identify small groups of peptides in each fraction rather than separate the complex digest all at once. For example, 1,484 unique proteins were identified from a total yeast cell lysate (Walker and Rigley, 2000), nearly all the proteins comprising the yeast 80S ribosomal complex could be identified (Tong et al., 1999), and the most comprehensive description of the human serum proteome at 490 identified proteins (Adkins et al., 2002) have all been described in a series of MudPit experimental reports. Despite the powerful descriptive power of multidimensional LC-MS/MS for identifying proteins, protein levels are difficult to quantify and compare between samples because of the variable nature of peak heights and retention times during liquid chromatography of complex protein digests.

Such concerns have been addressed in some measure by a new LC-MS/MS technique termed isotope-coded affinity tagging or ICAT (Tables 12-1 and 12-2) that permits comparative quantitation between samples and is sufficiently sensitive to identify low-expression proteins. The original method (Gygi et al., 1999) used heavy (deuterium) and light (hydrogen) bifunctional linking reagents to label native proteins from the control (i.e., $[^1H]_8$-linker) sample and the treated (i.e., $[^2H]_8$-linker) sample by a cysteine-reactive group at one end of the linker. Labeled peptides were collected by affinity chromatography, analyzed by LC-MS for relative quantitation of the isotopes on identical peptides, and finally identified individually by LC-MS/MS to obtain sequence tags of each peptide.

A related method using differential lysine guanidation is called mass-coded abundance tagging (MCAT), which is also useful for relative peptide quantitation and de novo peptide sequencing (Cagney and Emili, 2002). The ICAT method has been adapted for other functional groups such as phospho-peptides for quantitative comparison of phosphorylation (Cagney and Emili, 2002). Improved second-generation reagents feature cleavable ICAT linkers that improve protein expression analysis (Zhou et al., 2002). An o-nitrobenzyl-based photocleavable linker is coupled to an aminopropyl glass bead. The new linkers also contain a leucine group labeled with deuterium as the heavy linker ([Leu-^{2}H]$_7$-linker) or hydrogen ([Leu-^{1}H]$_7$-linker) with an attached iodoacetate moiety for reaction with cysteine-containing peptides. Samples are proteolytically digested, reduced, and captured on glass bead supports for separation from excess linker reagent, and the peptides are finally released by photocleavage for analysis by LC-ESI-MS/MS.

In contrast to the original method, the solid-phase reagent labels peptides after enzymatic digestion proteins, so it is more suitable for LC-MS analysis. Following a similar theme, differentially labeled proteins synthesized in cells cultured in normal ^{14}N and ^{15}N stable isotopes can be separated by 2D gel electrophoresis (Oda et al., 1999) or multidimensional chromatography (Washburn et al., 2002) for quantification by mass spectrometry. Quantification and identification with dual isotopes provide a large amount of information in a comparative manner. However, the requirements for sophisticated equipment and nonparallel analysis of samples limit the usefulness for this method in large multisample evaluations. Although 2D-MS and ICAT-MS have distinct advantages, a recent evaluation concludes that neither method provides comprehensive coverage of all proteins in model subcellular systems such as the mitochrondria or endoplasmic reticulum (Patton et al., 2002).

RETENTATE CHROMATOGRAPHY MASS SPECTROMETRY

Retentate chromatography mass spectrometry (RC-MS; Tables 12-1 and 12-2) stems from the observation that sensitive mass spectra of peptides and proteins can be generated by laser activation of chromatographic supports placed on MALDI targets. The product application of RC-MS, called SELDI-ToF for surface-enhanced laser desorption ionization–time of flight, refers to a mass spectrometry method for measuring full-length proteins retained on various adsorptive surfaces. Although the concept of laser desorption of intact proteins is well established, it has been commercially exploited (Hutchens and Yip, 1993) only recently as a means of rapid protein profiling by combining mass spectrometry and protein chromatography (Issaq et al., 2002). A key feature of the SELDI-ToF instrument is its ability to provide a rapid protein expression profile from a variety of biological and clinical samples. In particular, protein profiling can be performed using only a few microliters of accessible biological fluids such as serum, plasma, and urine. It has been used for biomarker identification in projects including prostate, bladder, breast, and ovarian cancers (Witzmann and Li, 2002).

The SELDI instrument is a specially adapted laser desorption ionization mass spectrometer capable of ionizing proteins from an adsorptive surface, or biochip. The instrument is equipped with a pulsed UV nitrogen laser with a mass accuracy range and specialized software to normalize, interpret, and compare spectra from several different groups. In an analogous manner to gel electrophoresis separation of denatured proteins according to size, a spectrum of protein masses is measured after laser energy ionizes and volatilizes proteins from the target surface in the presence of an overlaying organic matrix. The system is most effective in profiling relatively low-molecular-weight proteins (<20 kD), although higher MW molecules can be detected by adjusting laser intensity and the organic matrix (α-cyano-4-hydroxycinnamic acid [CHCA] for low MW range; 3,5-dimethoxy-4-hydroxycinnamic acid or SPA [sinapinnic acid] for high-protein MW range).

The analysis or reading of each sample is relatively rapid and can be automated so that 8 to 16 surfaces can be read on a single biochip in about 1–2 hours. Each 2 mm surface on a biochip (10 mm × 80 mm) adsorbs proteins from a complex protein mixture by chemical affinity (anionic, cationic, hydrophobic, hydrophilic, metal) or by more selective biochemical attraction (antibody, receptor, oligonucleotide, peptide). Selectivity on the binding surface can be varied by pH, ionic strength, detergent, organic solvent, and other modifiers. Nonspecific bound proteins are removed by repeated washing steps.

The capabilities of the SELDI technology (Table 12-2) are selectivity of proteins for specific chemical or biochemical surfaces; sensitivity for displaying subsets of proteins from complex mixtures; rapid analysis by reading hundreds of samples per day; automation by use of liquid handling workstations for accurate microsample processing; measurement of native proteins without enzymatic digestion; and production of biomarker patterns to define the presence of disease without exact knowledge of protein identities. Limitations of this technology are the nonglobal assessment of protein mixtures; the limited mass range that SELDI-Tof effectively measures; the ion suppression effect, which can affect protein detection; and the inability to directly identify proteins. A recent exciting development for protein identification from biochips is the construction of a biochip interface on a tandem quadrupole ToF MS instrument that permits sequence tag generation directly off the chip (Weinberger et al., 2002).

Innovative approaches to sample preparation and data analysis have led to some major advances in the field of biomarker identification of disease using SELDI analysis despite some limitations in definitive protein identification like that achieved by protein fingerprinting and sequence tags by traditional

MS applications. In terms of sample preparation, complex samples such as serum can be analyzed by direct application of 1–3 μl serum upon a chemical surface, or subsets of serum proteins can be analyzed by multidimensional chromatography.

Adsorption of proteins to strong anion exchange (SAX) resin and sequential pH elution of proteins results in fractionated subsets of proteins that can be further subdivided by adsorption onto the different SELDI surfaces. The approach produces a composite protein profile that can isolate abundant serum proteins like serum albumin to one or two fractions and improve detection of other low-abundance proteins that may be more informative of the disease state or chemical exposure. Another global protein biochip approach has been termed tagless extraction-retentate chromatography, in which methionine groups from digested protein fragments are selectively extracted by covalent attachment to bromoacetyl reactive groups, washed, and reductively released from solid supports for subsequent profiling on biochips for direct MS/MS analysis (Weinberger et al., 2002).

Sophisticated analysis of SELDI data examines each spectrum as a series of 1,500 data points (Petricoin et al., 2002). The use of serum samples from well-characterized patients as training sets can result in the development of a critical set of protein ions diagnostic of disease (Petricoin et al., 2002). When clinical serum samples were analyzed by SELDI-ToF for these critical protein ions from blinded serum samples, sophisticated algorithms were able to select normal from diseased patients with a high degree of accuracy, reaching nearly 100% in some cases for ovarian cancer (Petricoin et al., 2002) and some types of breast cancer (Li et al., 2002).

PROTEIN ARRAYS AND ALTERNATE MICROARRAYS

Protein arrays permit miniaturized, highly parallel assays for protein identification, protein-protein interactions, enzyme activities, and affinity studies for a great number of ligands and substrates. Proteins can be arrayed on glass or filter surfaces or can also be organized in capillary systems to form microfluidic arrays (Walter et al., 2002). The primary challenge is to preserve proteins in their biologically active form.

Development of high-throughput detection of proteins or specific protein features started with formatting in 96-well arrays, although multiplexed Western blotting has also been a viable means for increased protein expression analysis. Validation of high-throughput protein expression is by Western blot or immunohistochemistry. Specialized functional assays such as kinase, gel shift, and electrophoresis mobility shift (EMSA) are also useful in validating phosphorylation or transcription factor activation. Multiplexed Western blotting such as the PowerBlot (Yoo et al., 2002) permits comparative analysis of hundreds of proteins by carefully mixing antibody sets designed to maximize resolvable signals after electrotransfers from gel-separated proteins.

Many assays that have been bead captured and gel based have been converted to 96-well polystyrene microtiter plates to standardize the format for automation and higher throughput. In many ways, plate formats are a bridge to the higher-density microarray formats and remain quite useful until full conversion to microarray or microfluidic formats can be made. Microtiter plate assays for panels of cytokines are widely available because of their wide use in clinical medicine.

The term *protein microarray* has often been used to collectively refer to immobilized peptides, enzymatic substrates, polypeptides, and antibodies (Fig. 12-3). Protein arrays hold great promise for revolutionizing protein profiling, functional proteomics, and proteoinformatics because of the massive parallel processing capabilities, quantitative data, and ability to integrate into

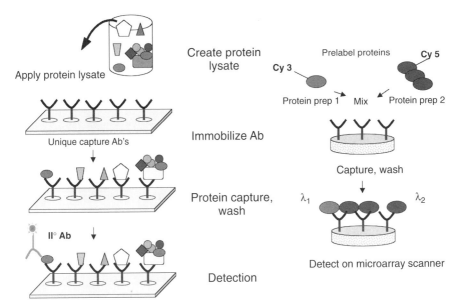

Figure 12-3 *Protein identification by antibody microarrays. Two schemes are shown for capture of proteins by specific antibodies immobilized on solid surfaces in microarray format. On the left, a protein lysate is applied to an array of unique antibodies that have been validated for selective capture of specific proteins. For detection, a second, signal-generating antibody must selectively bind to the captured protein, creating a "sandwich" assay. The scheme on the right of the figure uses a different approach. Proteins from each lysate (protein preps 1 and 2) are prelabeled with different fluorescent dyes, then mixed and applied to the collection of immobilized capture antibodies. After incubation and washing to remove unbound protein, the array is scanned at wavelengths ($\lambda 1$ and $\lambda 2$) to detect fluorescent signal, and a ratio is created from the intensities. If more protein is present in one sample, a greater signal will result. Note that in the sandwich assay approach the labeled secondary antibody can detect its free target protein directly or as part of a complex (far right Ab complex), which can lead to difficulties in interpretation. In the protein prelabeling scheme on the right, it is sometimes difficult to label all proteins stoichiometrically in each sample with fluorescently tagged cross-linking reagents.*

established hardware and software systems already developed for DNA microarrays (Mitchell, 2002).

The yeast two-hybrid system is widely used as a molecular biology–based screening method to study protein-protein interactions. However, this system cannot readily account for multiple components under particular conditions, cofactors, and posttranslational modifications. Microarrays of expressed proteins have been used to study protein-protein interactions, kinase activity, and small molecule targets on a targeted scale (MacBeath and Schreiber, 2000) and a genomic scale in yeast (Zhu et al., 2000, 2001). A considerable advantage is that the conditions of the experiment such as pH, temperature, ionic strength, presence of cofactors, and protein modification state can be controlled for proteins under study (MacBeath, 2002).

The biggest challenge in protein profiling using antibody microarrays is selection of validated antibodies that are useful in the desired sample environment (Table 12-2). Many of the initial reports using antibody arrays assayed for cytokines, since serum presents a relatively simple sample assay environment compared to tissue and there are numerous validated antibodies available for this clinically important set of proteins. Tissue and cell lysates present more complex assay environments with more opportunities for antibody cross-reactivity and other interferences that erode the biological meaningfulness of the data.

Antibody microarrays (MacBeath, 2002) are constructed for antigen capture in small sample volumes with detection by sandwich immunoassay or antigen labeling (Fig. 12-3). In sandwich immunoassay, capture antibodies are arrayed and immobilized to select specific proteins, which are then found by a second labeled detection antibody. In antigen labeling, all proteins in the sample are prelabeled (i.e., fluorescent dyes) prior to capture by immobilized antibody arrays. In direct assay systems, sample proteins are directly immobilized onto surfaces in much the same way that a dot blot is conducted. Selected proteins are then detected by prelabeled specific antibodies or signal amplified secondary antibodies, like the ultrasensitive PCR-based detection method of rolling circle amplification (Schweitzer et al., 2000) or the highly sensitive resonance light scattering reagents (Yguerabide and Yguerabide, 2001).

The capabilities of antibody microarray technology (Table 12-2) are similar to those for DNA array methods: selectivity of immunoreagents in complex protein lysates; rapid, massively parallel analysis of proteins; small sample volume requirement; automation and compatibility with DNA microarray technologies in hardware, software, and bioinformatics; and analysis of native proteins, which affords information on specific structure and protein-protein interactions. Limitations of this technology are the requirement for a multitude of validated antibodies; potential cross-reactivity with related proteins, which complicates biological interpretation; nonglobal assessment of protein expression defined by selection of antibody arrays; protein-protein interactions that can mask desired epitopes and reduce the signal or can alternately inflate the target protein signal by detection of attached nontarget proteins bound to the target (Table 12-2, Fig. 12-3).

Phage displayed antibodies that express a short-chain variable fragment of immunoglobulins (scFv) can provide an alternative source to antibody arrays (Jenkins and Pennington, 2001). In addition, small affinity proteins or affibodies selected from combinatorial libraries have been engineered to contain a common scaffold protein for increased chemical stability for use in array-based assays with wide affinity for target proteins (Nord et al., 1997). Perhaps one of the most novel protein capture array technologies does not involve antibodies or affinity ligands at all, but instead uses molecular imprinting (Bruggemann, 2002; Zimmerman et al., 2002). Target molecules such as peptides or polypeptides are bound to a surface and surrounded by functional monomers that assemble and polymerize around the peptide (Fig. 12-4). To manufacture a protein capture array, peptides that correspond to a unique sequence on protein surfaces can be spotted onto glass slides and coated with hydrophilic monomers and cross-linkers that polymerize to form a hydrogel layer. When the reversibly bound peptide is removed, it leaves a unique

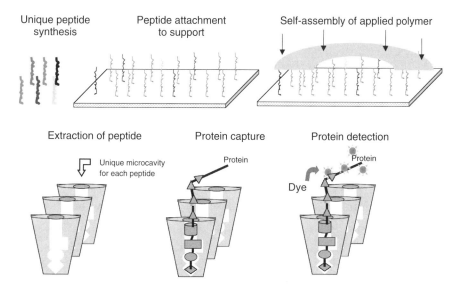

Figure 12-4 *Protein capture microarrays formed by molecular imprinting technology. A scheme for creating selective protein capture microarrays is shown that does not use ligand or antibody affinity reagents. Peptides are carefully chosen to correspond to unique surface epitopes of native proteins, then synthesized and reversibly immobilized onto glass surfaces in microarray format. A microthin layer of a special self-assembling polymer is applied to the surface of the peptide microarray that forms a molecular imprint or mold for each peptide. After removal of the peptides, a unique microcavity remains in the matrix that can selectively capture native protein from an applied lysate containing the corresponding peptide sequence. After washing to remove unbound protein, captured proteins are labeled with universal protein detection reagents and measured for intensity at each spot.*

molecular imprint or microcavity in the polymer surface that only fits like molecules containing the peptide. Large-scale protein capture microarrays are under intense commercial development.

Aptamer Arrays

In addition to antibodies, the other high-affinity reagents capable of binding to sets of diverse biomolecular targets like the proteome would be nucleic acid–binding species or apatmers (Cox and Ellington, 2001). Like immunoaffinity reagents, aptamer binding sets are in rapid development. Aptamers (Table 12-1) are generally described as single-stranded nucleic acids that, in contrast to antisense oligonucleotides or ribozymes, become biologically active by folding to a specific globular structure that imparts high-affinity binding to a variety of biological targets (Cerchia et al., 2002). Aptamer candidates are screened from large libraries of unique nucleic acids ($>10^{14}$) by a procedure called systematic evolution of ligands by exponential enrichment (SELEX).

Photo-SELEX (Golden et al., 2000) is a next-generation method whereby incorporation of a modified nucleotide into the oligonucleotide library aptamers can permit binding tightly to their target proteins and photocross-linking to these targets. The selection of aptamers on the basis of affinity and ability to photocross-link the target protein provides a specificity akin to the ELISA sandwich assay. Limitations of aptamers are their relatively limited range of physicochemical properties, having no equivalents to the hydrophobic and basic residues of some amino acids (Joyce, 1998), and large-scale screening is required to match aptamers to a suitably large collection of purified target proteins. However, once suitable aptamers have been identified, the advantages of aptamer arrays are that they can be printed using the same technologies as mRNA or oligoexpression arrays. High specificity and sensitivity can be achieved because aggressive washing removes much of the background binding. Proponents envision a photo-SELEX diagnostic system that could eventually quantify thousands of analytes from patient serum, urine, and tissue (Golden et al., 2000).

Peptide Arrays

The technology for microarraying presynthesized peptides onto glass supports by covalent attachment has only recently been developed in a manner that minimizes background (Table 12-1; Reimer et al., 2002). Initial peptide microarray experiments for determining kinase activity were conducted on chips coated with a monolayer of cross-linked bovine serum albumin that contained a functional group to anchor the substrate peptides by their amino groups (MacBeath and Schreiber, 2000).

Another elegant study reported attachment of c-Src kinase substrates to a chip on a self-assembled monolayer of alkanethiolates on gold (Houseman et al., 2002). Phosphorylation of the immobilized peptides was characterized by surface plasmon resonance (SPR-BIA), fluorescence, and phosphorimaging. Three inhibitors of the enzyme were quantitatively evaluated in an array format on a single, homogeneous substrate. Another application has shown the use of peptide arrays in cell adhesion assays in which surface idiotype of WEHI-231 cells could discriminate among arrayed peptides for a specific binding peptide (Falsey et al., 2001). These proof-of-principle studies suggest a wide diversity of applications for peptide microarrays in drug development, biochemistry, and toxicoproteomics studies.

SUMMARY

A description of the complex nature of protein structures, actions, and integration of organizational hierarchies in health and disease has been described as systems biology. Multiple high-throughput technologies and international cooperative strategies will likely be needed to meet the challenge of defining clinical and experimental proteomes and their response to toxic reagents (Merrick, 2003). High-throughput strategies in protein expression will produce large volumes of information but little scientific advancement unless these technologies are accompanied by validation of high-throughput results, informed hypothesis development, and rigorous experimentation. The outstanding progress in miniaturized, automated, and comprehensive protein measurements, particularly in microarray formats, ensures that these tools for protein expression will rapidly increase their reach from the experimental laboratory into drug development, diagnostics, and the clinic in the very near future.

REFERENCES

Adkins JN, Varnum SM, Auberry KJ, Moore RJ, Angell NH, Smith RD, et al. (2002). Toward a human blood serum proteome: analysis by multidimensional separation coupled with mass spectrometry. *Mol Cell Proteomics* **1**:947–955.

Bruggemann O (2002). Molecularly imprinted materials—receptors more durable than nature can provide. *Adv Biochem Eng Biotechnol* **76**:127–163.

Cagney G, Emili A (2002). De novo peptide sequencing and quantitative profiling of complex protein mixtures using mass-coded abundance tagging. *Nat Biotechnol* **20**: 163–170.

Cerchia L, Hamm J, Libri D, Tavitian B, de Franciscis V (2002). Nucleic acid aptamers in cancer medicine. *FEBS Lett* **528**:12–16.

Cox JC, Ellington AD (2001). Automated selection of anti-protein aptamers. *Bioorg Med Chem* **9**:2525–2531.

Falsey JR, Renil M, Park S, Li S, Lam KS (2001). Peptide and small molecule microarray for high throughput cell adhesion and functional assays. *Bioconjug Chem* **12**:346–353.

Figeys D (2002). Adapting arrays and lab-on-a-chip technology for proteomics. *Proteomics* **2**:373–382.

Garrels JI (1979). Two dimensional gel electrophoresis and computer analysis of proteins synthesized by clonal cell lines. *J Biol Chem* **254**:7961–7977.

Golden MC, Collins BD, Willis MC, Koch TH (2000). Diagnostic potential of PhotoSELEX-evolved ssDNA aptamers. *J Biotechnol* **81**:167–178.

Gygi SP, Rist B, Gerber SA, Turecek F, Gelb MH, Aebersold R (1999). Quantitative analysis of complex protein mixtures using isotope-coded affinity tags. *Nat Biotechnol* **17**:994–999.

Houseman BT, Huh JH, Kron SJ, Mrksich M (2002). Peptide chips for the quantitative evaluation of protein kinase activity. *Nat Biotechnol* **20**:270–274.

Huang L, Baldwin MA, Maltby DA, Medzihradszky KF, Baker PR, Allen N, et al. (2002). The identification of protein-protein interactions of the nuclear pore complex of *Saccharomyces cerevisiae* using high throughput matrix-assisted laser desorption ionization time-of-flight tandem mass spectrometry. *Mol Cell Proteomics* **1**:434–450.

Hunter TC, Andon NL, Koller A, Yates JR, Haynes PA (2002). The functional proteomics toolbox: methods and applications. *J Chromatogr B Analyt Technol Biomed Life Sci* **782**:165–181.

Hutchens TW, Yip TT (1993). New desorption strategies for the mass spectrometric analysis of macromolecules. *Rapid Commun Mass Spectrom* **7**:576–580.

Issaq HJ, Veenstra TD, Conrads TP, Felschow D (2002). The SELDI-TOF MS approach to proteomics: protein profiling and biomarker identification. *Biochem Biophys Res Commun* **292**:587–592.

Jenkins RE, Pennington SR (2001). Arrays for protein expression profiling: towards a viable alternative to two-dimensional gel electrophoresis? *Proteomics* **1**:13–29.

Joyce GF (1998). Nucleic acid enzymes: playing with a fuller deck. *Proc Natl Acad Sci USA* **95**:5845–5847.

Li J, Zhang Z, Rosenzweig J, Wang YY, Chan DW (2002). Proteomics and bioinformatics approaches for identification of serum biomarkers to detect breast cancer. *Clin Chem* **48**:1296–1304.

Link AJ, Eng J, Schieltz DM, Carmack E, Mize GJ, Morris DR, et al. (1999). Direct analysis of protein complexes using mass spectrometry. *Nat Biotechnol* **17**:676–682.

MacBeath G (2002). Protein microarrays and proteomics. *Nat Genet* 32(suppl **2**):526–532.

MacBeath G, Schreiber SL (2000). Printing proteins as microarrays for high-throughput function determination. *Science* **289**:1760–1763.

Merrick BA (2003). The Human Proteome Organization (HUPO) and environmental health. *Environ Health Perspect Toxicogenomics* **111**:1–5.

Mitchell P (2002). A perspective on protein microarrays. *Nat Biotechnol* **20**:225–229.

Nakai K (2001). Review: prediction of in vivo fates of proteins in the era of genomics and proteomics. *J Struct Biol* **134**:103–116.

Natsume T, Nakayama H, Isobe T (2001). BIA-MS-MS: biomolecular interaction analysis for functional proteomics. *Trends Biotechnol* **19**:S28–S33.

Nord K, Gunneriusson E, Ringdahl J, Stahl S, Uhlen M, Nygren PA (1997). Binding proteins selected from combinatorial libraries of an alpha-helical bacterial receptor domain. *Nat Biotechnol* **15**:772–777.

Oda Y, Huang K, Cross FR, Cowburn D, Chait BT (1999). Accurate quantitation of protein expression and site-specific phosphorylation. *Proc Natl Acad Sci USA* **96**: 6591–6596.

Patton WF, Schulenberg B, Steinberg TH (2002). Two-dimensional gel electrophoresis; better than a poke in the ICAT? *Curr Opin Biotechnol* **13**:321–328.

Petricoin EF, Ardekani AM, Hitt BA, Levine PJ, Fusaro VA, Steinberg SM, et al. (2002). Use of proteomic patterns in serum to identify ovarian cancer. *Lancet* **359**:572–577.

Reimer U, Reineke U, Schneider-Mergener J (2002). Peptide arrays: from macro to micro. *Curr Opin Biotechnol* **13**:315–320.

Schweitzer B, Wiltshire S, Lambert J, O'Malley S, Kukanskis K, Zhu Z, et al. (2000). Inaugural article: immunoassays with rolling circle DNA amplification: a versatile platform for ultrasensitive antigen detection. *Proc Natl Acad Sci USA* **97**: 10113–10119.

Tong W, Link A, Eng JK, Yates JR III (1999). Identification of proteins in complexes by solid-phase microextraction/multistep elution/capillary electrophoresis/tandem mass spectrometry. *Anal Chem* **71**:2270–2278.

Walker J, Rigley K (2000). Gene expression profiling in human peripheral blood mononuclear cells using high-density filter-based cDNA microarrays. *J Immunol Methods* **239**:167–179.

Walter G, Bussow K, Lueking A, Glokler J (2002). High-throughput protein arrays: prospects for molecular diagnostics. *Trends Mol Med* **8**:250–253.

Washburn MP, Ulaszek R, Deciu C, Schieltz DM, Yates JR III (2002). Analysis of quantitative proteomic data generated via multidimensional protein identification technology. *Anal Chem* **74**:1650–1657.

Weinberger SR, Viner RI, Ho P (2002). Tagless extraction-retentate chromatography: a new global protein digestion strategy for monitoring differential protein expression. *Electrophoresis* **23**:3182–3192.

Witzmann FA, Li J (2002). Cutting-edge technology. II. Proteomics: core technologies and applications in physiology. *Am J Physiol Gastrointest Liver Physiol* **282**: G735–G741.

Yguerabide J, Yguerabide EE (2001). Resonance light scattering particles as ultrasensitive labels for detection of analytes in a wide range of applications. *J Cell Biochem Suppl* **37**:71–81.

Yoo GH, Piechocki MP, Ensley JF, Nguyen T, Oliver J, Meng H, et al. (2002). Docetaxel induced gene expression patterns in head and neck squamous cell carcinoma using cDNA microarray and PowerBlot. *Clin Cancer Res* **8**:3910–3921.

Zhou H, Ranish JA, Watts JD, Aebersold R (2002). Quantitative proteome analysis by solid-phase isotope tagging and mass spectrometry. *Nat Biotechnol* **20**:512–515.

Zhu H, Klemic JF, Chang S, Bertone P, Casamayor A, Klemic KG, et al. (2000). Analysis of yeast protein kinases using protein chips. *Nat Genet* **26**:283–289.

Zhu H, Bilgin M, Bangham R, Hall D, Casamayor A, Bertone P, et al. (2001). Global analysis of protein activities using proteome chips. *Science* **293**:2101–2105.

Zimmerman SC, Wendland MS, Rakow NA, Zharov I, Suslick KS (2002). Synthetic hosts by monomolecular imprinting inside dendrimers. *Nature* **418**:399–403.

13

Analytical Proteomics Approaches to Analysis of Protein Modifications: Tools For Studying Proteome-Environment Interactions

Daniel C. Liebler

INTRODUCTION

Gene-environment interactions are thought to be critical determinants of human susceptibility to disease (Chen et al., 2002; Becker and Chan-Yeung, 2002; Ritenbaugh, 2000; Schaid et al., 1999; Shields and Harris, 2000; Barnes, 1999; Blumenthal and Blumenthal, 2002). The concept of gene-environment interactions is perhaps best defined in the context of genetics and epidemiology (Clayton and McKeigue, 2001). At the molecular-chemical level, gene-environment interactions do not necessarily result from direct interaction of environmental agents with DNA but rather from the interactions of chemical

Toxicogenomics: Principles and Applications. Edited by Hamadeh and Afshari
ISBN 0-471-43417-5   Copyright © 2004 John Wiley & Sons, Inc.

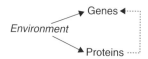

Figure 13-1 *Relationship between gene-environment interactions and gene-protein-environment interactions.*

and physical agents with proteins. Proteins are a principal point of contact with chemical and physical environmental agents, and effects on proteins and proteomes affect protein signaling networks and transcription factors, which regulate the expression of genes. Environmental factors that modify protein functions thus affect gene expression (Fig. 13-1).

DNA microarrays have made possible the simultaneous analysis of expression of thousands of genes and facilitate the study of cells as systems, in which the totality of expression of all genes dictates the state of the cell. Of course, gene expression changes are often implicitly assumed to reflect downstream changes in proteomes. However, this relationship between gene expression and the proteome is probably not that simple for two reasons. First, studies of gene expression and protein expression indicate that the correlation between mRNA levels and protein levels is poor, especially for the great majority of proteins, which are expressed at relatively low abundance (Gygi et al., 1999b; Griffin et al., 2002; Chen et al., 2002). Second, any protein product of a single gene may exist in multiple forms due to posttranslational modifications, the existence of mutants, and the formation of splice variants. Thus critical aspects of stress responses can be observed only at the level of the proteome.

The dynamic modification of proteins is an essential control mechanism in living systems. Diverse endogenous modifications, such as phosphorylation, acetylation, glycosylation, and ubiquitination, govern many protein functions, interactions, and turnover (Pawson, 2002; Johnson and Lapadat, 2002; Freiman and Tjian, 2003; Glickman and Ciechanover, 2002; Hay, 2001). Environmental agents can affect the modification status of proteins in several ways (Fig. 13-2). Some environmental chemicals and drugs are metabolized to reactive electrophiles, which are thought to be causative agents in drug and chemical toxicities (Miller and Miller, 1981; Nelson, 1995; Guengerich and Liebler, 1985). Other environmental stresses can lead to the formation of oxidants and endogenous electrophiles that can also modify proteins (Burcham, 1998). Electrophile modifications apparently play essential, although poorly understood, roles in toxicity and cancer (Cohen et al., 1997; Guengerich, 2001). Moreover, endogenous regulatory protein modifications are sensitive to modulation by environmental factors (Pawson, 2002; Johnson and Lapadat, 2002; Freiman and Tjian, 2003; Glickman and Ciechanover, 2002; Mao et al., 2000; Saitoh and Hinchey, 2000; Hong et al., 2001; Muller et al., 2001).

The recent development of powerful analytical methodology, gene and protein sequence databases, and bioinformatics tools has driven the growth of

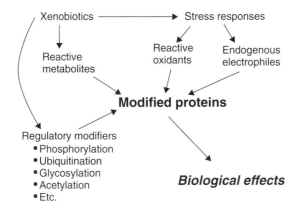

Figure 13-2 *Environmental influences on protein modifications.*

proteomics (Liebler, 2001; Griffin and Aebersold, 2001; Mann et al., 2001). Proteomics approaches offer new tools to study the effects of environmental factors on proteins and proteomes. A particularly important application of proteomics approaches is in the study of protein modifications and how they are affected by environmental factors. Recent advances in proteomics methods enable detailed studies of protein modifications and the effects of environmental agents on modifications and functions of modified proteins (Liebler, 2002). The focus of this chapter is on the development and application of methods for identifying protein targets for modification and mapping modifications at the level of amino acid sequence. These methods and approaches will enable detailed mechanistic studies of protein-environment interactions that underlie gene-environment interactions.

ANALYTICAL PLATFORMS FOR PROTEOMICS ANALYSES

Recent advances in MS instrumentation and protein and peptide separations, together with new data mining and database search algorithms and software, have enabled modern proteomics (Mann et al., 2001; Pandey and Mann, 2000; Yates, 2000; Griffin and Aebersold, 2001). The greatest challenge in proteomic analyses is the exceptionally wide dynamic range for protein expression— there is at least a million-fold difference in concentration between the least abundant and most abundant proteins in cells. Moreover, modified protein forms, including xenobiotic- and posttranslationally modified variants often are present at low stoichiometry compared to unmodified forms. Thus, the challenge in analyses of modified proteins is the detection of both high-abundance and lower-abundance components.

A classical approach to proteome analysis is based on initial protein separation, typically by 2D-SDS-PAGE, followed by selection of protein spots of interest for identification by MS. A 2D gel-based approach can be a productive means of detecting and comparing modifications in relatively abundant proteins, particularly because some modifications alter the mobility of the proteins in isoelectric focusing separations, and the forms are resolved on the gels (Wilkins et al., 1999). Nevertheless, detection of low-abundance species is often limited by the narrow dynamic range for protein staining. The dynamic range for stain intensities is several orders of magnitude less than that for protein abundances, and thus the lower-abundance species may not be detected. This problem is compounded in analyses of modified and unmodified proteins, as the modified forms are often present at much lower abundance. These factors have limited the utility of 2D gel-based approaches and have led investigators to employ "shotgun" proteomics strategies to detect low-abundance protein modifications.

In "shotgun" analysis, protein mixtures are proteolytically digested (generally without prior protein fractionation), then the resulting peptide mixtures are subjected to multidimensional chromatographic separations and LC-MS-MS (Link et al., 1999; Washburn et al., 2001). High-throughput correlation of MS-MS spectra to database sequences with parallel computing and the Sequest algorithm (Eng et al., 1994) enables rapid protein identification from the thousands to millions of spectra that such analyses generate (Sadygov et al., 2002). Yates and colleagues have used the term *MudPIT* (multidimensional protein identification technique) to describe this type of analysis (Washburn et al., 2001; Wolters et al., 2001). The advantage of this approach is that the multidimensional peptide separations resolve complex peptide mixtures into simplified subfractions, which allows the mass spectrometer to acquire MS-MS spectra of a greater number of peptides than is possible in single-dimension reverse-phase LC-MS-MS analyses (Link et al., 1999; Wolters et al., 2001). This approach is applicable not only to the identification of proteins but also to mapping certain modifications that can easily be predicted to occur at specific targets (e.g., phosphorylation of serine; MacCoss et al., 2002).

Analysis of protein modifications is best done with tandem MS instruments, which yield MS-MS spectra of the peptides analyzed. MS-MS spectra often provide a definitive location of the sequence position of modifications (see below). Matrix-assisted laser desorption ionization–time of flight (MALDI-ToF) instruments typically provide mass measurements of intact peptide ions, although peptide fragmentation information can be obtained by analysis in postsource decay mode (Spengler et al., 1992; Roepstorff, 2000). Thus, most MALDI-ToF instruments can be used to detect modified peptides but not to map the modifications to specific sequence locations.

The most widely used tandem mass spectrometers are ion trap instruments, which are (relatively) inexpensive, robust, and display good sensitivity for peptide MS-MS analyses (Jonscher and Yates, 1997). Their only significant drawbacks are relatively poor mass accuracy (approximately $\pm 0.5\,m/z$ unit for precursor and product ions in typical LC-MS-MS analyses). This limitation generally does not preclude reliable identifications of native and modified peptides from MS-MS spectra, except for different modifications of similar nominal mass (Jonscher and Yates, 1997; Yates, 1998). Quadrupole–time of flight (Q-TOF) MS instruments operated with either MALDI or electrospray sources provide higher resolution and mass accuracy, which can be valuable in discriminating between apparent modifications of similar nominal mass (Morris et al., 1997; Chalkley and Burlingame, 2001). Recently developed MALDI-ToF-ToF instruments are capable of high-throughput MS-MS analyses with very high resolution and mass accuracy for both precursor and product ions (Medzihradsky et al., 2000; Bienvenut et al., 2002).

MAPPING PROTEIN MODIFICATIONS FROM MS-MS DATA: THE SALSA ALGORITHM

Analyses of MS-MS data with tools such as Sequest are directed primarily at protein identification. Sequest can correlate MS-MS spectra of modified peptides with the correct database sequences when modifications at specific sites can be anticipated (e.g., phosphorylation of tyrosine). However, modifications that are unanticipated in either sequence specificity or mass are not generally detected with Sequest. To address this problem, we developed the SALSA algorithm and software. SALSA is a tool that enables rapid identification of spectra corresponding not only to unmodified peptides but also to variant and modified forms of those same sequences (Hansen et al., 2001; Liebler et al., 2002). This enables the user to mine large data sets for spectra of peptide adducts without necessarily having knowledge of the site specificity of adduct formation or even the mass of the adduct.

SALSA evaluates MS-MS spectra for specific, user-defined features, including product ions at specific m/z values, neutral or charged losses from singly or doubly charged precursors, and ion pairs. The MS-MS scans containing the specified features are then scored on the basis of the intensities of the scored ion signals. The scores are independent of the concentrations of the analytes in the mixture. Peptide adducts formed with the electrophiles 1,4-benzoquinone (Mason and Liebler, 2000), 2-chloroacetyl chloride, and S-(2-chloroacetyl)glutathione (Jones and Liebler, 2000), 7,8-dihydroxy-9,10-epoxy-7,8,9,10-tetrahydrobenzo[a]pyrene (Harriman et al., 1998), and dehydromonocrotaline pyrrole (Hansen et al., 2001) all have been shown to display characteristic combinations of product ions, neutral or charged losses, or ion pairs that would enable detection of their MS-MS spectra with SALSA.

Application of the SALSA algorithm to the detection of peptide MS-MS ion series enables identification of MS-MS spectra displaying characteristics of specific peptide sequences. SALSA analysis scores MS-MS spectra based on correspondence between theoretical ion series for peptide sequence motifs and actual MS-MS product ion series, regardless of their absolute positions on the *m/z* axis. Analyses of tryptic digests of bovine serum albumin (BSA) by LC-MS-MS followed by SALSA analysis detected MS-MS spectra for both unmodified and multiple modified forms of several BSA tryptic peptides (Liebler et al., 2002). Application of SALSA to sequence-specific mapping of xenobiotic adducts on proteins is described below.

QUANTITATIVE ANALYSIS OF MODIFIED PROTEINS

LC-MS-MS has recently been applied to quantitative comparisons of proteome changes. Gygi and Aebersold reported an innovative approach to quantitative proteome analysis with the use of isotope-coded affinity tags (ICAT; Gygi et al., 1999a). The ICAT approach lends itself to *relative* protein quantitation of proteins between two samples. (This is analogous to two-color fluorescence labeling for analysis of relative gene expression levels by DNA microarrays.) Both high- and low-abundance proteins are detected by LC-MS-MS with data-dependent scanning (Gygi et al., 2000). In addition, MS-MS data provide unambiguous protein identification via sequence information. The ICAT reagents label proteins by means of thiol-reactive electrophiles, which restricts their application to analysis of cysteine-containing proteins and peptides. Recent improvements on this approach include the development of a photocleavable, solid-phase labeling reagent that provides improved labeling efficiency and MS-MS fragmentation characteristics (Zhou et al., 2002).

Although the ICAT strategy is generally applicable to relative protein quantitation, it is not well suited to quantitation of modified proteins and peptides. This is because modifications (particularly electrophile modifications) often occur on cysteine residues, thus blocking their subsequent reaction with the ICAT reagent. For analysis of modifications on other amino acid targets, use of ICAT reagents would require that the modified peptides also had available cysteines for ICAT labeling, which is generally unlikely. To apply a stable isotope tagging approach to the analysis of adducted peptides and proteins, we used d_0- and d_5-phenyl isocyanate (PIC) as N-terminal reactive tags for essentially all peptides in proteolytic digests (Mason and Liebler, 2003). PIC reacts quantitatively with peptide N-terminal amines within minutes at neutral pH, and the PIC-labeled peptides undergo informative MS-MS fragmentation. Ratios of d_0- and d_5-PIC-labeled derivatives of several model peptides were linear across a 10,000-fold range of peptide concentration ratios, thus indicating a wide dynamic range for quantitation. Application of PIC labeling enabled relative quantitation of several styrene oxide adducts of human hemoglobin in LC-MS-MS analyses (see below).

SEQUENCE-SPECIFIC MAPPING OF ELECTROPHILE ADDUCTS ON PROTEINS

Sequence Mapping of Aliphatic Epoxide Adducts of Human Hemoglobin

Hemoglobin (Hb) adducts formed by epoxides of styrene, ethylene, and butadiene are among the most widely studied markers of exposure to these chemicals (for reviews, see Skipper et al., 1994; Farmer, 1995; Ehrenberg et al., 1996). Indirect methods of analysis suggested that adducts with cysteine and glutamate are formed with aliphatic epoxides (Sepai et al., 1993; Rappaport et al., 1993; Yeowell-O'Connell et al., 1996). However, only the N-terminal valine adducts have been unambiguously measured by a modified Edman degradation procedure (Tornqvist et al., 1986; Pauwels et al., 1997). Kaur and colleagues employed fast atom bombardment tandem MS to identify a histidine adduct of styrene oxide in human Hb (Kaur et al., 1989). More recently, Moll et al. (Moll et al., 2000) employed electrospray MS to identify peptides adducted with butadiene monoxide in mouse erythrocytes in vitro. This was the first systematic application of MS to comprehensive mapping of Hb adducts. However, the analyses did not include tandem MS to provide sequence-specific mapping of adducts.

We recently applied LC-MS-MS and SALSA to analyze aliphatic epoxide adducts on human Hb (Badghisi and Liebler, 2002). Hb was incubated with styrene oxide, ethylene oxide, and butadiene dioxide (40 μM to 40 mM) to form adducts, digested with trypsin, and analyzed by LC-MS-MS on a ThermoFinnigan LCQ ion trap MS instrument. Data-dependent scanning was used for automated acquisition of peptide MS-MS spectra. The SALSA algorithm was used to detect MS-MS spectra of native and modified Hb peptides.

The adducted sites identified at 40 mM are the N-terminal valines of both Hbα and Hbβ, Glu7, Cys93, and His77, 97, and 143 of the β chain and His45 of the α chain. Specific shifts in the b- and y-ion series in MS-MS spectra confirmed the locations of each adduct. At 40 μM styrene oxide, only the N-terminal valines of the Hbα and Hbβ chains and Cys93 of the Hbβ chain were adducted at detectable levels. Figure 13-3 illustrates the sequence selectivity of Hb adduction by styrene oxide. The right-hand panel depicts all histidine, cysteine, and N-terminal amine nucleophiles in human Hb, whereas the center and right-hand panels depict amino acids adducted by styrene oxide at 40 mM and 40 μM, respectively. Our results suggest that alkylation displays considerable selectivity, even among nucleophiles that would appear to have similar steric accessibility to the electrophile. This study also demonstrates the utility of LC-MS-MS and SALSA for mapping modifications on proteins, particularly when the target selectivity of the modifying agent is not known beforehand.

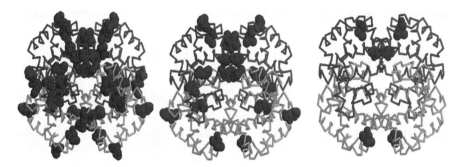

Figure 13-3   *Sequence selectivity of adduction of human Hb by styrene oxide. Left image highlights all Cys, His, and N-termini. Center image highlights residues adducted by styrene oxide at 40 mM (α N-terminus, His 20α, His 58α, β N-terminus, Cys 93β, Cys 112β, His 63β, His 77β, His 97β, His 143β). Right image depicts residues adducted by styrene oxide at 40 μM (α N-terminus, β N-terminus, Cys 93β).*

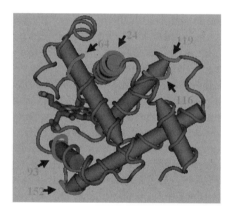

Figure 13-4   *Histidine residues in bovine myoglobin labeled by HNE. See text for discussion.* (Source: Reprinted from Alderton et al., 2003, by permission of the American Chemical Society.)

Sequence Mapping of Chemical-Protein Adducts on Myoglobin and Glutathione-S-Transferases

A recent collaboration with Dr. Cameron Faustman (University of Connecticut) focused on identifying the mechanisms by which lipid oxidation led to the oxidation of myoglobin in meat, which is a significant process in spoilage (Alderton et al., 2003). The work focused on mapping myoglobin adducts of 4-hydroxy-2-nonenal (HNE), a reactive α,β-unsaturated aldehyde derived from the lipid peroxidation, which accelerates the oxidation of oxymyoglobin to metmyoglobin. HNE-adducted myoglobin was analyzed by LC-MS-MS and SALSA, which identified six His residues that were readily adducted, including the proximal (His93) and distal (His64) ligands of the heme group (Fig. 13-4). Mb tertiary structure was altered in a manner consistent with destabilization. These results suggest that HNE accelerates bovine skeletal muscle OxyMb oxidation by covalent modification at histidine residues, possibly by modifying the stereoelectronic environment of the heme group.

Other work in collaboration with the laboratory of Dr. M. W. Anders has applied LC-MS-MS and SALSA to study the inactivation of glutathione transferase zeta (GSTZ1-1), which catalyzes the *cis-trans* isomerization of maleylacetone (MA) to fumarylacetone (FA), respectively. LC-MS-MS and SALSA analyses of tryptic digests of hGSTZ1 polymorphic variants revealed that the active site (SSC*SWR) and C-terminal (LLVLEAFQVSHPC*R) cysteine residues of hGSTZ1-1 were covalently modified by MA and FA. These data indicate that MA and FA are substrate and product inactivators of hGSTZ1-1 and covalently modify hGSTZ1-1 at the active-site cysteine residue in the absence of glutathione.

UNRESOLVED PROBLEMS IN DISCOVERING PROTEIN TARGETS OF REACTIVE INTERMEDIATES

Despite our recent progress and that of others in analytical proteomics, the problem of comprehensive analysis of protein-xenobiotic adducts remains formidable. The central problem remains the tremendous complexity of proteomes and the very wide dynamic range for protein expression, which is about a million-fold. The difference in abundance between highly abundant unadducted proteins or peptides and their lower-abundance adducted counterparts generally precludes reliable detection of the latter. Automated analysis modes in which LC-MS-MS instruments are operated primarily results in acquisition of MS-MS spectra of the most abundant species. The value of multidimensional LC separations prior to MS-MS is that they "spread out" the peptide mixture, thus giving the MS instrument more opportunity to acquire spectra of less abundant species.

Unfortunately, in analysis of complex samples containing relevant levels of adducts, virtually all instrument time is "wasted" on unmodified species. One solution to this problem is to use increasingly complex multidimensional LC separations prior to MS. However, this ultimately is impractical and inefficient—too much instrument time is still wasted on unmodified peptides. In principle, affinity enrichment of adducted proteins and peptides can be incorporated into the analysis.

Antibodies have been developed against a number of xenobiotic adducts (Satoh et al., 1985; Hargus et al., 1994; Halmes et al., 1996; Bartolone et al., 1992; Bulera et al., 1995; Rombach and Hanzlik, 1997; Kleiner et al., 1998) and posttranslational modifications (Marcus et al., 2000; Gronborg et al., 2002; Stancato and Petricoin, 2001). These antibodies directed against small chemical epitopes generally are used for Western blot analysis, but the affinity of the antibodies for their targets may vary considerably, which can limit their practical utility for immunoaffinity enrichment.

Nucleic acid aptamers could also offer a useful affinity enrichment tool (Brody and Gold, 2000). Aptamers can recognize not only proteins but small molecule structures and with higher affinities than many antibodies. Although development of specific aptamers against xenobiotic adducts has not yet been reported, this approach may prove useful for future work.

A particularly innovative approach to proteome-wide analyses termed activity-based protein profiling has been developed by Cravatt and colleagues, who have used bifunctional probes to profile functionally related classes of proteins in complex mixtures (Adam et al., 2002a, b, c). These probes combine an electrophilic reactive group directed toward a specific enzyme active site class, a linker, and an affinity tag, which allows selective isolation of labeled proteins. Variation of the chemistry of the reactive group and its stereoelectronic environment enable selective capture of different enzyme classes (Adam et al., 2002b).

Although these investigators have pursued this strategy to identify catalytically functional proteins, the general approach can be broadened and generalized to identify subgroups of proteins that display virtually any characteristic, including susceptibility to modification by a class of electrophiles. Thus model electrophiles containing affinity tags may be useful probes to identify representative classes and functional domains of proteins susceptible to modification. This offers a powerful approach to profile proteomes for susceptible molecular targets of reactive electrophiles.

ACKNOWLEDGMENTS

I gratefully acknowledge the contributions of present and past members of the Liebler laboratory, including Dan Mason, Ph.D., Julie Jones, Ph.D., Beau Hansen, Sean Davey, Laura Tiscareno, Hamid Badghisi, and Jeanne Burr. I also thank Drs. Cameron Faustman and Amy Alderton (University of Connecticut) and Drs. M. W. Anders, Hoffman Lantum, and Wayne Anderson (University of Rochester) for productive collaborations described herein. Work in my laboratory is supported by NIH grants ES10056, ES11811, ES06694, and ES07091 and a grant from ThermoFinnigan Corporation.

REFERENCES

Adam GC, Sorensen EJ, Cravatt BF (2002a). Chemical strategies for functional proteomics. *Mol Cell Proteomics* **1**:781–790.

Adam GC, Sorensen EJ, Cravatt BF (2002b). Proteomic profiling of mechanistically distinct enzyme classes using a common chemotype. *Nat Biotechnol* **20**:805–809.

Adam GC, Sorensen EJ, Cravatt BF (2002c). Trifunctional chemical probes for the consolidated detection and identification of enzyme activities from complex proteomes. *Mol Cell Proteomics* **1**:828–835.

Alderton AL, Faustman C, Liebler DC, Hill DW (2003). Induction of redox instability of bovine myoglobin by adduction with 4-hydroxy-2-nonenal. *Biochemistry.*

Badghisi H, Liebler DC (2002). Sequence mapping of epoxide adducts in human hemoglobin with LC-tandem MS and the SALSA algorithm. *Chem Res Toxicol* **15**: 799–805.

Barnes KC (1999). Gene-environment and gene-gene interaction studies in the molecular genetic analysis of asthma and atopy. *Clin Exp Allergy* **29**(suppl 4):47–51.

Bartolone JB, Birge RB, Bulera SJ, Bruno MK, Nishanian EV, Cohen SD, Khairallah EA (1992). Purification, antibody production, and partial amino acid sequence of the 58-kDa acetaminophen-binding liver proteins. *Toxicol Appl Pharmacol* **113**: 19–29.

Becker AB, Chan-Yeung M (2002). Primary prevention of asthma. *Curr Opin Pulm Med* **8**:16–24.

Bienvenut WV, Deon C, Pasquarello C, Campbell JM, Sanchez JC, Vestal ML, Hochstrasser DF (2002). Matrix-assisted laser desorption/ionization-tandem mass spectrometry with high resolution and sensitivity for identification and characterization of proteins. *Proteomics* **2**:868–876.

Blumenthal JB, Blumenthal MN (2002). Genetics of asthma. *Med Clin North Am* **86**: 937–950.

Brody EN, Gold L (2000). Aptamers as therapeutic and diagnostic agents. *J Biotechnol* **74**:5–13.

Bulera SJ, Birge RB, Cohen SD, Khairallah EA (1995). Identification of the mouse liver 44-kDa acetaminophen-binding protein as a subunit of glutamine synthetase. *Toxicol Appl Pharmacol* **134**:313–320.

Burcham PC (1998). Genotoxic lipid peroxidation products: their DNA damaging properties and role in formation of endogenous DNA adducts. *Mutagenesis* **13**: 287–305.

Chalkley RJ, Burlingame AL (2001). Identification of GlcNAcylation sites of peptides and alpha-crystallin using Q-TOF mass spectrometry. *J Am Soc Mass Spectrom* **12**: 1106–1113.

Chen G, Gharib TG, Huang CC, Taylor JM, Misek DE, Kardia SL, Giordano TJ, Iannettoni MD, Orringer MB, Hanash SM, Beer DG (2002). Discordant protein and mRNA expression in lung adenocarcinomas. *Mol Cell Proteomics* **1**:304–313.

Clayton D, McKeigue PM (2001). Epidemiological methods for studying genes and environmental factors in complex diseases. *Lancet* **358**:1356–1360.

Cohen SD, Pumford NR, Khairallah EA, Boekelheide K, Pohl LR, Amouzadeh HR, Hinson JA (1997). Selective protein covalent binding and target organ toxicity. *Toxicol Appl Pharmacol* **143**:1–12.

Ehrenberg L, Granath F, Tornqvist M (1996). Macromolecule adducts as biomarkers of exposure to environmental mutagens in human populations. *Environ Health Perspect* **104**(suppl 3):423–428.

Eng JK, McCormack AL, Yates JR (1994). An approach to correlate tandem mass-spectral data of peptides with amino-acid-sequences in a protein database. *J Am Soc Mass Spectrom* **5**:976–989.

Farmer PB (1995). Monitoring of human exposure to carcinogens through DNA and protein adduct determination. *Toxicol Lett* **82–83**:757–762.

Freiman RN, Tjian R (2003). Regulating the regulators: lysine modifications make their mark. *Cell* **112**:11–17.

Glickman MH, Ciechanover A (2002). The ubiquitin-proteasome proteolytic pathway: destruction for the sake of construction. *Physiol Rev* **82**:373–428.

Griffin TJ, Aebersold R (2001). Advances in proteome analysis by mass spectrometry. *J Biol Chem* **276**:45497–45500.

Griffin TJ, Gygi SP, Ideker T, Rist B, Eng J, Hood L, Aebersold R (2002). Complementary profiling of gene expression at the transcriptome and proteome levels in *Saccharomyces cerevisiae. Mol Cell Proteomics* **1**:323–333.

Gronborg M, Kristiansen TZ, Stensballe A, Andersen JS, Ohara O, Mann M, Jensen ON, Pandey A (2002). A mass spectrometry-based proteomic approach for identification of serine/threonine-phosphorylated proteins by enrichment with phospho-specific antibodies: identification of a novel protein, Frigg, as a protein kinase A substrate. *Mol Cell Proteomics* **1**:517–527.

Guengerich FP (2001). Forging the links between metabolism and carcinogenesis. *Mutat Res* **488**:195–209.

Guengerich FP, Liebler DC (1985). Enzymatic activation of chemicals to toxic metabolites. *Crit Rev Toxicol* **14**:259–307.

Gygi SP, Rist B, Gerber SA, Turecek F, Gelb MH, Aebersold R (1999a). Quantitative analysis of complex protein mixtures using isotope-coded affinity tags. *Nat Biotechnol* **17**:994–999.

Gygi SP, Rochon Y, Franza BR, Aebersold R (1999b). Correlation between protein and mRNA abundance in yeast. *Mol Cell Biol* **19**:1720–1730.

Gygi SP, Corthals GL, Zhang Y, Rochon Y, Aebersold R (2000). Evaluation of two-dimensional gel electrophoresis–based proteome analysis technology. *Proc Natl Acad Sci USA* **97**:9390–9395.

Halmes NC, McMillan DC, Oatis JE Jr., Pumford NR (1996). Immunochemical detection of protein adducts in mice treated with trichloroethylene. *Chem Res Toxicol* **9**:451–456.

Hansen BT, Jones JA, Mason DE, Liebler DC (2001). SALSA: a pattern recognition algorithm to detect electrophile-adducted peptides by automated evaluation of CID spectra in LC-MS-MS analyses. *Anal Chem* **73**:1676–1683.

Hargus SJ, Amouzedeh HR, Pumford NR, Myers TG, McCoy SC, Pohl LR (1994). Metabolic activation and immunochemical localization of liver protein adducts of the nonsteroidal anti-inflammatory drug diclofenac. *Chem Res Toxicol* **7**:575–582.

Harriman SP, Hill JA, Tannenbaum SR, Wishnok JS (1998). Detection and identification of carcinogen-peptide adducts by nanoelectrospray tandem mass spectrometry. *J Am Soc Mass Spectrom* **9**:202–207.

Hay RT (2001). Protein modification by SUMO. *Trends Biochem Sci* **26**:332–333.

Hong Y, Rogers R, Matunis MJ, Mayhew CN, Goodson ML, Park-Sarge OK, Sarge KD (2001). Regulation of HSF1 by stress-induced SUMO-1 modification. *J Biol Chem.*

Johnson GL, Lapadat R (2002). Mitogen-activated protein kinase pathways mediated by ERK, JNK, and p38 protein kinases. *Science* **298**:1911–1912.

Jones JA, Liebler DC (2000). Tandem MS analysis of model peptide adducts from reactive metabolites of the hepatotoxin 1,1-dichloroethylene. *Chem Res Toxicol* **13**:1302–1312.

Jonscher KR, Yates JR (1997). The quadrupole ion trap mass spectrometer—a small solution to a big challenge. *Anal Biochem* **244**:1–15.

Kaur S, Hollander D, Haas R, Burlingame AL (1989). Characterization of structural xenobiotic modifications in proteins by high sensitivity tandem mass spectrometry. Human hemoglobin treated in vitro with styrene 7,8-oxide. *J Biol Chem* **264**: 16981–16984.

Kleiner HE, Rivera MI, Pumford NR, Monks TJ, Lau SS (1998). Immunochemical detection of quinol—thioether-derived protein adducts. *Chem Res Toxicol* **11**:1283–1290.

Liebler DC (2001). *Introduction to Proteomics: Tools for the New Biology*. Totowa, NJ: Humana Press.

Liebler DC (2002). Proteomic approaches to characterize protein modifications: new tools to study the effects of environmental exposures. *Environ Health Perspect* **110**(suppl 1):3–9.

Liebler DC, Hansen BT, Davey SW, Tiscareno L, Mason DE (2002). Peptide sequence motif analysis of tandem MS data with the SALSA algorithm. *Anal Chem* **74**: 203–210.

Link AJ, Eng J, Schieltz DM, Carmack E, Mize GJ, Morris DR, Garvik BM, Yates JR III (1999). Direct analysis of protein complexes using mass spectrometry. *Nat Biotechnol* **17**:676–682.

MacCoss MJ, McDonald WH, Saraf A, Sadygov R, Clark JM, Tasto JJ, Gould KL, Wolters D, Washburn M, Weiss A, Clark JI, Yates JR III (2002). Shotgun identification of protein modifications from protein complexes and lens tissue. *Proc Natl Acad Sci USA* **99**:7900–7905.

Mann M, Hendrickson RC, Pandey A (2001). Analysis of proteins and proteomes by mass spectrometry. *Annu Rev Biochem* **70**:437–473.

Mao Y, Sun M, Desai SD, Liu LF (2000). SUMO-1 conjugation to topoisomerase I: a possible repair response to topoisomerase-mediated DNA damage. *Proc Natl Acad Sci USA* **97**:4046–4051.

Marcus K, Immler D, Sternberger J, Meyer HE (2000). Identification of platelet proteins separated by two-dimensional gel electrophoresis and analyzed by matrix assisted laser desorption/ionization–time of flight–mass spectrometry and detection of tyrosine-phosphorylated proteins. *Electrophoresis* **21**:2622–2636.

Mason DE, Liebler DC (2000). Characterization of benzoquinone-peptide adducts by electrospray mass spectrometry. *Chem Res Toxicol* **13**:976–982.

Mason DE, Liebler DC (2003). Quantitative analysis of modified proteins by LC-MS-MS of peptides labeled with phenyl isocyanate. *J Proteome Res*.

Medzihradsky KF, Campbell JM, Baldwin MA, Falick AM, Juhasz P, Vestal ML, Burlingame AL (2000). The characteristics of peptide collision-induced dissociation using a high performance MALDI-ToF/ToF tandem mass spectrometer. *Anal Chem* **72**:552–558.

Miller EC, Miller JA (1981). Searches for ultimate chemical carcinogens and their reactions with cellular macromolecules. *Cancer* **47**:2327–2345.

Moll TS, Harms AC, Elfarra AA (2000). A comprehensive structural analysis of hemoglobin adducts formed after in vitro exposure of erythrocytes to butadiene monoxide. *Chem Res Toxicol* **13**:1103–1113.

Morris HR, Paxton T, Panico M, McDowell R, Dell A (1997). A novel geometry mass spectrometer, the Q-TOF, for low-femtomole/attomole-range biopolymer sequencing. *J Protein Chem* **16**:469–479.

Muller S, Hoege C, Pyrowolakis G, Jentsch S (2001). SUMO, ubiquitin's mysterious cousin. *Nat Rev Mol Cell Biol* **2**:202–210.

Nelson SD (1995). Mechanisms of the formation and disposition of reactive metabolites that can cause acute liver injury. *Drug Metab Rev* **27**:147–177.

Pandey A, Mann M (2000). Proteomics to study genes and genomes. *Nature* **405**: 837–846.

Pauwels W, Farmer PB, Osterman-Golkar S, Severi M, Cordero R, Bailey E, Veulemans H (1997). Ring test for the determination of N-terminal valine adducts of styrene 7,8-oxide with haemoglobin by the modified Edman degradation technique. *J Chromatogr B Biomed Sci Appl* **702**:77–83.

Pawson T (2002). Regulation and targets of receptor tyrosine kinases. *Eur J Cancer* **38**(suppl 5):S3–10.

Rappaport SM, Ting D, Jin Z, Yeowell-O'Connell K, Waidyanatha S, McDonald T (1993). Application of Raney nickel to measure adducts of styrene oxide with hemoglobin and albumin. *Chem Res Toxicol* **6**:238–244.

Ritenbaugh C (2000). Diet and prevention of colorectal cancer. *Curr Oncol Rep* **2**: 225–233.

Roepstorff P (2000). MALDI-TOF mass spectrometry in protein chemistry. *EXS* **88**: 81–97.

Rombach EM, Hanzlik RP (1997). Detection of benzoquinone adducts to rat liver protein sulfhydryl groups using specific antibodies. *Chem Res Toxicol* **10**:1407–1411.

Sadygov RG, Eng J, Durr E, Saraf A, McDonald H, MacCoss MJ, Yates JR (2002). Code developments to improve the efficiency of automated MS/MS spectra interpretation. *J Proteome Res* **1**:211–215.

Saitoh H, Hinchey J (2000). Functional heterogeneity of small ubiquitin-related protein modifiers SUMO-1 versus SUMO-2/3. *J Biol Chem* **275**:6252–6258.

Satoh H, Fukuda Y, Anderson DK, Ferrans VJ, Gillette JR, Pohl LR (1985). Immunological studies on the mechanism of halothane-induced hepatotoxicity: immunohistochemical evidence of trifluoroacetylated hepatocytes. *J Pharmacol Exp Ther* **233**: 857–862.

Schaid DJ, Buetow K, Weeks DE, Wijsman E, Guo SW, Ott J, Dahl C (1999). Discovery of cancer susceptibility genes: study designs, analytic approaches, and trends in technology. *J Natl Cancer Inst Monogr*, pp 1–16.

Sepai O, Anderson D, Street B, Bird I, Farmer PB, Bailey E (1993). Monitoring of exposure to styrene oxide by GC-MS analysis of phenylhydroxyethyl esters in hemoglobin. *Arch Toxicol* **67**:28–33.

Shields PG, Harris CC (2000). Cancer risk and low-penetrance susceptibility genes in gene-environment interactions. *J Clin Oncol* **18**:2309–2315.

Skipper PL, Peng X, Soohoo CK, Tannenbaum SR (1994). Protein adducts as biomarkers of human carcinogen exposure. *Drug Metab Rev* **26**:111–124.

Spengler B, Kirsch D, Kaufmann R, Jaeger E (1992). Peptide sequencing by matrix-assisted laser-desorption mass spectrometry. *Rapid Commun Mass Spectrom* **6**:105–108.

Stancato LF, Petricoin EF III (2001). Fingerprinting of signal transduction pathways using a combination of anti-phosphotyrosine immunoprecipitations and two-dimensional polyacrylamide gel electrophoresis. *Electrophoresis* **22**:2120–2124.

Tornqvist M, Mowrer J, Jensen S, Ehrenberg L (1986). Monitoring of environmental cancer initiators through hemoglobin adducts by a modified Edman degradation method. *Anal Biochem* **154**:255–266.

Washburn MP, Wolters D, Yates JR (2001). Large-scale analysis of the yeast proteome by multidimensional protein identification technology. *Nat Biotechnol* **19**:242–247.

Wilkins MR, Gasteiger E, Gooley AA, Herbert BR, Molloy MP, Binz PA, Ou K, Sanchez JC, Bairoch A, Williams KL, Hochstrasser DF (1999). High-throughput mass spectrometric discovery of protein post-translational modifications. *J Mol Biol* **289**:645–657.

Wolters DA, Washburn MP, Yates JR (2001). An automated multidimensional protein identification technology for shotgun proteomics. *Anal Chem* **73**:5683–5690.

Yates JR (1998). Database searching using mass spectrometry data. *Electrophoresis* **19**:893–900.

Yates JR (2000). Mass spectrometry. From genomics to proteomics. *Trends Genet* **16**: 5–8.

Yeowell-O'Connell K, Jin Z, Rappaport SM (1996). Determination of albumin and hemoglobin adducts in workers exposed to styrene and styrene oxide. *Cancer Epidemiol Biomarkers Prev* **5**:205–215.

Zhou H, Ranish JA, Watts JD, Aebersold R (2002). Quantitative proteome analysis by solid-phase isotope tagging and mass spectrometry. *Nat Biotechnol* **20**:512–515.

14

Introduction to Metabolomics and Metabolic Profiling

Robert E. London and David R. Houck

INTRODUCTION: WHAT IS METABOLOMICS?

In the 1990s, the ability to study biological phenomena in a high-throughput and parallel manner gave birth to the various "omics" fields; *metabolomics* is the most recent (but not the last?) word to be injected into the modern vernacular of the biological sciences. The "omics" can be considered a collection of technologies aimed at accelerating our understanding of biology: high-throughput and parallel approaches to fundamental understanding of biological systems. Ultimately, uncovering the intricate relationships and interactions between genes, proteins, and metabolism will provide humankind with a type of roadmap for the treatment and diagnosis of disease. From the perspective of metabolism, there are increasing efforts to comprehensively measure metabolite pools and to relate changes in those pools to genotypes, as well as environmental and xenobiotic challenges. The intent of this chapter is to introduce the reader to the concept of the metabolome and provide a guide to modern analytical technologies, particularly nuclear magnetic resonance (NMR), which are used in toxicological applications of metabolomics.

Toxicogenomics: Principles and Applications. Edited by Hamadeh and Afshari
ISBN 0-471-43417-5 Copyright © 2004 John Wiley & Sons, Inc.

Although the word *metabolomics* is relatively new, the study of metabolic processes has been a longstanding focus of biochemical and physiological research. Metabolism is the sum of all enzyme-catalyzed reactions in living cells that transform organic molecules, or the sum of all processes to acquire and use energy in a living organism. Metabolic pathways therefore produce or consume energy by breaking and making bonds in low-molecular-weight biochemicals (<1,500 daltons). Thus, as proteome is a word to describe the full complement of endogenous proteins, metabolome represents all the small molecules, or metabolites, in an organelle, cell, tissue, organ, or organism. A simplistic relationship that connects genes to metabolites is shown in Figure 14-1.

Expression of the endogenous genetic code (genome) generates messages (transcriptome) that determine the composition of the proteome, which in turn establishes the catalytic factories of metabolism. The relationship is a regulatory cycle, because fundamentally the metabolome regulates the genome through specific metabolite-protein interactions that either directly or indirectly control gene transcription. There are many examples of genes in biosynthetic pathways that are feedback regulated by intermediates or end-product metabolites. Perhaps the most studied and characterized regulation systems in mammals are those involved in cholesterol homeostasis (Emery et al., 2001; Osborne, 2000) and glucose metabolism (Tappy and Minehira, 2001). Nitric oxide and oxygen, probably the smallest of metabolites, are also regulators of

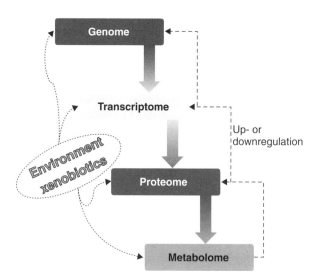

Figure 14-1 *The relationship from genome to metabolome. The genome determines the potential composition of the metabolome, and the metabolome will influence (positively or negatively) expression of that potential. The environment and foreign chemistry will also impact the make up of the "omes."*

gene transcription (Wenger 2002; Pineda-Molina and Lamas, 2001). Whereas the genome and proteome determine the potential expression of the metabolome, the environmental and nutritional inputs will ultimately modulate the actual metabolic state of a biological system; those inputs include xenobiotics and toxic molecules that interact with DNA, mRNA, and proteins.

Interestingly, the actual definition and usage of the word *metabolomics* and other "omics" terms are apparently unsettled concepts within the practicing community. In fact, the "omics" jargon is a bit confusing in the metabolite analysis literature. The word *metabolomics* was apparently first published by Tweeddale et al. in 1998: "Effect of Slow Growth on Metabolism of *Escherichia coli*, as Revealed by Global Metabolite Pool ('metabolome') Analysis." Metabolome aptly describes the comprehensive metabolite pool in any cell, tissue, organism, or biological fluid; therefore, metabolomics can be considered a group of technologies aimed at the elucidation and study of metabolomes. Like genomes and proteomes, there are multiple metabolomes, from unicellular organisms and mammalian cells in culture to the overall metabolome of a tissue. Of course, you can also refer to the metabolome of biofluids such as urine, cerebral spinal fluid (CSF), and plasma. In fact, metabolites found in biofluids of higher organisms can be indicative of the overall metabolic state of an organism and the response to xenobiotic challenges (Videen and Ross, 1994).

Recently the word metabo*n*omics has been coined to label the study of metabolite patterns in biological fluids and the correlation of metabolite patterns to xenobiotic challenge and particular disease states: "metabonomic approach to toxicological evaluation—the study of the multivariate time-resolved metabolic changes in biofluids, cells, and tissues to pathophysiological insult or genetic modification" (Nicholson et al., 1999). The literature covering the application of NMR to metabolite analysis can generally be considered to fall into two categories: (1) detailed spectroscopic characterization of endogenous compounds or xenobiotic metabolites using many of the tools of modern NMR spectroscopy, or (2) low-resolution, high-throughput spectroscopic analysis, which seeks to identify spectral characteristics and to arrive at broad categorizations of spectral patterns as being normal or indicative of various pathological conditions. The former studies can be grouped under the general heading of metabolomics—that is, the qualitative and quantitative determination of the molecular composition of a biological sample using a broad range of methodologies. The latter approach, developed primarily by Jeremy Nicholson and coworkers, has been dubbed metabonomics.

Although in some cases the line between these two approaches may be poorly defined, a hallmark of the metabonomic approach is, in the evaluation of the present reviewers, the fundamentally different treatment of spectral data, which is evaluated by a broad range of pattern recognition techniques (Lindon et al., 2001) rather than by more traditional spectroscopic analysis. Another apparent distinction between metabolomics and metabonomics is the study of cell and tissue metabolism (metabo*l*omics) versus a systems, or organism, approach to metabolism using biofluids as indicators (metabo*n*omics) (Nicholson et al., 2002). Certainly in the case of toxicological research, these approaches are not mutually exclusive. A good illustration of the interrelationship is provided by the metabonomic study of paracetamol-treated rats indicating a major, unidentified urinary metabolite subsequently identified as pyroglutamate (5-oxoproline) using more exhaustive spectroscopic methods (Ghauri et al., 1993). Ultimately, both approaches must address the key toxicological question, "What is the critical biochemical lesion in the cell and how does this relate mechanically to the exhibited pathological condition?" (Nicholson and Wilson, 1989)."

Metabolic profiling and *biochemical profiling* are also commonly used terms to describe metabolite analysis. Metabolite profiling may be considered as the analytical processes used to measure the quality and/or quantity of small molecules in a biological sample. The process could involve one or more techniques including liquid chromatography (LC), gas chromatography (GC), mass spectrometry (MS), and nuclear magnetic resonance (NMR). Historically, profiling endogenous metabolites was restricted to specific compounds, or compound types. For example, the organic acids in urine or the flavenoids found in blood following ingestion of red wine can be "profiled" by any of several techniques such as GC-MS, LC-MS, or NMR; the profile can be thought of as a snapshot of the levels of metabolites at a certain time and state.

At present, a metabolite profile can be viewed as a subset of a metabolome; for a variety of reasons to be discussed later, current analytical technologies are unlikely to completely elucidate (quantitatively and/or qualitatively) the entire population of small molecules in any biological system (organelle or organism). Simply put, the complexity of the problem is enormous. Even so, efforts are underway in academia and industry to "profile" the entire complement of endogenous metabolites in various cell types, tissues, and biofluids. Herein is the challenge: to develop integrated analytical methods and data-reduction technologies to deal with the inherent complexity of biological systems. The term *metabolic profiling* has also been used in the drug metabolism literature to capture xenobiotic metabolism in mammals: the reactions and intermediates involved in degradation/excretion of xenobiotics. While this is an appropriate use of the term, this chapter does not discuss analysis of drug metabolites.

OBJECTIVES AND APPLICATIONS OF METABOLOMICS

Metabolomics is being applied to a wide variety of applications, from human nutrition and toxicology to the development of pharmaceuticals and agricultural products. The numerous applications of metabolomics in fact share several underlying objectives: (1) discovering metabolite signatures as prognostic indicators, diagnostics, or biomarkers of disease states, (2) establishing toxicological markers for drug development and environmental toxicology, (3) understanding mechanisms of metabolic diseases, and (4) correlating metabolite phenotypes (metabotype) with genotype and environmental input (e.g., nutrition). One of the more developed diagnostic applications that existed for decades before "omics" is screening of neonates for genetic disorders in intermediary metabolism. The recently reviewed analytical approaches to clinical diagnosis of metabolic disorders (Chace, 2001; Chace et al., 2002; Sim et al., 2002) provide excellent methods for metabolite analysis and are good examples of the application of MS to metabolomics.

One of the most successful implementations of NMR screening of biofluids to date involves the characterization of blood plasma lipoproteins, based on a measurement and analysis of the lipid methyl resonances (Otvos, 1999). In this application, the NMR analysis is able to go beyond the usual separation of lipoproteins into HDL, LDL, and VLDL by characterizing lipoprotein subclasses within these categories. These subclasses have unequal associations with coronary heart disease. Nuclear magnetic resonance distinguishes among the subclasses on the basis of slight differences in the spectral properties of the lipids carried within the particles, which vary according to the diameter of the phospholipid shell. These studies have shown that individuals with elevated triglycerides are likely to have higher-risk lipoprotein subclass profiles. Triglyceride-rich lipoproteins drive the metabolic reactions that produce LDL of abnormal size and cholesterol content. The quantities of these abnormal LDL particles and the associated risk of coronary heart disease are underestimated by conventional cholesterol measurements. Nuclear magnetic resonance spectroscopy measures lipoprotein subclasses directly and efficiently, and produces information that may improve the assessment and management of cardiovascular disease risk.

As already mentioned, with respect to the study of the metabolome, metabolic profiling does not refer to elucidation of xenobiotic (drug) metabolism. However, metabolomic analyses have been used to understand how xenobiotics influence primary and intermediary metabolism in humans. For example, NMR-based analyses of metabolites in urine and plasma have uncovered particular metabolite signatures of toxic events induced by numerous compounds such as mercuric chloride, hydrazine, and antibiotics (Shockcor and Holmes, 2002). Correlation of metabolite profiles to genotype in plants has demonstrated that establishing metabolic phenotypes is an important approach to the study of gene function (Fiehn et al., 2000; Fiehn, 2002). Finally, metabolic flux analysis, especially with stable isotopes, has provided insight into many metabolic pathways, from biosynthesis of coenzymes in microbes (Houck et al., 1991) and the control of glucose metabolism in mammals (Jones et al., 1998) to hepatic methyl transfer reactions (London et al., 1987). This chapter focuses on the techniques of metabolomics and metabolite analysis as applied to toxicological studies in mammalian systems.

DEALING WITH COMPLEXITY

It is probably safe to state that no metabolic profile generated by one or more analytical techniques comprehensively represents the composition of a metabolome in a prokaryote or eukaryote. The overall complexity of the problem derives from genetic variation, tissue and cell-type diversity, nutritional variation, and the huge number of diverse small molecules that comprise the metabolome. Analytical sciences have historically dealt with specific analyses of individual species (e.g., histamine or ATP) or classes of molecules (e.g., amino acids or nucleotides), but the goal of metabolomics is comprehensive analysis of the entire population of small molecules. At the end of 2002, the KEGG database (http://www.genome.ad.jp/kegg/) had over 10,000 entries for "compounds," and greater than 8,000 of them had a molecular weight within a range of 30–1,500 Daltons (Goto et al., 1998, 2002).

Compounds between 1,000 and 1,500 Da are mostly peptides and coenzyme A conjugates. The vast majority (>4,000) of compounds less than 1,500 Da fall between 100 and 600 Da. In addition, many macromolecular structures (e.g., membranes) consist of tightly but dynamically associated small molecules. The analytical challenge therefore is to have broadly applicable methods to measure qualitative and at least semiquantitative changes in metabolites of different molecular weight, polarity, charge state, compartmentation, ability and stability. Another source of complexity is the disparate concentration ranges of metabolites (e.g., glucose and ATP versus steroid hormones); thus the dynamic range of any method will be pushed to the limits.

The dynamic range of detection and quantitative analysis depend on both the molecule and the detection technique. The integral of a resonance in NMR is directly proportional to the quantity of a compound within receiver space of the NMR probe. Yet generally dynamic range in ^{1}H-NMR is limited by the fact that the analysis is done in H_2O so that the water protons are present at ~100M concentration. The dynamic range characteristics of modern NMR spectrometers then limit the analysis to metabolites present at concentrations ≥0.1 mM—that is about 6 orders of magnitude. Some investigators have lyophilized the samples and redissolved them in deuterium oxide (D_2O; e.g., Feng et al., 2002), reducing the water proton concentration to about 100 mM and the detection sensitivity to ~1 µM. Although the dynamic range then becomes less of a limitation, the inherent low sensitivity of the NMR experiment requires multihour signal averaging to achieve this sensitivity. Further, such analysis requires that the D_2O contain no impurities at these concentrations that could interfere with the analysis and that the low-molecular-weight metabolites exhibit at least some resonances that do not overlap the resonances of more concentrated compounds. Finally, solvent replacement significantly increases the time required to process samples.

The dynamic range of mass spectrometry (MS) will depend on the particular ionization and ion analysis techniques. LC-MS is a concentration-dependent technique, and modern nanoflow systems can detect subpicomolar quantities of molecules in microliter samples; however, the quantitative dynamic range could be 10^2 to 10^5, depending on the molecule and instrument type. Another limiting factor is ionization in MS methods that utilize atmospheric ionization for molecules in aqueous solution. The quantity and quality of ionization for a particular molecule of interest is likely to vary across different samples. First of all, salts and buffers will influence the type of ion observed; for example, sodium readily forms adducts with oxygen functionalities, causing a "shift" in a molecular ion from M + H (M + 1) to M + Na (M + 23). In addition, the ionization of a molecule can be suppressed by other ions that are simultaneously introduced into the ion source. All of these analytical variables must be addressed and controlled to precisely measure biological perturbations across samples.

SAMPLE PREPARATION AND HANDLING

General Considerations

Compartmentation and structural complexity of cells and tissues present barriers to the release of metabolites for analysis. A generic representation of the sample preparation process for metabolite analysis, from tissue treatment to data analysis, is given in Figure 14-2. Metabolites in a biological sample must

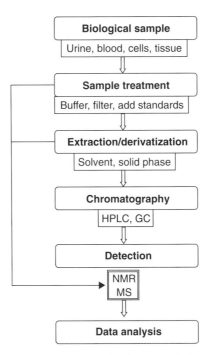

Figure 14-2 *The processes of metabolomics, from sample handling to data analysis.*

be made available for analysis. Many sample types will require disruption of cells and tissue, followed by centrifugation, extraction, and concentration. For example, detailed lipid profiling, usually carried out by gas chromatography (GC), involves solvent extraction, saponification, methylation of carboxylic acids, and concentration prior to analysis. The extraction and analytical methods for fatty acid profiling have essentially remained unchanged for almost five decades (Folch et al., 1957).

MS approaches to analysis of newborn blood samples usually begin with extraction of dried blood from filter paper (Chace, 2001); the extract can be directly infused into the MS, or further derivatized and/or chromatographed. As indicated in the scheme shown in Figure 14-2, certain steps of metabolome analysis could be circumvented or shortened, depending on the sample and method of analysis. In the case of biological fluids (CSF or urine), the sample preparation may be as simple as adding a D_2O buffer and internal standards for NMR. Filtered fluids could also be directly infused into an electrospray MS (Pitt et al., 2002; Vreken et al., 1999). In any case, comprehensive determination of the quality and quantity of metabolites in any one sample will require parallel implementation of several processing and detection methods.

NMR Spectroscopy

Experimental issues related to the optimization of biofluid analysis by NMR spectroscopy have been discussed by Nicholson and Wilson (1989), Tate et al. (2001), and Deprez et al. (2002). Some of the physical characteristics of various biofluids that influence the specific approaches utilized for NMR studies are summarized by Lindon et al. (2000). Issues such as sample viscosity, protein, salts, divalent cations, and lipid content will significantly affect the NMR spectrum. High viscosity can lead to significant broadening of the resonances. The presence of protein and lipids typically give rise to broad resonances, which have the effect of producing an uneven baseline, interfering with the simple quantitation of small metabolite concentrations. Binding interactions can significantly broaden the resonances of small metabolites in the presence of macromolecules due to chemical exchange (e.g., fig. 1 of Deprez et al., 2002).

There are currently many approaches to baseline smoothing—for example, polynomial fitting and subtraction—which appear to be sufficient to deal with such problems for most samples. Nevertheless, it is important to note that in general the protein/lipid composition of a given biofluid may depend significantly on whether the animal has been treated with xenobiotics. For example, urine derived from rats treated with toxic substances may contain large concentrations of excreted proteins. Such differences mean that the baseline correction methods used for the control sample may be inadequate for samples derived from treated animals. Deprez et al. (2002) have utilized a protocol involving acetonitrile precipitation of proteins and applied this approach to NMR studies of blood plasma.

Probably the most significant difference between the analysis of biofluids and the analysis of mixtures of organic compounds obtained in a chemical reaction is the need to deal with water, which is present at ~55 M (110 M protons). Since even the most abundant metabolites are typically at millimolar concentrations, it is important to effectively suppress the large water resonance and to utilize a spectrometer with high dynamic range characteristics. There are a large number of methods for suppressing water/solvent resonances involving nonexcitation of the water, separation based on relaxation rate differences, the use of field gradients, and the use of selective periods of presaturation. The latter approach, used in combination with the nuclear Overhauser experiment, appears to be most widely used at present for the analysis of biofluids (e.g., Deprez et al., 2002).

The use of spin-echo sequences such as the CPMG experiment works well for the elimination of broad resonances arising from macromolecules. In many of the reported high-throughput metabonomic studies, sample preparation has involved the addition of 10–20% D_2O (v/v) for the field/frequency lock. Although lyophilizing and redissolving samples in D_2O adds considerable time to the analysis, this can be set up as a batch process and could substantially improve the dynamic range characteristics of the experiment, as well as allowing more complete analysis of metabolites with resonances near water. If such a procedure is followed, concentrations of volatile compounds such as acetic acid and TMAO may be reduced.

The comparative analysis of biofluids that is typically the objective of metabonomic studies requires that they be at approximately the same pH. This is a particular issue for urine, for which the pH may vary widely; but even for other types of samples, pH variation will significantly affect resonance shifts. Further, toxicological studies are likely to produce significant pH variation, particularly in urine, resulting from the perturbation of mechanisms that regulate pH. One approach for dealing with this type of variability has been the addition of a high concentration of a proton-free buffer such as sodium phosphate to overwhelm sample pH variation. Physiological samples typically contain high salt concentrations, and in order to standardize the pH value, an additional 300–400 mM phosphate buffer is frequently added. These high salt concentrations reduce the sensitivity of the NMR experiment by lowering the Q of the NMR rf coil. In addition, a sample containing high concentrations of phosphate and organic compounds presents an ideal bacterial growth medium, so antibacterial agents should also be added. Sodium azide is often the agent of choice, again because it lacks proton resonances.

Following the above discussion, the instrumentation requirements for metabolic analysis are generally dependent on the objectives of the studies, particularly whether a complete characterization of the metabolite compositions of a given sample is desired, or whether a comparative, metabonomic analysis is the goal of the study. In the first case, it is optimal to take full advantage of the most sensitive—that is, highest field spectrometer and sophisticated spectroscopy—experiments that have been developed. LC-NMR methods may also be critical in some cases for the characterization of low-abundance metabolites, as well as two-dimensional methods for resonance assignment, potentially involving multinuclear characterization. Alternatively, metabonomic analysis is at present largely limited to low-resolution studies that facilitate comparisons among different samples in which salt concentrations, pH, protein concentrations, and other physical parameters may vary to a limited extent. These experiments can be performed adequately on low-field spectrometers, although due to the lower sensitivity, the total time required to analyze a given data set will increase. For example, at 300 MHz the total experimental time required to achieve a given signal/noise level will increase by a factor of about 4.6 relative to a study performed at 500 MHz.

In addition to studies of urine and blood plasma, metabolite analysis has often involved the study of cell extracts. A variety of extraction protocols are used, depending primarily on whether the analysis is focused on water-soluble or lipophilic components of the cell. Of course, extraction minimizes the water/solvent interference problem, since the sample can then be prepared in a deuterated solvent. Metabolite extraction procedures are eliminated if solid-state NMR methods are used for the analysis of intact tissue samples.

Such an approach has recently been used in a study of the effects of α-naphthylisothiocyanate (ANIT; Waters et al., 2002). This approach utilizes high-speed "magic angle" sample spinning (MAS) to remove the effects of chemical shift anisotropy, achieving relatively narrow line widths for solid samples. ANIT treatment was found to produce increased levels of nucleosides, nucleoside bases, and nucleoside monophosphates. The solid-state spectra showed a decrease in triglyceride resonances to below control values, while the spectra obtained from chloroform/methanol extracts showed a rise in triglycerides. Additionally, ^{1}H NMR analysis of the extracts identified a number of signals from unknown metabolites that were not observed in the ^{1}H-MAS-NMR analysis. In general, more research is needed to validate the applicability of the solid-state approach to metabolite analysis.

MASS SPECTROMETRY AND OTHER CHROMATOGRAPHY-COUPLED TECHNIQUES

Liquid and gas chromatography are probably the most widely used techniques for separating metabolites prior to analytical detection. Hundreds of chromatographic methods for analysis of specific metabolites or metabolite classes (e.g., sterols and lipids) have been validated and published over the years. Comprehensive analysis of hundreds or thousands of metabolites using one or two chromatographic methods is both a practical and a theoretical problem. In addition to factors such as method selectivity, chromatographic-coupled methods will be limited by the intrinsic analytical peak capacity of the system, which includes the column and detector. For any given system, the maximum number of detectable metabolites will be determined by the peak capacities of both the column and the detector.

Davis and Giddings (1983) presented an excellent statistical analysis of chromatographic peak capacity. They demonstrated that when the number of peaks in a sample (n) equals the peak capacity of a column (n_c), the maximum number of observable peaks is $0.37n_c$. In fact, peak overlap will occur when $n > 0.18n_c$. Utilizing detectors such as mass spectrometers that have high analytical peak capacities increases the overall peak capacity of the system (Table 14-1); the system's peak capacity can be estimated as the multiple of n_c and the peak capacity of the detector (n_d).

At unit mass resolution, a mass spectrometer can detect 1,400 mass peaks between 100 and 1,500 amu, but the actual peak capacity is one-third to one-quarter less than 1,400 because each molecule will exhibit three to four isotopomer peaks. Therefore, applying the assessment used for chromatographic peaks, the analytical peak capacity of the MS detector (n_d) will actually be ~80–90 ($1,400/3 \times 0.18$). For example, assuming an HPLC column with a peak capacity of 100 (with some peak overlap) is coupled to a single-quadrupole ESI-MS that has a unit mass resolution between 100 and 1,000 amu, the maximum peak capacity of an LC-MS system is approximately 9,000 (100×90).

Of all the chromatography detectors available to researchers today, mass spectrometers offer the greatest potential for comprehensive metabolite analysis. This is in part due to the rapid evolution in MS technology during the past decade. The advent of electrospray ionization extended the power of MS to broad applications within the biological sciences; mass spectra of large and highly polar molecules can now be obtained either directly from biological samples, or via infusion from, for example, HPLC. In addition, the instrumentation has become less cumbersome, and software is becoming especially user-friendly.

With a few exceptions, the operation of commercial mass spectrometers no longer requires MS specialists; in fact, electrospray mass spectrometers are now important bench-top components of many chemistry and biology laboratories. The power of MS for metabolite analysis derives from both its high sensitivity (pg) and the fact that metabolites can be separated by the mass spectrometer according to molecular weight at a minimum resolution of 1 atomic mass unit (amu). Thus, depending on the particular resolution of the mass analyzer, the detector peak capacity of MS could range from 100 to potentially 50,000. Overlap of metabolite peaks in a mass spectrum will occur when compounds are isobaric (having the same molecular mass); therefore, metabolites having the same molecular formula—for example, various hexose sugars (mass 180.16)—will not be distinguished by molecular mass alone; such compounds must be either separated by chromatography prior to MS or subjected to tandem MS to observe product MS spectra. There are now numerous types of mass spectrometers, but the most useful instruments for metabolomic analyses are listed in Table 14-1.

TABLE 14-1 Summary of the various mass analyzers used for metabolite profiling[a]

Mass Analyzer	Resolution $(m/\Delta m)$	Mass Accuracy (amu)	Quantitative Analysis	Spectrum Acquisition Rate (Hz)	Tandem MS	Peak Capacity (n_d)
Quadrupole triple quad	1,000–2,000	0.1	Best	1–2	No MS^2	100
Time of flight (TOF)	5,000–10,000	0.001–0.01	Poor	>100	No	500–1,000
Quadrupole-TOF	5,000–15,000	0.001–0.01	Fair	>100	MS^2	500–1,500
Ion trap	1,000–2,000	0.1	Fair	1–2	MS^n	100
FT-MS	10,000–500,000	<0.001	Fair	>100	MS^n	1,000–50,000

[a] These are general parameters for a mass range of 50–1,500 amu and will vary depending on the specific instrument configurations and design. MS^2 and MS^n indicate the ability to perform either a single-tandem MS experiment for any given precursor ion (MS^2) or a sequence of tandem MS experiments on ions trapped in a collision cell (MS^n). If the mass spectrometer is coupled to a chromatography system, the total system peak capacity will be increased by the peak capacity of the column ($n_s = n_c \times n_d$).

Although a detailed description of the various MS technologies (see Willoughby et al., 2002; Rossi and Sinz, 2002) is outside the scope of this chapter, it is important to highlight the technologies that provide the highest throughput and peak capacity for metabolite analysis. Atmospheric pressure ionization (API) techniques such as electrospray ionization (ESI) are commonly used methods for transforming polar metabolites from aqueous solutes into gas-phase ions in a mass spectrometer. It is important to note API techniques, known as soft ionization methods, do not yield uniform ionization across all classes of biomolecules. As a general rule of thumb, the more heteroatoms and functional groups (e.g., COOH and NH_2) a molecule contains, the greater the tendency to ionize via API. Ionization can be enhanced by the addition of some volatile buffers such as acetic acid, formic acid, trifluoroacetic acid, and ammonium acetate (0.5%) to the infusion solvent.

Electron impact MS, a higher-energy technique, is most useful for analysis of metabolites separated by gas chromatography. Many nonvolatile compounds such as organic acids can be converted to volatile derivatives for GC analysis (Chace, 2001; Wolfe, 1992). ESI-MS, probably the most useful ionization method for analysis of biomolecules, can be accomplished with different mass analyzers, such as quadrupole (Q), time of flight (ToF), and Fourier transform ion-cyclotron resonance (FT) systems (Table 14-1). Bench-top ToF-MS systems have become relatively affordable and user-friendly. Most importantly, ToF offers fast spectrum acquisition (e.g., >10 Hz), good mass accuracy, and superior mass resolution. As a result of the increased resolution, coupling a high-resolution ToF-MS to HPLC can achieve analytical peak capacities of 10^6. However, the bench-top ToF-MS can have a somewhat limited dynamic range, which is highly compound dependent.

Triple-quadrupole instruments, in contrast, are now the gold standard for sensitivity and quantitative dynamic range, because of the ability to monitor product ions using tandem MS (also known as MS/MS and MS^2 techniques). Briefly, tandem MS essentially isolates and fragments a specific molecular ion (precursor ion)—for example the dipeptide ala-val (m/z ~189)—to yield a product ion—for example, valine at m/z ~117; triple quadrapole, quadrapole-ToF, ion trap, and FT mass spectrometers have the ability to isolate ions for MS/MS experiments. The result is a substantial (>10-fold) increase in signal to noise.

Getting back to dynamic range, a bench-top ESI-ToF might have a lower limit of detection of 1 pmole glutamate, but the response may not be linear above 100 pmole. For other molecules the quantitative window could be in the nmole range, and/or the dynamic range could be less than 10^3. However, tandem MS using a triple-quadrupole MS can easily deliver 4–5 orders of magnitude in quantitative dynamic range. On the downside, quadrupole instruments do not generally have the high spectrum acquisition rate and mass resolution that are important for detection of a multitude of metabolites in a single chromatographic method.

At the high end of instrumentation, FT-MS provides the greatest resolving power and therefore unmatched analytical peak capacity. However, FT instrumentation is complex and not as widely implemented; it is only a matter of time and development before FT-MS becomes a tool of choice for metabolite analysis. Because of the high resolving power of both ESI-ToF and FT-MS, it is possible to analyze complex mixtures without the intervention of chromatography. Flow injection or direct infusion of samples into the MS can provide high-throughput analysis, possibly up 6–12 samples per minute; the rate-limiting step is the automatic sample introduction, usually via a commercial autosampler.

Numerous studies have utilized HPLC-coulometric methods to profile red-ox active metabolites in biological samples, including neurotransmitters (Acworth and Bowaers, 1997), flavonoids in plasma (Nurmi and Adlercreutz, 1999), endogenous antioxidants in plasma (Finckh et al., 1995), and the metabolomes of rat plasma and mitochondria (Kristal et al., 1998, 2002). Electrochemical detection offers high sensitivity and a quantitative dynamic range of 10^5 to 10^6 for most red-ox active metabolites. Sensitivity is on the order of 5–10 pg, depending on the structure of a particular metabolite.

Modern coulometric arrays have up to 16 channels that detect oxidation potentials from 0 to 900 mV, usually in increments of 60 mV. In a first approximation, the peak capacity of an HPLC-coulometric system will be 16 times the peak capacity of the column(s). Most oxidizable low-molecular-weight compounds in a sample can be analyzed by coupling HPLC to a coulometric array detector. The response of each compound is monitored on three different channels, and because the relative response across those channels is compound dependent, the analysis provides qualitative information. Approximately 600–1,200 metabolites have been detected in serum and plasma by coupling gradient chromatographic systems to coulometric arrays (Matson et al., 1990; Milbury, 1997; Kristal, 2002). Most recently, the technique has been used to identify changes in the rat plasma metabolome in response to dramatic changes in dietary intake (Shi et al., 2002). From approximately 300 analytically reproducible metabolites, subsets of 37 and 63 metabolites could reliably distinguish rats fed ad libitum (AL) from dietary restricted (DR) cohorts. Examples of 16 channel array traces from analysis of AL and DR sera are shown in Figure 14-3.

NMR METHODS: DATA PRESENTATION AND INTERPRETATION

For more detailed analysis of the components present in a biofluid, the two most useful NMR approaches for dealing with the greater complexity and density of resonances are two-dimensional NMR spectroscopy (Croasmun and Carlson, 1994) and LC-NMR (Albert, 2002). For the benefit of the nonspecialist, a very brief introduction to the application of these approaches is summarized below. Two-dimensional NMR spectra provide correlation information that reveals the relationships of the resonances in a particular sample. Analysis of a contour plot, the most common format for 2D data analysis,

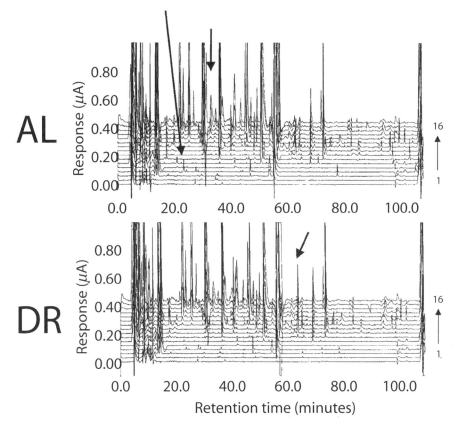

Figure 14-3 *HPLC-coulometric array analysis of plasma samples from dietary-restricted (DR) rats and ad libitum (AL) consumption rats. The 16 traces in each chromatographic experiment are the responses at 60 mV increments over a total potential of 0–900 mV. Numerous differences are apparent from inspection (arrows), and 63 metabolites were determined to be diagnostic for the cohorts via principal components analysis and hierarchical cluster analysis (Shi et al., 2002; Vigneau-Callahan et al., 2001). (Source: Reprinted with permission from American Society of Nutritional Sciences © 2001.)*

provides information such as: Do the resonances correspond to protons on a single molecule or on different molecules? Are the protons located near each other in the molecular structure (typically <3 bonds apart)? Are the protons located near each other in space (typically <5 Å)? Can we assign a particular resonance or multiplet of resonances to a particular proton in the compound?

There are many different 2D NMR experiments that can be performed, and the information content of the contour plot depends on which experiment has been used to generate it. One of the most common 2D NMR experiments is the COrrelated SpectrospY or COSY experiment, which generates off-diagonal or cross peaks for pairs of nuclei that are involved in scalar coupling interactions. For protons, this typically requires pairs of ¹H nuclei separated by two or three bonds in the molecular structure. Thus, a cross peak implies that the two protons are on the same molecule and are separated by two or three bonds. This type of information is extremely useful for assigning the resonances—that is, determining which protons in the molecule correspond to which resonances. This information supplements the more familiar chemical shift and coupling constant data available from 1D NMR spectra.

Another common type of 2D NMR experiment is the nuclear Overhauser Effect Spectroscopy, or NOESY experiment. In NOESY contour plots, cross-peak intensities depend on the distance of the pair of interacting nuclei in space, rather than on the molecular framework of the molecule. For this reason, the NOESY experiment is most useful for the assignment of isolated protons or methyl groups that are >3 bonds from other protons in the mole-cule. The NOESY experiment also provides conformational information that can be related to the rotamer populations about single bonds. Hence, this experiment is a central feature of structural NMR work used, for example, to determine the structure of proteins in solution.

A third type of 2D NMR experiment correlates different nuclear species—for example, a ^{13}C resonance with a ^{1}H resonance. While in the other exam-ples given above the two axes of the 2D NMR experiment are generally both ^{1}H, in the third type of experiment the two axes are different, corresponding to ^{13}C shifts and ^{1}H shifts. The most commonly used example of this type of experiment is the HSQC (heteronuclear single-quantum coherence) experi-ment (Croasmun and Carlson, 1994), which correlates ^{1}H with the ^{13}C shift of a scaler-coupled nucleus.

NMR flow methods can be used either to process large numbers of samples or, following the LC-NMR approach, to generate a set of spectra, ideally cor-responding to individual or a limited number of the components of a mixture (depending on how well the chromatrographic separation works). It is impor-tant to emphasize here that the NMR approach inherently achieves a spectral separation, which largely eliminates the need for physical separation, so the analysis of a mixture generally does not require any additional chromatogra-phy. However, as the number of components becomes sufficiently large, or if the resonances corresponding to some metabolites of interest are over-whelmed by the signals from more abundant compounds, it may become desirable to perform an LC-NMR experiment.

In general, LC-NMR has proven to be a powerful method for the identifi-cation of drug metabolites, and more recently, for the analysis of biofluids. Figure 14-4A illustrates a section of the pseudo-2D NMR spectrum generated using an HPLC NMR instrument applied to the analysis of whole-rat urine obtained 56 h after a hydrazine dose of 120 mg/kg (Nicholls et al., 2001). Figure 14-4B–E correspond to the 1D ^{1}H NMR spectra at each of the retention times indicated. Spectrum 14-4B contains a mixture of 2-aminoadipic acid and taurine, spectrum 14-4D contains creatinine, and spectrum 14-4E contains Nα-acetyl-L-citrulline. Spectrum 14-4C contains unidentified urinary metabolite(s). The ability to obtain separate spectra for compounds

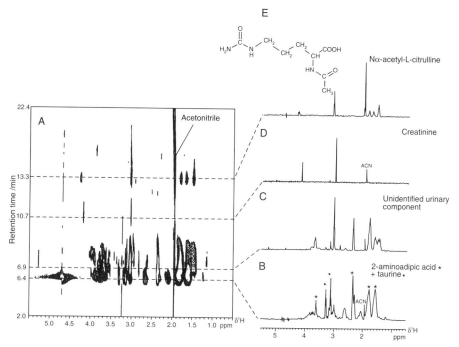

Figure 14-4 *Application of HPLC NMR to the identification of urinary metabolites. The sample was obtained from whole-rat urine 56 h after a 120 mg/kg dose of hydrazine. (A) On-flow HPLC NMR pseudo-2D ¹H NMR spectrum (500 MHz); (B) 2-aminoadipic acid plus unidentified coeluting metabolites; (C) unidentified urinary component(s); (D) creatinine; (E) Na-acetyl-citrulline (Nicholls et al., 2001). (Source: Reprinted with permission from American Chemical Society, copyright 2001.)*

such as Nα-acetyl-L-citrulline is clearly of great utility for the identification of unknown metabolites, particularly if there is significant spectral overlap with more abundant compounds.

METABOLIC MARKERS

One approach to evaluating the procedures that have been introduced is to consider the metabolite markers that have been identified following various technologies. Some of the best examples of using MS for metabolite analysis are found within clinical applications such as newborn screening. Hundreds of metabolites from human blood can be analyzed from dried blood samples collected from newborns; this includes the now validated tandem-MS analysis for several inborn errors of metabolism (Chace, 2001; Chace et al., 2002; Pitt et al., 2002).

As part of the routine screening for inborn errors of metabolism (IEM), small blood samples from newborns usually are spotted on filter paper cards known as Gutherie cards. The samples are submitted to screens for specific metabolites that are diagnostic of inherited disorders in amino metabolism such as phenylketonuria (PKU), maple syrup urine disease (MSUD), and homocysteinuria. Prior to the advances in MS technology, several different analyses, including a bacterial inhibition assay, were required to screen these diseases; in general, the methods were prone to high false positive rates, causing unnecessary and stressful treatment regimes.

Rigorous analysis for all three disorders can now be accomplished by the single tandem-MS method without the use of chromatography. For example, amino acids and other small-molecule metabolites in the spots are first extracted with methanol containing stable isotopically labeled standards; the endogenous amino acids and standards are then converted to butyl esters. The sample can be directly infused into a triple quadrupole MS; the MS/MS mode is set up to scan for the ions derived from the loss of 102 amu, butylformate (Fig. 14-5).

Defects in fatty acid beta-oxidation represent some of the most common IEMs, and at least 21 enzyme deficiencies have been correlated to clinical IEM. The increased levels of acylcarnitines that are indicative of beta-oxidation disorders are readily determined by plasma analysis using tandem-MS methods. These MS-based analyses are illustrative of metabolite screening methods that use clinically accepted metabolic markers for the diagnosis of disease. Rigorous analysis of 62 urinary metabolites, including organic acids and amino acids, was accomplished using simple sample preparation and tandem-MS (Pitt et al., 2002). The method provided quantitative results (variability <15%) for most metabolites and reliably identified 105 of the 108 individuals with IEM in the study. Similar quantitative MS methods could be utilized to correlate changes in specific plasma urinary metabolites (metabolic markers) with exposure to toxic substances.

As previously mentioned, metabolomic analysis has been applied to the investigation of toxicological responses to drug therapy. As an example, perturbations in lipid metabolism are believed to cause adverse responses to certain therapeutic agents for type 2 diabetes, particularly thiazolidinedione agonists of PPARγ (peroxisome proliferator-activated receptor γ). A study of the structural lipid metabolome in obese male mice followed phenotypic, metabolic, and gene expression changes resulting from chronic exposure to the PPARγ agonist rosiglitazone (Watkins et al., 2002). The study utilized standard GC methods to quantitatively measure lipids in plasma, liver, and heart tissue. The comprehensive lipid analysis provided detailed information on drug-induced perturbations in lipid metabolism. For example, several fatty acids were metabolic markers for increased fatty acid synthesis in liver and adipose tissue; the markers correlated well with accumulation of triglycerides in the liver and increased fatty acids in adipose tissue (Fig. 14-6). In this case, researchers discovered metabolic markers for a previously known pharmacodynamic response to rosiglitazone. However, to realize the true utility of metabolite analysis in the assessment of drug toxicity, metabolomics

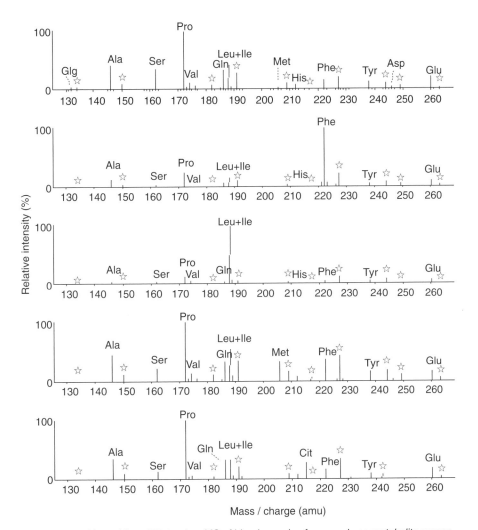

Figure 14-5 *Neutral-loss 102 tandem MS of blood samples from newborn metabolite screening. Top: control. Second: PKU. Third: MSUD. Fourth: homocystinuria. Bottom: acute citrullinemia. The stars indicate internal stable-isotope standards (Chace et al., 2002).* (Source: Reprinted with permission from the Annual Review of Genomics and Human Genetics, *Volume 3 © 2002 by Annual Reviews; www.annualreviews.org.)*

must demonstrate the ability to foretell specific tissue toxicity and/or perturbations in metabolic pathways caused by administration of new agents in development.

Toxicological studies of rodents, for which genetic and dietary factors can be fully controlled, provide the most ideal conditions for comparative biofluid analysis and the identification of metabolic markers. Table 14-2, based on a similar table compiled by Shockcor and Holmes (2002), presents a tabulation

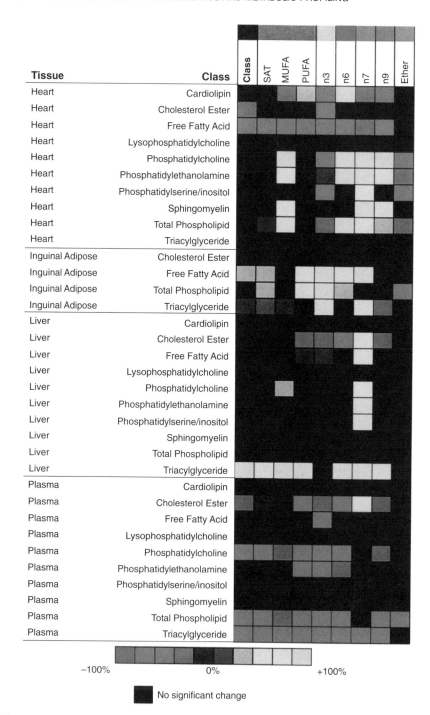

Figure 14-6 *Metabolite heat map summarizes the influence of rosiglitazone on tissue-specific lipid metabolomes (Watkins et al., 2002). (*Source: *Reprinted with permission from* Journal of Lipid Research *© 2001, Lipid Research, Inc.)*

TABLE 14-2 Metabolic markers associated with various toxins as determined via metabonomic analysis[a]

Chemical Agent	Target Organ/Toxicity Type	Associated Biomarkers
Adriamycin	Heart and kidney (glomerulus)	↑creatine, taurine, ↓citrate ↑↓αKG
Allyl alcohol	Liver (periportal)	↑creatine, lactate, phenylacetylglycine, N-methyl nicotinamide, taurine ↓citrate, αKG
Amiodarone	Phospholipidosis (lung)	↑ phenylacetylglycine, DMG
α-Naphthylisothiocyanate (ANIT)	Liver (cholestasis)	↑acetate, bile acids, glucose ↓citrate, hippurate, αKG, succinate
2-Bromoethanamine	Kidney (papilla) and mitochondrial dysfunction	↑adipic acid, DMG, glutaric acid, N-acetylglycine ↓↑succinate, TMAO
Butylated hydroxytoluene	Liver	↑glucose, taurine
Cadmium chloride	Testicular	↑creatine, glucose ↓citrate
Carbon tetrachloride	Liver	↑taurine, creatine ↓citrate, αKG, succinate
2-Chloroethanamine	Kidney (papilla) and mitochondrial dysfunction	↑adipic acid, DMG, glutaric acid, N-acetylglycine ↓↑succinate, TMAO
Chloroquine	Phospholipidosis and liver necrosis	↑phenylacetylglycine, DMG
S-(1,2-dichlorovinyl)-L-cysteine (DCVC) z4	Kidney (S2/3 proximal tubular)	↑citrate, succinate
S-(1,2-dichlorovinyl)-L-homocysteine (DCVHC)	Kidney (S2/3 proximal tubular)	↑acetate, amino acids, glucose, organic acids ↓citrate, creatinine, hippurate, αKG, succinate
Ethionine	Liver	↑glucose, taurine ↓αKG
Galactosamine	Liver (hepatitis-like lesion)	↑acetate, betaine, bile acids, creatine, organic acids, taurine, urocanic acid ↓hippurate, αKG, succinate
Haxachlorobutadiene	Kidney (S3 proximal tubular)	↑acetate, amino acids, glucose, organic acids ↓citrate, creatinine, hippurate, αKG, succinate
Hydrazine	Liver (steatosis)	↑2-amino adipate, β-alanine, creatine, N-acetyl-citrulline ↓creatinine, fumarate, hippurate, TMAO
Lanthanum nitrate		↑acetoacetate, alanine, aromatic amino acids, DMA, ethanol, glucose, hippurate, αKG, lactate, succinate, taurine, TMAO ↓allantoin, citrate, creatinine, glucose, urea

TABLE 14-2   *Continued*

Chemical Agent	Target Organ/Toxicity Type	Associated Biomarkers
Lead acetate	Liver, lung, and kidney	↑acetate, creatine, glucose, lactate, N-acetyls, taurine ↓citrate, hippurate, αKG, succinate
Mercuric chloride	Kidney (S3 proximal tubular)	↑acetate, amino acids, glucose, organic acids ↓citrate, creatinine, hippurate, αKG, succinate
Paracetamol (N-acetyl-*p*-aminophenol)		↑pyroglutamate
p-aminophenol	Kidney (S3 proximal tubular)	↑acetate, amino acids, glucose, organic acids ↓citrate, creatinine, hippurate, αKG, succinate
Paraquat	Kidney and lung	↑amino acids, glucose, lactate, organic acids ↓citrate, creatinine, hippurate, valine
Propyleneimine	Kidney (papilla)	↑N-acetylglycine ↓↑αKG, TMAO
Puromycin aminonucleoside	Kidney (glomerulus and proximal tubular)	↑acetate, alanine, creatine, formate, glucose, macromolecules (proteins and lipids), taurine, TMAO ↓citrate, αKG
Sodium chromate	Kidney (S1 proximal tubular)	↑glucose ↓citrate, hippurate, αKG
Sodium fluoride	Kidney (proximal tubular)	↑acetate, amino acids, glucose, organic acids, threonine ↓citrate, creatinine, hippurate, αKG, succinate
TCTFP	Kidney (S3 proximal tubular)	↑acetate, amino acids, glucose, organic acids ↓citrate, creatinine, hippurate, αKG, succinate
Thioacetamide	Liver and kidney	↑acetate, amino acids, creatine, glucose, organic acids, taurine, TMAO ↓citrate, hippurate, αKG, succinate
Tris(2,3-dibromopropyl) phosphate	Kidney (proximal tubular)	↑glucose, lactate
Uranyl nitrate	Kidney (S3 proximal tubular)	↑acetate, amino acids, glucose, organic acids, urea ↓citrate, creatinine, hippurate, αKG, succinate

[a] This table is based on table 1 from Shockcor and Holmes, 2002. We dropped some entries corresponding to species other than rats, and included data reported for paracetamol (Ghauri et al., 1997), La(NO₃)₃ (Feng et al., 2002), and tris(2,3-dibromopropyl)phosphate (Fukuoka et al., 1987).
αKG, α-ketoglutarate; DMA, dimethylamine; DMG, dimethylglycine; TMAO, trimethylamine oxide.

of some of the metabolic markers that have been identified based on NMR studies of urine and in some cases other biofluid samples. This table illustrates the emergence of some patterns as well as some of the limitations of urinary biomarker analysis.

The TCA intermediates citrate, alpha-ketoglutarate (αKG), succinate, and in a few cases fumarate, constitute approximately 25% of the identified bio-markers listed in table 1 of Shockcor and Holmes (2002). Measurements of these compounds have proven to be exceedingly variable. For example, in a study of the toxicity of $HgCl_2$, Nicholson et al. (1985) initially observed con-siderable elevations of succinic and acetic acids, while a subsequent study did not observe elevations of these metabolites (Gartland et al., 1989a). It was suggested that differences in the fasting protocol might be responsible for this discrepancy. However, the urinary ^1H spectra from both studies exhibit large acetate and lactate resonances (Fig. 14-7). The Shockcor and Holmes table lists

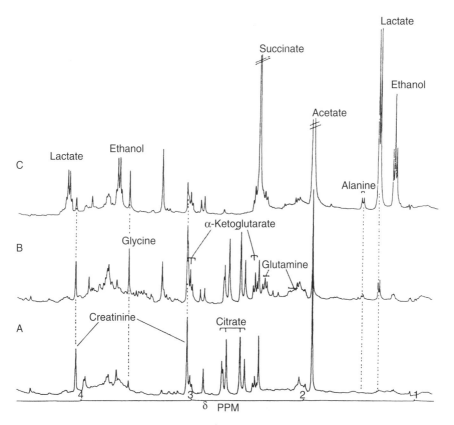

Figure 14-7 *Upfield region of the 400 MHz ^1H NMR spectra of urine obtained from fasted rats exposed to $HgCl_2$. (A) control; (B) 0–24 h after 1 mg/kg $HgCl_2$, an acute nephrotoxin damaging the pars recta of the proximal tubule; (C) 0–24 h after 2 mg/kg $HgCl_2$. (Source: Reprinted with permission from Nicholson et al., 1985.)*

acetate as a biomarker but does not specifically mention lactate, despite its apparent presence as a dominant metabolite in both studies.

Another important factor that can influence the excretion of TCA intermediates is renal carbonic anhydrase activity. Nicholson and coworkers discuss the possibility of false positive readings, citing the example of the carbonic anhydrase inhibitor acetazolamide (Nicholson et al., 2002). Administration of this inhibitor massively reduces the excretion of TCA intermediates, apparently by interfering with renal carbonic anhydrase. Since as noted in Table 14-2, altered secretion of TCA intermediates represents an important biomarker for many of the toxicants that have been studied, Nicholson et al. (2002) propose using supervised methods of analysis that include models of renal acidosis to avoid misinterpreting these data. They also discuss false negative readings that can occur when toxicants cause a significant metabolic perturbation without producing much tissue damage. Such effects are more common for low-potency toxicants.

Beyond the variability of TCA metabolites already noted, we unexpectedly found many apparent discrepancies between the biomarker patterns listed in the review of Shockcor and Holmes and spectra or conclusions presented in the primary literature. For example, the biomarkers listed for amiodarone and chloroquine toxicity in the Shockor and Holmes table differ qualitatively from those given in table 1 of Espina et al. (2001). In a study of uranyl nitrate, Anthony et al. (1994) noted the striking appearance of high concentrations of 3-D-hydroxybutyrate and parallel absence of acetoacetate and acetone in rat urine samples, and suggested this pattern could provide a useful biomarker for proximal tubular damage. However, the tabulation of Shockcor and Holmes does not specifically mention 3-D-hydroxybutyrate as a biomarker for uranyl nitrate toxicity, although this might be included as one of the organic acid markers. Effects of a $20\,\mu l/kg$ dose of propylene imine on rat urine illustrated in Figure 14-8 (Gartland et al., 1989a) do not correspond closely with the data in Table 14-2. The principal metabolites observed after long exposure times that are apparent from the spectra (glucose/sugars, succinate, acetate, lactate, alanine) are not specifically mentioned in the Shockcor/Holmes table. Ethanol, a compound that appears in the ^{1}H NMR spectra of urine treated with a number of different compounds such as $HgCl_2$ (Fig. 14-7) and $La(NO_3)_3$ (Table 14-2), is also not listed in their table, perhaps because it originates from gut bacteria. The presence of ethanol in the urine is apparently indicative of a loss of alcohol dehydrogenase (Nicholson et al., 1985) and perhaps cytochrome P450 activity, since the latter is also involved in ethanol metabolism (McGehee et al., 1994). The study of Bairaktari et al. (1998) appears to indicate that valine rises rather than falls in response to paraquat. Finally, from the recent review by Lindon et al. (2000), the discriminatory metabolites for Hg toxicity include valine, taurine, TMAO, and glucose; for bromoethanamine, the discriminatory metabolites include acetate, methylamine, dimethylamine, lactate, and creatine. A comparison of Lindon's discriminatory metabolites

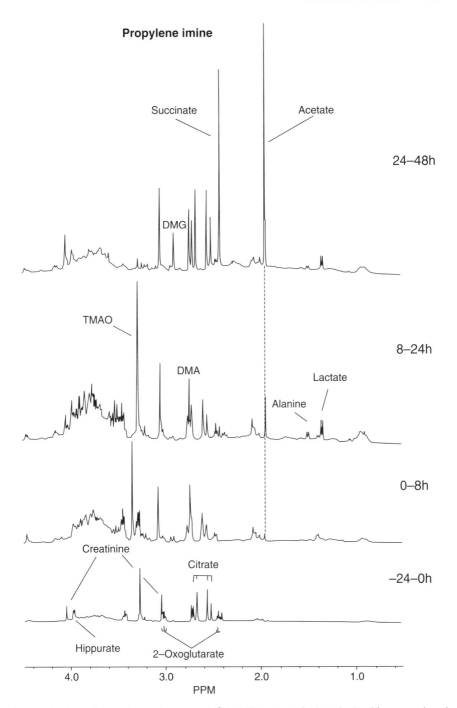

Figure 14-8 *Upfield region of the 400MHz ¹H NMR spectra of urine obtained from rats dosed with 20μl/kg propylene imine. The lower spectrum corresponds to a predose control, and the upper spectra to samples obtained over the times indicated. (*Source: *Reprinted with permission from Gartland et al., 1989a.)*

with the listings in Table 14-2 (Table 1 of Shockcor and Holmes, 2002) reveals significant differences.

These apparent inconsistencies may arise because more recent studies based on larger samples and/or different protocols have found metabolic correlations that differ from those observed in the earlier literature. However, recent presentations are sparse in spectral and quantitative data, choosing instead to emphasize principal component analysis (PCA) that demonstrates data clustering behavior for chemically dosed animals (Holmes et al., 1998). In addition, it may be that that the metabolic markers identified in table 1 of Shockcor and Holmes are derived from studies other than urine, such as the more comprehensive tissue analysis of the effects of alpha-naphthylisothio-cyanate (ANIT; Waters et al., 2002). The latter conclusion has significant implications for the use of urine NMR spectra only as a basis for the analysis of organ toxicity.

In contrast to the problems already noted using TCA intermediates as metabolic markers, Ghauri et al. (1993) have identified pyroglutamate (5-oxo-proline) as a marker for the toxicity of high doses of paracetamol (N-acetyl-p-aminophenol). The accumulation of this compound is apparently related to an interference with the metabolism of sulfur-containing amino acids, since it can be prevented by coadministration of methionine (Joo, 1993). Thus, it was proposed that the 5-oxoprolinuria resulted from the lack of sufficient dietary sulfur-containing amino acids to compensate for the loss of sulfur as sulfate and cysteine conjugates of paracetamol. The resonances for both, as well as the paracetamol glucuronide, were also observed in the ^{1}H spectrum of the urine.

In any case, these observations suggest that 5-oxoprolinuria may similarly prove to be a biomarker for other chemical agents that have the effect of depleting the levels of sulfur-containing amino acids. Recent NMR studies of hydrazine-treated rats have also found increases in 2-aminoadipate, β-alanine, citrulline, N-α-acetylcitrulline, and argininosuccinate, as well as the more common metabolites (Nicholls et al., 2001). As discussed previously, the identification of these biomarkers was based on the use of LC-NMR methodology (see Fig. 14-4). Hence, advances in technology may expand the ability of this approach to identify new and more unique biomarkers.

METABOLITE CONTRIBUTIONS OF DIET AND INTESTINAL FLORA

Considering the organism as a collection of interacting, metabolically active organs, it is necessary to include the intestinal flora as well. Inclusion of the microbial metabolic pool is inherently more problematic, since there is a greater possibility for significant variation. Further, intestinal bacteria may exhibit variable sensitivity to drugs or toxins that are under study. In general, the contributions of intestinal flora to urinary metabolites will be dependent on the particular metabolite under consideration.

As one example, recent metabonomic studies indicate that trimethylamine oxide (TMAO) is a common biomarker for renal toxicity (e.g., see Table 1 of Shockcor and Holmes, 2002). Significant variations in urinary TMA and TMAO levels also have been reported to allow differentiation between two strains of mice (Gavaghan et al., 2000). As illustrated in Figure 14-9 (Smith et al., 1994), both dietary choline levels and intestinal bacteria play a major role in methylamine metabolism. The formation of TMA from choline appears to occur primarily in the gut bacteria (Smith et al., 1994), while the subsequent oxidation to TMAO probably occurs predominantly in the liver. The formation of TMAO thus provides a particularly good illustration for the complex metabolic interrelationships that must be interpreted in order to make the most effective use of metabolomic analysis.

Phenylacetate, a product of phenylalanine and phenethylamine catabolism, has also been observed in urine and proposed to be correlated with depression and other psychiatric illnesses (Gonzalez-Sastre et al., 1988; Faull et al., 1989; McGregor et al., 1996). The levels of phenylacetate and various phenylacetate conjugates in urine have been shown to be strongly affected by oral antibiotic administration (Hryhorczuk et al., 1984; Brugues and Sastre, 1986), again highlighting the role that intestinal bacteria can play in phenylalanine metabolism. Phenylacetylglycine has also been identified as a biomarker for the toxicity resulting from chloroquine (Espina et al., 2001). Hence, an evaluation of the contributions of intestinal bacterial to phenylacetate metabolism

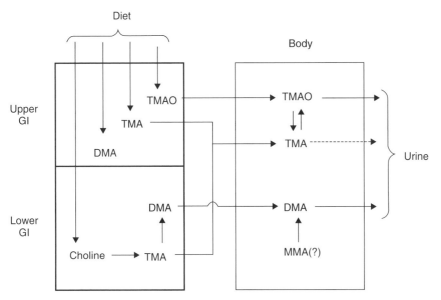

Figure 14-9 *Summary of key processes in the metabolism and excretion of methylamines illustrating contributions of flora in the upper and lower GI tracts, as well as the animal. The possible synthesis of dimethylamine (DMA) from monomethylamine (MMA) is suggested but not demonstrated at present.* (Source: *Reprinted from Smith et al., 1994.*)

is important for the interpretation of levels appearing in various biofluids. Contributions of gut bacteria to hippurate and other aromatic compounds found in the urine have also been noted (Phipps et al., 1998). In general, understanding the complex metabolic interrelationships between the host organism and gut bacteria is central to the analysis of metabolites in various body fluids.

Small-bowel bacterial overgrowth (SBBO) is frequently encountered in patients with advanced chronic kidney failure (Dunn et al., 1998). This condition contributes to the pathophysiology and decreased nutritional status of chronic dialysis patients. It is associated with elevated levels of dimethylamine (DMA) and its carcinogenic metabolite, nitrosodimethylamine (NDMA). Hence, elevated urinary DMA and NDMA levels may serve as markers for this condition. In general, a renal toxicant might produce SBBO and its attendant metabolic consequences over a period of time, or the toxicant might interfere with the growth of intestinal bacteria, having the opposite effect. Hence, it is in general important to determine how a particular compound under study is affecting intestinal flora.

As noted, levels of DMA provide one marker for the metabolic activity of intestinal bacteria, but these levels are also strongly dependent on dietary choline and potentially on other compounds that could serve as precursors to methylamine synthesis. In the metabonomic literature there has been relatively little discussion of the extent to which SBBO may influence the metabolite compositions. The observation of elevated DMA and TMAO levels in rats treated with $La(NO_3)_3$ over a period of 3 months (Feng et al., 2002) is consistent with an SBBO effect. Further, urine samples derived from studies of rats treated with $HgCl_2$ (Nicholson et al., 1985) and $La(NO_3)_3$ (Feng et al., 2002) exhibit significant concentrations of ethanol (Fig. 14-7). It has been proposed that this elevation results from a loss of alcohol dehydrogenase activity. However, SBBO secondary to loss of renal function may also be a contributory factor in such studies.

ISOTOPIC ENRICHMENT AND METABOLIC FLUX

Although characterization of the metabolite concentrations is the typical objective of metabolic analysis, studies designed to determine metabolic flux can be particularly informative. Indeed, the contributions of various pathways may change dramatically under conditions in which concentrations of particular metabolites remain relatively unperturbed. Such effects occur due to the great redundancy of most complex biological systems. Metabolic flux analysis involves the use of isotopically labeled precursors and analysis of the labeling pattern of particular metabolic products. In general, isotopic labels are distributed at different positions in the metabolite under study. The set of compounds that are isostructural but have variable isotopic composition are called isotopomers.

We emphasize that the use of the stable ^{13}C isotope in metabolic tracer studies differs conceptually from the use of ^{14}C in several important ways: (1) for the stable isotope experiment, the information of greatest interest is typically the label distribution, rather than the specific activity of a labeled compound or a labeled position in the molecule; and (2) while the ^{14}C background is essentially nonexistent, the natural abundance of 1.1% for ^{13}C must often be taken into consideration. NMR provides an extremely useful approach for the determination of the isotope distribution in a given compound, since each isotopomer exhibits a unique NMR spectrum, although in general some isotopomers are more easily resolved than others. While isotopomer composition is typically obtained by direct one-dimensional ^{13}C NMR spectroscopy, a two-dimensional J-resolved HSQC spectrum provides a considerably more sensitive and rapid means of obtaining this information (Burgess et al., 2001).

MS, also widely applied to isotopomer analysis, has been used to study lipid, carbohydrate, amino acid, and steroid metabolism (Szafanek et al., 1974; Tserng et al., 1984; Katz et al., 1989; Lee et al., 1991). A recent study of citric acid cycle flux in perfused rat hearts provided a detailed comparison of three techniques for isotopomer analysis: full-scan MS, tandem MS, and ^{13}C NMR (Jeffrey et al., 2002). The flux of carbon units from three ^{13}C-labeled precursors, [2-^{13}C]acetate, [1-^{13}C]glucose, and [3-^{13}C]pyruvate, was followed by determining the isotopomer distribution of glutamate, a transamination product of the citric acid cycle intermediate α-ketoglutarate. Tandem MS of a glutamate derivative proved to be at least as accurate as NMR for the analysis of the flux through the citric acid cycle; full-scan MS data were more prone to significant error.

Given its greater sensitivity with respect to NMR, tandem MS could potentially be used to follow flux through a pathway by isotopomer analysis of a single low-abundance biomolecule. Applying a traditional GC-MS approach to isotopomer analysis in a modern gene knockout model, Kurland and colleagues (Xu et al., 2002) measured the flux of murine glucose metabolism during and after infusion of ^{13}C-labeled precursors into wild-type and PPARα-KO mice. The isotopomer flux approach clearly revealed previously unobserved defects in the glucose metabolism of the PPARα-KO mouse, especially in the fasted state; lactate production and conversion of lactate to glucose were both dramatically impaired. Simply measuring metabolite levels in response to feeding and/or insulin treatment would provide much less insight into the mechanisms of metabolic deficiencies.

As an illustration of the NMR approach, Carvalho et al. (2002) have studied the effects of CCl_4 on metabolic flux parameters in the liver of Sprague-Dawley rats. These studies utilized [U-^{13}C]propionate as a labeling precursor, which has been exploited for studies of gluconeogenesis and Krebs cycle activity in both rats and humans (Jones et al., 1998, 2001). This compound functions as an anaplerotic precursor that is converted into [1,3,4-$^{13}C_3$]succinyl-CoA, a TCA intermediate. Plasma glucose becomes labeled with ^{13}C as a consequence of gluconeogenic outflow via phosphoenolpyruvate. A simplified diagram of hepatic metabolism showing anaplerotic and gluconeogenic pathways is shown in Figure 14-10.

Analysis of the ^{13}C labeling pattern of metabolites in a hepatic extract yielded information on the effects of CCl_4 on the flux parameters. These flux parameters can be converted to absolute values when integrated with an independent measurement of absolute flux through any one of the pathways. Using the approach outlined, Carvalho et al. found that in both fed and 24 h fasted male Sprague-Dawley rats, CCl_4 administration caused a significant increase in relative gluconeogenic flux, although net hepatic glucose output was reduced. Absolute citrate synthase flux dropped by 47%, indicating that oxidative Krebs cycle flux was highly susceptible to CCl_4 injury. The reduction in absolute fluxes indicates a significant loss of hepatic metabolic capacity, while the relative maintenance of gluconeogenic fluxes indicates a reorganization of metabolic activity toward preserving hepatic glucose output. Recycling of phosphoenolpyruvate (PEP) via pyruvate and oxaloacetate was extensive under all conditions and was not significantly altered by CCl_4 injury.

In the above study, most of the information on hepatic metabolism was derived from NMR analysis of liver extracts. Chemical biopsy approaches represent an alternative and less invasive approach for obtaining similar information (Beylot et al., 1996). In a chemical biopsy a xenobiotic capable of forming an adduct with a cellular metabolite is administered subsequent to

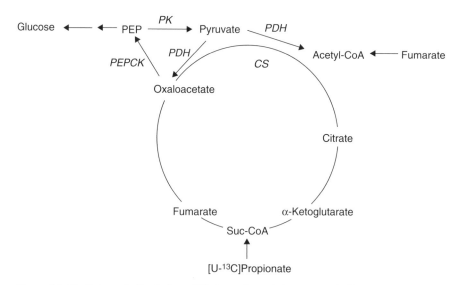

Figure 14-10 *Schematic illustration of Krebs cycle and gluconeogenic fluxes. The enzymatic fluxes indicated for CS (citrate synthase), PDH (pyruvate dehydrogenase), PEPCK (phospho-enolpyruvate carboxykinase), and PK (pyruvate kinase) represent effective, and in some cases multistep conversions. Isotopomer distributions derived from observations of the labeling in glutamate and glucose are used to determine these parameters.*

the addition of a labeled precursor. The adduct is subsequently excreted, and the labeling pattern in the metabolite can then be analyzed. For example, paracetamol (Tylenol) forms a glucuronide conjugate in the liver, which is subsequently found in the urine. The labeling of the excreted glucuronide then reports on the labeling of the hepatic glucose pool (Jones et al., 1998). Similarly, the labeling of hepatic glutamine can be monitored by administration of phenylacetate, which forms a glutamine conjugate. The labeling of cytosolic acetyl-CoA has been monitored by administration of sulfamethoxazole. Of course, in contrast with in vivo spectroscopic methods, only a time-averaged result is obtained.

Deuterium is a popular labeling tool for metabolic studies and is most often detected by mass spectrometry. The ^{2}H isotope has been less frequently used for metabolic flux analysis than ^{13}C, since as a spin 1 nucleus it is subject to resonance broadening by the quadrupolar relaxation mechanism. However, in a selected number of instances, the metabolism of deuterated substrates has been followed directly using ^{2}H NMR. An attractive aspect of this approach is the relative ease of such studies and the low natural abundance of deuterium (1.56×10^{-2}% compared, for example, with 1.1% for ^{13}C).

The metabolism of [methyl-^{2}H$_3$]methionine provides an interesting application of the analysis of hepatic metabolism by ^{2}H NMR. In this case, the low natural abundance and rapid relaxation behavior of the deuterium make the experiments particularly easy to implement. In vivo ^{2}H NMR studies of the metabolism of both L- and D-[methyl-^{2}H$_3$]methionine have been reported in which the biotransformations were monitored using a surface coil placed over the liver of an anesthetized rat (London et al., 1987; London and Gabel, 1988). In the study of L-methionine, the rapid transmethylation reaction leading to the production of [methyl-^{2}H$_3$]sarcosine was observed (Fig. 14-11). Such a study is in effect monitoring the activities of methionine adenosyl transferase and glycine N-methyl transferase.

Over longer time periods, a number of other metabolites become deuterated at rates related to the activities of the enzymes involved in these processes. Some of the deuterated sarcosine produced is oxidized in the mitochondrion via sarcosine dehydrogenase, with the label ultimately ending up as deuterated water, while some may be excreted. Interestingly, an analogous study using the deuterated D-methionine precursor gave essentially equivalent results, demonstrating the rapid conversion of the D- to the L-isomer (London and Gabel, 1988). A comparison of the metabolism of L- and D-[methyl-^{2}H$_3$]methionine showed that the results could be influenced by treatment with sodium benzoate, a known inhibitor of D-amino acid oxidase.

The conversion of D- to L-methionine involves this enzyme followed by reamination catalyzed by the enzyme glutaminase II. Thus the effects of a specific chemical agent on these pathways could be studied using this technique. A reduction in the rates of L-methionine clearance by benzoate was also noted, possibly resulting from a reduction in glycine levels as the latter was used for formation of hippurate from benzoate. This would reduce the

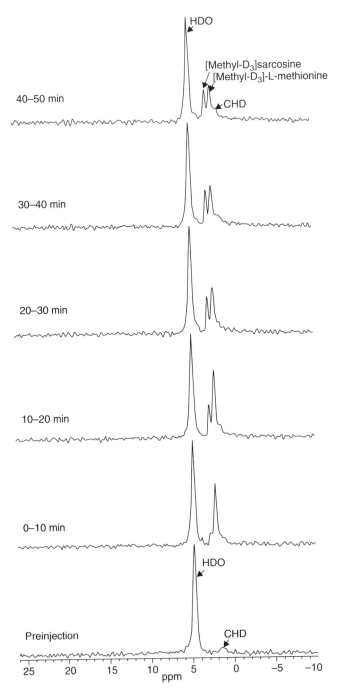

Figure 14-11   *A series of ^{2}H NMR spectra illustrating the initial metabolism [methyl-^{2}H$_3$]methionine in the liver of an anesthetized rat. The spectra were obtained using a 1.3 cm surface coil placed above the liver (London et al., 1987). (*Source: Reprinted with permission from American Chemical Society, copyright 1987.)*

availability of glycine for the glycine N-methyl transferase reaction. Although the reported in vivo studies utilized a surface coil approach, deuterated metabolites were observed in a hepatic extract and presumably could also be monitored in urine by ^{2}H NMR. Deuterium NMR spectroscopy has also been used to monitor the metabolism of deuterated drugs (e.g. Farrant et al., 1993; Akira et al., 1994).

As a final example, we consider the use of deuterated water to evaluate the metabolic flux leading to the production of blood glucose. In this type of approach, ^{2}H NMR is used to determine the incorporation of deuterium derived from tracer levels of deuterated water into the various positions of glucose. This incorporation arises not from the direct conversion of a labeled precursor but from enzymatic or tautomeric equilibria, such as the triose phosphate isomerase reaction that introduces a deuteron into the C-2 position of glyceraldehyde-3-phosphate:

Human studies have utilized 2.5 g D_2O/kg body water, followed by 0.5% ^2H_2O *ad libitum* until completion of the study (Jones et al., 2002). In this type of study, blood samples are obtained from the subject and ^{2}H NMR spectra such as those shown in Figure 14-12 are obtained for a derivatized form of the glucose (monoacetone glucose), which has been shown to provide excellent resolution of the deuterium NMR signals. The relative intensities of the deuterium resonances at positions 2, 5, and 6_S are used to determine the contributions of glycogen, glycerol, and PEP to the glucose pool as illustrated in Figure 14-12, according to the relations

$$\text{Glucose fraction from Glycogen} = 1 - (H5/H2)$$

$$\text{Glucose fraction from glycerol} = (H5 - H6_S)/H2$$

$$\text{Glucose fraction from Krebs cycle} = H6_S/H2.$$

As in most metabolic studies, this interpretation is predicated on a number of assumptions, particularly complete equilibration of glucose-6-phosphate isomerase, triose phosphate isomerase, and fumarase reactions, and the absence of glycogen cycling. Although the emphasis of such studies to date has been for clinical evaluation, toxins that affect glycogen mobilization or other points in the gluconeogenic process can be evaluated using this type of experiment.

DATA ANALYSIS

Classically, analytical chemists utilize the concentration-dependent response of authentic standards to measure levels of metabolites in biological systems. However, the standard curve approach is not practically applicable to comprehensive analysis of the inherently complex metabolome. In order to facilitate a high-throughput approach to metabolic characterization, a substantial effort has been directed toward automated data analysis approaches that

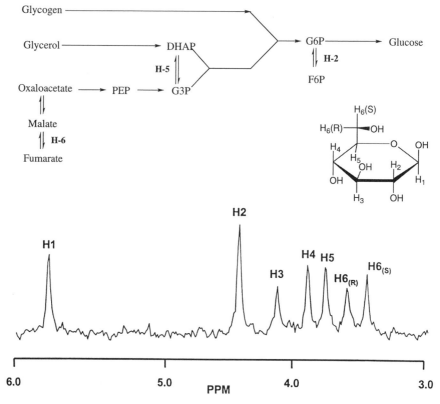

Figure 14-12 Schematic illustration of gluconeogenesis illustrating the steps at which deuterium is introduced into the precursor molecules. A ^2H NMR spectrum of monoacetone glucose illustrating the use of this approach for the analysis of ^2H incorporation at the various positions of the glucose molecule is shown. (Source: Spectra and diagram derived from the "Symposium: Understanding Metabolism in the Mouse," Thursday, May 2, 2002, presented by the Southwestern Biomedical Magnetic Resonance Facility, Southwestern Medical School.)

utilize pattern recognition software (e.g., Lindon et al., 2001). The output from the measurements is a multidimensional data set: for example, a set of intensities as a function of chromatographic retention time, mass, or NMR chemical shift.

Study of temporal changes in metabolite pools will add a time dimension, as well as a substantial amount of experimental time to the project. The scale of the data reduction challenge can be exemplified by HPLC-MS analyses. A chromatographic run time of 20 minutes, at an acquisition rate of 5 spectra per second in a TOF instrument, would produce 6,000 individual MS spectra, comprising 50,000 points each; the challenge is to extract all the real MS responses of potentially >1,000 metabolites in each chromatogram. The amplitudes of those responses must then be compared across the set of biological samples. One way to visualize and analyze HPLC-MS data is to create contour plots of run time versus m/z versus amplitude. Differences in the contours from sample to sample would reflect changes in metabolite levels.

In general, the acute effects of a chemical agent may be very different from the chronic effects, so in some cases short-term changes in metabolite levels are opposite those observed over long time periods. Differences in the time-dependent response may allow separations that cannot be made based on fixed-term studies only. A general problem that characterizes these data sets is that small differences in conditions—for example, sample pH or salt content—make comparisons among individual subjects difficult. In the case of NMR, this analytical problem has been approached by binning the data (Lindon et al., 2001).

Although some alternative approaches, such as the partial linear fit approach of Vogels et al. (1996), have been applied to the analysis of complex mixtures, this approach has not be widely applied to the comparative analysis of biofluids. The binning approach effectively transforms a high-resolution spectrum into a low-resolution data set; it is equivalent to directly obtaining a low-resolution spectrum (a typical resolution used for this type of analysis is 0.04 ppm = 24 Hz at an operating frequency of 600 MHz) but has the advantage that the raw data actually contain more information that can be accessed at a later point in the analysis. This procedure results in a great loss of information content but greatly facilitates data comparisons among samples using pattern recognition programs (Lindon et al., 2001).

In some situations, the loss of information may lead to serious inaccuracies. As an extreme example, a perturbation that alters the ratio of two similar metabolites with similar subspectra will be lost if the two metabolites are not resolved in the binned spectra. Indeed, the loss of fine structure is in many ways equivalent to the difficulties that have been encountered from in vivo metabolic analysis by NMR, in which poor resolution inherent in the analysis of intact tissues significantly limits the information content of the spectra.

For the purpose of classifying samples and detecting a toxic response, principal component analysis (PCA) as applied by Nicholson and coworkers (Lindon et al., 2001) has been the most popular approach. PCA transforms the original variables into a set of uncorrelated variables called the principal components (PCs). The PCs are linear combinations of the original variables and do not correspond to particular metabolites. The first PC contains the largest part of the variance for a given data set, with subsequent components characterized by progressively smaller degrees of variance. A multidimensional scores plot with axes PC1, PC2, PC3, and so forth, is typically presented to highlight the differences among various groups of animals. Such presentations ideally exhibit clusters of data points that theoretically can be used to classify the type of toxic response resulting from a particular treatment protocol.

Time-dependent presentations are often developed, since the progress of a pathological response may be characterized by a complex, time-dependent set of metabolic changes. It is possible to retrieve more intuitive spectral information with a loadings plot, which identifies spectral regions responsible for differentiation between treated and control populations. This plot is somewhat analogous to a subtraction of spectra for treated and untreated animals but contains the statistical results for the entire group. A more extensive discussion of data-handling approaches is provided by Lindon et al. (2001).

SUMMARY

The screening of biofluids to obtain metabolic profiles has been a central feature of efforts to detect inborn errors of metabolism for many years. These clinical applications have historically necessitated quantitative analysis of specific metabolites or classes of metabolites. Modern analytical tools are now being used to discover metabolic markers of toxic and therapeutic responses. These recent attempts to develop a more comprehensive, high-throughput approach for the characterization of large numbers of metabolites have utilized mass spectrometry and NMR spectroscopy as well as modifications of these approaches and alternative approaches such as HPLC-coulometric array detectors. Characterization of the data generated by these studies increasingly relies on pattern recognition approaches to generate visual representations that ideally can provide a clear indication of how significantly the sample deviates from normal controls and can allow classification of a particular type of abnormality or toxicity.

The NMR literature in particular has tended to evolve toward more graphic presentations of data and away from direct information quantifying the specific metabolite changes that underlie these studies. This change is manifest in many of the recent papers in this area, as well as in the apparent inconsistencies that we have found between some of the biomarkers tabulated in a recent review and the available literature presenting ^{1}H NMR spectral data for biofluids—usually urine—for the same chemical agents. These discrepancies may arise because the database goes beyond the published literature on urinary metabolites.

In any case, newer instrumental capabilities, such as use of HPLC-NMR to separate and characterize the metabolites in biofluids, will continue to provide greater insight into toxicological response to industrial and environmentally significant chemicals and drugs. Several toxicological studies have noted that metabolic perturbations preceded histological changes, indicating the promise of such approaches for monitoring effects such as the early response to drug therapy. Clearly, the possibility of identifying unique individual responses before a significant and potentially irreversible toxic response develops offers great biomedical potential.

In particular, metabolomic studies are now being applied to the drug development process. As evidence for this, numerous small companies (at least 10) have been formed to provide a variety of metabolomic services to larger drug development organizations. One goal of these companies is to rapidly obtain early metabolic markers of drug toxicity (or safety) that can be used to prioritize or eliminate compounds prior to expensive safety assessment in animals and humans.

Another objective is to discover metabolic markers of therapeutic responses in humans; potentially metabolic signatures could be used as pharmacodynamic end points or for distinguishing responders from nonresponders. Essentially, these companies are expecting that metabolomics will accelerate drug discovery and ultimately improve chances of clinical success. A challenge for these applications will be clearly distinguishing "normal" metabolite profiles from perturbed profiles.

In clinical settings, metabolic patterns presented by PCA analyses have yet to be proven as rigorously predictive of events that are manifested as true therapeutic or toxic effects. Moreover, use of metabolic patterns in support of drug safety or efficacy will require educating regulatory agencies before clinical use is widely accepted. From a more fundamental perspective, a more extensive understanding of metabolomes ultimately complements the genomic and proteomic data becoming available so that a more complete understanding of the interactive nature of cellular response to chemical agents can be developed.

REFERENCES

Acworth IN, Bowaers M (1997). In: IN Acworth, M Naoi, H Parvez, S Parvez (eds), *Progress in HPLC-HPCE, Coulometric Electrode Array Detectors for HPLC*, vol 6. Utrecht, The Netherlands: VSP, pp 169–290.

Akira K, Farrant RD, Lindon JC, Caddick ST, Nicholls AW, Nicholson JK (1994). High-field deuterium nuclear magnetic resonance spectroscopic monitoring of the pharmacokinetics of selectively deuterated benzoic acid in man. *Anal Biochem* **221**:297–302.

Albert K (ed) (2002). *On-line LC-NMR and Related Techniques.* The Atrium, Southern Gate, Chichester, West Sussex, England: John Wiley & Sons.

Anthony ML, Gartland KPR, Beddell CR, Lindon JC, Nicholson JK (1994). Studies of the biochemical toxicology of uranyl-nitrate in the rat. *Arch Toxicol* **68**:43–53.

Bairaktari E, Katopodis K, Siamopoulos KC, Tsolas O (1998). Paraquat-induced renal injury studied by ^{1}H NMR spectroscopy of urine. *Clin Chem* **44**:1256–1261.

Beylot M, Peroni O, Diraison F, Large V (1996). New methods for in vivo studies of hepatic metabolism. *Reprod Nutr Develop* **36**:363–373.

Brugues JM, Gonzalez-Sastre F (1986). Influence of intestinal flora on the elimination of phenylacetic acid in urine. *Clin Chem* **32**:223.

Burgess SC, Carvalho RA, Merritt ME, Jones JG, Malloy CR, Sherry AD (2001). C-13 isotopomer analysis of glutamate by J-resolved heteronuclear single quantum coherence spectroscopy. *Anal Biochem* **289**:187–195.

Carvalho RA, Jones JG, McGuirk C, Sherry AD, Malloy CR (2002). Hepatic gluconeogenesis and Krebs cycle fluxes in a CCl4 model of acute liver failure. *NMR Biomed* **15**:45–51.

Chace DH (2001). Mass spectrometry in the clinical laboratory. *Chem Rev* **101**:445–477.

Chace DH, Kalas TA, Naylor EW (2002). The application of tandem mass spectrometry to neonatal screening for inherited disorders of intermediary metabolism. *Annu Rev Genomics Hum Genet* **3**:17–45.

Croasmun WR, Carlson RMK (1994). *Two-Dimensional NMR Spectroscopy: Applications for Chemists and Biochemists.* New York: VCH Publishers.

Cutler P, Bell DJ, Birrell HC, Connelly JC, Connor SC, Holmes E, Mitchell BC, Monte SY, Neville BA, Pickford R, Polley S, Schneider K, Skehel JM (1999). An integrated proteomic approach to studying glomerular nephrotoxicity. *Electrophoresis* **20**: 3647–3658.

Davis JM, Giddings JC (1983). Statistical theory of component overlap in multicomponent chromatograms *Anal Chem* **55**:418–424.

Deprez S, Sweatman BC, Connor SC, Haselden JN, Waterfield CJ (2002). Optimisation of collection, storage and preparation of rat plasma for H-1 NMR spectroscopic analysis in toxicology studies to determine inherent variation in biochemical profiles. *J Pharmaceut Biomed Anal* **30**:1297–1310.

Dunn SR, Simenhoff ML, Ahmed KE, Gaughan WJ, Eltayeb BO, Fitzpatrick MED, Emery SM, Ayres JW, Holt KE (1998). Effect of oral administration of freeze-dried *Lactobacillus acidophilus* on small bowel bacterial overgrowth in patients with end stage kidney disease: reducing uremic toxins and improving nutrition. *Int Dairy J* **8**:545–553.

Emery JG, Ohlstein EH, Jaye M (2001). Therapeutic modulation of transcription factor activity. *Trends Pharmacol Sci* **22**:233–240.

Espina JR, Shockcor JP, Herron WJ, Car BD, Contel NR, Ciaccio PJ, Lindon JC, Holmes E, Nicholson JK (2001). Detection of in vivo biomarkers of phospholipidosis using NMR-based metabonomic approaches. *Magn Res Chem* **39**:559–565.

Farrant RD, Salman SR, Lindon JC, Cupid BC, Nicholson JK (1993). Deuterium NMR spectroscopy of biofluids for the identification of drug metabolites—application to N, N-dimethylformamide. *J Pharm Biomed Anal* **11**:687–692.

Faull KF, King RJ, Barchas JD, Czernansky JG (1989). CSF phenylacetic acid and hostility in paranoid schizophrenia. *Psychiatry Res* **30**:111–118.

Feng J, Li X, Pei F, Chen X, Li S, Nie Y (2002). ^{1}H NMR analysis for metabolites in serum and urine from rats administered chronically with $La(NO_3)_3$ *Anal Biochem* **301**:1–7.

Fiehn O (2002). Metabolomics—the link between genotypes and phenotypes. *Plant Mol Biol* **48**:155–171.

Fiehn O, Kopka J, Dormann P, Altmann T, Trethewey RN, Willmitzer L (2000). Metabolite profiling for plant functional genomics. *Nat Biotechnol* **18**:1157–1161.

Finckh B, Kontush A, Commentz J, Hubner C, Burdelski M, Kohlschutter A (1995). Monitoring of ubiquinol-10, ubiquinone-10, carotenoids, and tocopherols in neonatal plasma microsamples using high-performance liquid chromatography with coulometric electrochemical detection. *Anal Biochem* **232**:210–216.

Folch J, Lees M, Sloane-Stanely GH (1957). A simple method for the isolation and purification of total lipids from animal tissues. *J Biol Chem* **226**:497–509.

Fukuoka M, Takahashi T, Tanaka A, Yamaha T, Naito K, Nakaji Y, Kobayashi K, Tobe M (1987). Nephrotoxic effect of tris(2,3-dibromopropyl)phosphate on rat urinary metabolites: assessment from ^{13}C-NMR spectra of urines and biochemical and histopathological examinations. *J Appl Toxicol* **7**:23–34.

Gartland KPR, Bonner FW, Nicholson JK (1989a). Investigations into the biochemical effects of region-specific nephrotoxins. *Mol Pharmacol* **35**:242–250.

Gartland KPR, Bonner FW, Timbrell JA, Nicholson JK (1989b). Biochemical characterization of para-aminophenol-induced nephrotoxic lesions in the F344 rat. *Arch Toxicol* **63**:97–106.

Gavaghan CL, Holmes E, Lenz E, Wilson ID, Nicholson JK (2000). An NMR-based metabonomic approach to investigate the biochemical consequences of genetic strain differences: application to the C57BL10J and Alpk : ApfCD mouse. *FEBS Lett* **484**:169–174.

Ghauri FYK, McLean AEM, Beales D, Wilson ID, Nicholson JK (1993). Induction of 5-oxoprolinuria in the rat following chronic feeding with N-acetyl 4-aminophenol (paracetamol). *Biochem Pharmacol* **46**:953–957.

Gonzalez-Sastre B, Mora J, Guillamat R, Queralto JM, Alvarez E, Udina C, Massana J (1988). Urinary phenylacetic acid excretion in depressive patients. *Acta Psychiatr Scand* **78**:208–210.

Goto S, Nishioka T, Kanehisa M (1998). LIGAND: chemical database for enzyme reactions. *Bioinformatics* **14**:591–599.

Goto S, Okuno Y, Hattori M, Nishioka T, Kanehisa M (2002). LIGAND: database of chemical compounds and reactions in biological pathways. Nucleic Acids Res **30**:402–404.

Holmes E, Nicholls AW, Lindon JC, Ramos S, Spraul M, Neidig P, Connor SC, Connelly J, Damment SJP, Haselden J, Nicholson JK (1998). Development of a model for classification of toxin-induced lesions using H-1 NMR spectroscopy of urine combined with pattern recognition. *NMR in Biomed* **11**:235–244.

Houck DR, Hanners JL, Unkefer CJ (1991). Biosynthesis of pyrroloquinoline quinone. 2. Biosynthetic assembly from glutamate and tyrosine. *J Am Chem Soc* **113**: 3162–3166.

Hryhorczuk LM, Novak EA, Gershon S (1984). Gut flora and urinary phenylacetic acid. *Science* **226**:996.

Jeffrey FM, Roach JS, Storey CJ, Sherry AD, Malloy CR (2002). ^{13}C isotopomer analysis of glutamate by tandem mass spectrometry. *Anal Biochem* **300**:192–205.

Jones JG, Solomon MA, Sherry AD, Jeffrey FM, Malloy CR (1998). ^{13}C NMR measurements of human gluconeogenic fluxes after ingestion of [U-^{13}C]propionate, phylacetate, and acetaminophen. *Am J Physiol* **275**:E843–E852.

Jones JG, Solomon MA, Cole SM, Sherry AD, Malloy CR (2001). An integrated ^{2}H and ^{13}C NMR study of gluconeogenesis and TCA cycle flux in humans. *Am J Physiol Endocrinol Metab* **281**:E848–E856.

Jones JG, Perdigoto R, Rodrigues TB, Geraldes CFGC (2002). Quantitation of absolute H-2 enrichment of plasma glucose by H-2 NMR analysis of its monoacetone derivative. *Magn Reson Med* **48**:535–539.

Joo F (1993). The blood-brain barrier in vitro—the second decade. *Neurochem Int* **23**: 499–521.

Katz J, Lee WN, Wals PA, Bergner EA (1989). Studies of glycogen synthesis and the Krebs cycle by mass isotopomer analysis with [U-13C]glucose in rats. *J Biol Chem* **264**:12994–13004.

Kristal BS, Vigneau-Callahan KE, Matson WR (1998). Simultaneous analysis of the majority of low-molecular-weight, redox-active compounds from mitochondria. *Anal Biochem* **263**:18–25.

Kristal BS, Vigneau-Callahan K, Matson WR (2002). Simultaneous analysis of multiple redox-active metabolites from biological matrices. *Methods Mol Biol* **186**: 185–194.

Lee W-NP, Edmond J, Byerley LO, Bergner EA (1991). Mass isotopomer analysis: theoretical and practical considerations. *Biol Mass Spectrom* **20**:451–458.

Lindon JC, Nicholson JK, Holmes E, Everett JR (2000). Metabonomics: metabolic processes studied by NMR spectroscopy of biofluids. *Concepts Magn Reson* **12**: 289–320.

Lindon JC, Holmes E, Nicholson JK (2001). Pattern recognition methods and applications in biomedical magnetic resonance. *Progr NMR Spectrosc* **39**:1–40.

London RE, Gabel SA (1988). A deuterium surface coil NMR study of the metabolism of D-methionine in the liver of the anesthetized rat. *Biochemistry* **27**:7864–7869.

London RE, Gabel SA, Funk A (1987). Metabolism of excess methionine in the liver of the intact rat: an in vivo ^{2}H NMR study. *Biochemistry* **26**:7166–7172.

Matson WR, Bouckoms A, Svedson C, Beal MF, Bird ED (1990). In: *Basic Clinical and Therapeutic Aspects of Alzheimer's and Parkinson's Diseases*, vol 2. New York: Plenum, pp 5512–5516.

McGehee RE, Ronis MJJ, Cowherd RM, Ingelmansundberg M, Badger TM (1994). Characterization of cytochrome-P450 2E1 induction in a rat hepatoma FGC-4 cell model by ethanol. *Biochem Pharmacol* **48**:1823–1833.

McGregor NR, Dunstan RH, Zerbes M, Butt HL, Roberts TK, Klineberg IJ (1996). Preliminary determination of the association between symptom expression and urinary metabolites in subjects with chronic fatigue syndrome. *Biochem Molec Med* **58**:85–92.

Milbury PE (1997). In: *Progress in HPLC, Coulometric Electrode Array Detectors for HPLC*. VSP International Science Publication, pp 125–141.

Nicholls AW, Holmes E, Lindon JC, Shockcor JP, Farrant RD, Haselden JN, Damment SJP, Waterfield CJ, Nicholson JK (2001). Metabonomic investigations into hydrazine toxicity in the rat. *Chem Res Toxicol* **14**:975–987.

Nicholson JK, Wilson ID (1989). High resolution proton magnetic resonance spectroscopy of biological fluids. **21**:449–501.

Nicholson JK, Timbrell JA, Sadler PJ (1985). Proton NMR spectra of urine as indicators of renal damage mercury-induced toxicity in rats. *Mol Pharm* **27**:644–651.

Nicholson JK, Lindon JC, Holmes E (1999). "Metabonomics": understanding the metabolic responses of living systems to pathophysiological stimuli via multivariate statistical analysis of biological NMR. *Xenobiotica* **11**:1181–1189.

Nicholson JK, Connelly J, Lindon JC, Holmes E (2002). Metabonomics: a platform for studying drug toxicity and gene function. *Nat Rev Drug Discov* **1**:153–161.

Nurmi T, Adlercreutz H (1999). Sensitive high-performance liquid chromatographic method for profiling phytoestrogens using coulometric electrode array detection: application to plasma analysis. *Anal Biochem* **274**:110–117.

Osborne TF (2000). Sterol regulatory element-binding proteins (SREBPs): key regulators of nutritional homeostasis and insulin action. *J Biol Chem* **275**:32379–32382.

Otvos J (1999). Measurement of triglyceride-rich lipoproteins by nuclear magnetic resonance spectroscopy. *Clin Cardiol* **22**(suppl 2):II21–II27.

Phipps AN, Stewart J, Wright B, Wilson ID (1998). Effect of diet on the urinary excretion of hippuric acid and other dietary-derived aromatics in rat. A complex interaction between diet, gut microflora and substrate specificity. *Xenobiotica* **28**:527–537.

Pineda-Molina E, Lamas S (2001). Nitric oxide as a regulator of gene expression: studies with the transcription factor proteins cJun and p50. *Biofactors* **15**:113–115.

Pitt JJ, Eggington M, Kahler SG (2002). Comprehensive screening of urine samples for inborn errors of metabolism by electrospray tandem mass spectrometry. *Clin Chem* **48**:1970–1980.

Rossi DT, Sinz MW (2002). In: Rossi and Sinz (eds), *Mass Spectrometry in Drug Discovery*. New York: Marcel Dekker.

Shi H, Vigneau-Callahan KE, Shestopalov AI, Milbury PE, Matson WR, Kristal BS (2002). Characterization of diet-dependent metabolic serotypes: primary validation of male and female serotypes in independent cohorts of rats. *J Nutr* **132**: 1039–1046.

Shockcor JP, Holmes E (2002). Metabonomic applications in toxicity screening and disease diagnosis. *Curr Topics Med Chem* **2**:35–51.

Sim KG, Hammond J, Wilcken B (2002). Strategies for the diagnosis of mitochondrial fatty acid beta-oxidation disorders. *Clin Chim Acta* **323**:37–58.

Smith JL, Wishnok JS, Deen WM (1994). Metabolism and excretion of methylamines in rats. *Toxicol Appl Pharmacol* **125**:296–308.

Szafranek J, Pfaffenberger DC, Horning EC (1974). The mass spectra of some per-O-acetylaldonitriles. *Carbohydr Res* **38**:97–105.

Tappy L, Minehira K (2001). New data and new concepts on the role of the liver in glucose homeostasis. *Curr Opin Clin Nutr Metab Care* **4**:273–277.

Tate AR, Damment SJP, and Lindon JC (2001). Investigation of the metabolite variation in control rat urine using ^{1}H NMR spectroscopy. *Anal Biochem* **291**:17–26.

Tserng KY, Gilfillan CA, Kalhan SC (1984). Determination of carbon-13 labeled lactate in blood by gas chromatography/mass spectrometry. *Anal Chem* **56**:517–523.

Tweeddale H, Notley-McRobb L, Ferenci T (1998). Effect of slow growth on metabolism of *Escherichia coli*, as revealed by global metabolite pool ("metabolome") analysis. *J Bacteriol* **180**:5109–5116.

Videen JS, Ross BD (1994). Proton nuclear magnetic resonance urinalysis: coming of age. *Kidney Int Suppl* **47**:S122–S128.

Vigneau-Callahan KE, Shestopalov AI, Milbury PE, Matson WR, Kristal BS (2001). Characterization of diet-dependent metabolic serotypes: analytical and biological variability issues in rats. *J Nutrition* **131**:924S–932S.

Vogels JTWE, Tas AC, Venekamp J, van der Greef J (1996). Partial linear fit: a new NMR spectroscopy preprocessing tool for pattern recognition applications. *J Chemometrics* **10**:425–438.

Vreken P, van lint AE, Bootsma AH, Overmars H, Wanders RJ, van Gennip AH (1999). Quantitative plasma acylcarnitine analysis using tandem mass spectrometry for the diagnosis of organic acidaemias and fatty acid oxidation defects. *J Inherit Metab Dis* **22**:302–306.

Waters NJ, Holmes E, Waterfield CJ, Farrant RD, Nicholson JK (2002). NMR and pattern recognition studies on liver extracts and intact livers from rats treated with a-naphthylisothiocyanate. *Biochem Pharmacol* **64**:67–77.

Watkins SM, Reifsnyder PR, Pan HJ, German JB, Leiter EH (2002). Lipid metabolome-wide effects of the PPARgamma agonist rosiglitazone. *J Lipid Res* **43**:1809–1817.

Wenger RH (2002). Cellular adaptation to hypoxia: O2-sensing protein hydroxylases, hypoxia-inducible transcription factors, and O2-regulated gene expression. *FASEB J* **16**:1151–1162.

Willoughby R, Sheehan E, Mitrovich S (2002). *A Global View of LC/MS*, 2nd ed. Pittsburgh, PA: Global View Publishing.

Wolfe RR (1992). *Radioactive and Stable Isotope Tracers in Biomedicine: Principles and Practice of Kinetic Analysis.* John Wiley & Sons.

Xu J, Xiao G, Trujillo C, Chang V, Blanco L, Joseph SB, Bassilian S, Saad MF, Tontonoz P, Lee WN, Kurland IJ (2002). Peroxisome proliferator-activated receptor alpha (PPARalpha) influences substrate utilization for hepatic glucose production. *J Biol Chem* **277**:50237–50244.

15

Toxicogenomics Resources

Hisham K. Hamadeh and Rupesh P. Amin

TOXICOLOGY INFORMATION/LINKS

TOXNET   http://toxnet.nlm.nih.gov/index.html

TOXNET provides searchable links to multiple databases containing information of toxicological significance. Included are links to the hazardous substances database (HSDB; containing information on human and animal toxicity, safety, etc.), GENETOX (information related to genetic toxicity/mutagenicity), Chemical Carcinogenesis Research Information System (carcinogenicity, mutagenicity, tumor promotion, and tumor inhibition, TOXLINE (literature references to biochemical, pharmacological, physiological, and toxicological effects of drugs and other chemicals), Developmental and Reproductive Toxicology and Environmental Teratology Information Center (current and older literature on developmental and reproductive toxicology), and others.

National Toxicology Program (NTP)   http://ntp-server.niehs.nih.gov/

Among other things, this site contains valuable information about potentially toxic chemicals Included are the *NTP Carcinogenesis and Toxicology Technical Reports* and *Short-Term Toxicity Reports* for numerous xenobiotics.

Toxicogenomics: Principles and Applications. Edited by Hamadeh and Afshari
ISBN 0-471-43417-5   Copyright © 2004 John Wiley & Sons, Inc.

Carcinogenic Potency Database (CPDB)
http://potency.berkeley.edu/cpdb.html

Information on experiments are reported, including all bioassays from the National Cancer Institute/National Toxicology Program (NCI/NTP) and experimental results from the general literature that meet a set of inclusion criteria. There is information on over 1,000 chemicals. Information consists of species, strain, and sex of test animal; features of experimental protocol such as route of administration, duration of dosing, dose level(s) in mg/kg body weight/day, and duration of experiment; histopathology and tumor incidence; carcinogenic potency and its statistical significance; shape of the dose-response curve; author's opinion regarding carcinogenicity; and literature citations.

HazDat, the Agency for Toxic Substances and Disease Registry's Hazardous Substance Release/Health Effects Database
http://www.atsdr.cdc.gov/hazdat.html

HazDat, the Agency for Toxic Substances and Disease Registry's Hazardous Substance Release/Health Effects Database, provides access to information on the release of hazardous substances from Superfund sites or from emergency events and information on the effects of hazardous substances on the health of human populations. Also included are substance-specific information such as the ATSDR Priority List of Hazardous Substances, health effects by route and duration of exposure, metabolites, interactions of substances, susceptible populations, and biomarkers of exposure and effects.

ANNOTATION

GeneCards   http://genome-www.stanford.edu/genecards/

GeneCards™ is a database of information related to human genes, their products and their role in various diseases. GeneCards offers information about the functions of all human genes that have an approved symbol, as well as selected others.

MedMiner   http://discover.nci.nih.gov/textmining/filters.html

Searches and organizes the literature on genes, gene-gene relationships, and gene-drug relationships. It uses GeneCards, PubMed, syntactic analysis, truncated keyword filtering of relationals, and user-controlled sculpting of Boolean queries to generate key sentences from pertinent abstracts [*BioTechniques* (1999) **27**:1210].

Online Mendelian Inheritance in Man (OMIM)
http://www.ncbi.nlm.nih.gov/entrez/query.fcgi?db=OMIM

OMIM is a database containing information and references on human genes and related genetic disorders. There are links to additional related resources at NCBI and elsewhere. The OMIM gene map presents the cytogenetic map location of disease genes and other expressed genes described in OMIM.

Tumor Gene Database   http://condor.bcm.tmc.edu/oncogene.html

The database contains information related to genes that are targets for cancer-causing mutations, protooncogenes, and tumor suppressor genes. Its goal is to provide a standard set of facts (e.g., protein size, biochemical activity, chromosomal location) about all known tumor genes.

Database Referencing of Array Genes ONline (DRAGON)
http://pevsnerlab.kennedykrieger.org/dragon.htm

Allows annotation of gene expression data and includes chromosome localization and tissue distribution information related to particular genes.

GENATLAS   http://www.dsi.univ-paris5.fr/genatlas/

GENATLAS compiles the information relevant to the mapping efforts of the Human Genome Project. This information is collected from original articles in the literature or from the proceedings of Human Gene Mapping and Single Chromosome Workshops. The site allows searching for genes associated with particular pathology or phenotype.

Atlas of Genetics and Cytogenetics in Oncology and Haematology
http://www.infobiogen.fr/services/chromcancer/

Peer reviewed on-line journal and database devoted to genes, cytogenetics, and clinical entities in cancer and cancer-prone diseases.

PATHWAYS

Cancer Genome Anatomy Program (CGAP)
http://cgap.nci.nih.gov/Pathways

Useful links to multiple pathways on the CGAP Web site have been obtained directly from BioCarta and KEGG (*Kyoto Encyclopedia of Genes and Genomes*). In addition, CGAP has linked each human gene in BioCarta and each human enzyme in KEGG to its CGAP Gene Info page, and each intermediary metabolite in KEGG to a CGAP Compound Info page.

Biological pathway maps can also be accessed from
http://cgap.nci.nih.gov/Pathways/Kegg_Standard_Pathways.

Expert Protein Analysis System (ExPASy)
http://www.expasy.ch/cgi-bin/search-biochem-index

Allows searches of biochemical/metabolic pathways.

GENE EXPRESSION DATA STANDARDS

Microarray Gene Expression Data Society   http://www.mged.org/

Provides information on ongoing initiatives to establish standards and recommendations for data annotation and exchange, facilitating the creation of
microarray databases and related software implementing these standards, and
promoting the sharing of high-quality, well-annotated data within the life sciences community. Includes links to Minimum Information About a Microarray Experiment (MIAME). In addition, authors, editors, and reviewers of
microarray gene expression papers can obtain guidance on the presence of
essential information in microarray publications.

GENOMIC ANALYSES (SEQUENCE)

Database of Expressed Sequence Tags (dbEST)
http://www.ncbi.nlm.nih.gov/dbEST/

dbEST is a division of GenBank that contains sequence data and other information on "single-pass" cDNA sequences, or expressed sequence tags (ESTs),
from a number of organisms.

Unigene Database   http://www.ncbi.nlm.nih.gov/UniGene/

UniGene is an experimental system for automatically partitioning GenBank
sequences into a nonredundant set of gene-oriented clusters.

TIGR Gene Index   http://www.tigr.org/tdb/tgi.html

Resource for structural, functional, and comparative analysis of genomes and
gene products from a wide variety of organisms.

LocusLink   http://www.ncbi.nlm.nih.gov/LocusLink/

LocusLink is a tool that combines descriptive and sequence information on
genetic loci through a single-query interface and covers information on offi-

cial nomenclature, aliases, sequence accessions, phenotypes, EC numbers, OMIM numbers, UniGene clusters, homology, map information, and related Web sites.

RefSeq   http://www.ncbi.nlm.nih.gov/LocusLink/refseq.html/

The Reference Sequence (RefSeq) collection aims to provide a comprehensive, integrated, nonredundant set of sequences, including genomic DNA, transcript RNA, and protein products, for major research organisms.

Spidey   http://www.ncbi.nlm.nih.gov/IEB/Research/Ostell/Spidey/

Spidey aligns one or more mRNA sequences to a single genomic sequence. Spidey will try to determine the exon/intron structure, returning one or more models of the genomic structure, including the genomic/mRNA alignments for each exon.

GENE EXPRESSION DATABASES/LIMS

GeneX   http://www.ncgr.org/genex/

A collaboration between the National Center for Genome Resources and the University of California, Irvine, which aims to deliver an Internet-available repository of gene expression data with an integrated tool set.

ArrayExpress   http://www.ebi.ac.uk/arrayexpress/

ArrayExpress, made available by the European Bioinformatics Institute (EBI), is a public repository for microarray data aimed at storing well-annotated data in accordance with MGED recommendations.

Maxd   http://bioinf.man.ac.uk/microarray/maxd/

Maxd is a data warehouse and visualization environment for genomic expression data developed by the University of Manchester.

Gene Expression Omnibus (GEO)   http://www.ncbi.nlm.nih.gov/geo/

Geo is NCBI's gene expression repository. It accepts data from nucleotide, antibody, and tissue arrays and serial analysis of gene expression (SAGE) platforms.

Another Microarray Database (AMAD)
http://www.microarrays.org/software.html

AMAD is a flat-file, Web-driven database system intended for use with microarray-generated data. It was written by Mike Eisen, Max Diehn, Paul Spellman, and Joseph DeRisi.

ExpressDB   http://arep.med.harvard.edu/ExpressDB/

ExpressDB is a relational database from Harvard University containing yeast and *Escherichia coli* RNA expression data.

Stanford Microarray Database (SMD)
http://genome-www5.stanford.edu/MicroArray/SMD/

SMD is Stanford University's solution for storing raw and normalized data from microarray experiments, as well as their corresponding image files. In addition, SMD provides interfaces for data retrieval, analysis, and visualization.

ChipDB   http://staffa.wi.mit.edu/chipdb/public/

ChipDB is MIT's database housing yeast gene expression data. It has some analysis capabilities.

GXD   http://www.informatics.jax.org/mgihome/GXD/aboutGXD.shtml

The Gene Expression Database (GXD), made available by the Jackson laboratory, is a community resource for gene expression information from the laboratory mouse.

HuGE Index   http://www.hugeindex.org/

The Human Gene Expression Index aims to provide a comprehensive database to understand the expression of human genes in normal human tissues by measuring mRNA expression levels of thousands of genes with high-density oligonucleotide array technology.

READ (RIKEN cDNA Expression Array Database)
http://read.gsc.riken.go.jp/

A database of gene expression data maintained by RIKEN, the Institute of Physical and Chemical Research in Japan.

RNA Abundance Database (RAD)   http://www.cbil.upenn.edu/RAD2/

RNA Abundance Database (RAD), by the University of Pennsylvania, is a public gene expression database designed to hold data from array-based and nonarray-based (SAGE) experiments. The ultimate goal is to allow comparative analysis of experiments performed by different laboratories using different platforms and investigating different biological systems.

SGD   http://genome-www4.stanford.edu/cgi-bin/SGD/expression/expressionConnection.pl

A gene expression database of *Saccharomyces* at Stanford University provides simultaneous search of several microarray studies for gene expression data for a given gene or ORF.

Yale Microarray Database (YMD)   http://info.med.yale.edu/microarray/

YMD is a gene expression database developed by several laboratories, mostly from Yale University.

BodyMap   http://bodymap.ims.u-tokyo.ac.jp/

BodyMap is a data bank of expression information of human and mouse genes in various tissues or cell types and various timings.

GENE EXPRESSION ANALYSIS

Acuity   http://www.axon.com/GN_Acuity.html

Contains various visualization/exploratory tools such as hierarchical, k-means and k-medians clustering, SOM, PCA, and gene shaving. It has a scripting engine for customizable analysis through VBScript, JavaScript, or ActiveX objects.

Expression Profiler   http://ep.ebi.ac.uk/EP/

EP has many useful modules with capabilities in clustering analysis and visualization of gene expression and sequence data.

GeneLinker   http://microarray.genelinker.com/

This software contains data preprocessing tools and analysis options such as clustering, SOMs, PCA, and visualization functions. It also provides various data annotation schemes as well as proprietary prediction and supervised classification algorithms.

Spotfire DecisionSite   http://www.spotfire.com/products/fg.asp

Spotfire DecisionSite is an enterprise solution that offers analysis capabilities such as hierarchical and k-means cluster analysis, PCA, profile search, coincidence testing, normalization, and a number of interactive plots for visualization of data. It also allows for access to GATC databases.

Partek Pro   http://www.partek.com/html/products/products.html

This package allows for comprehensive processing, analysis, and visualization of data. Capabilities include exploratory data analysis, robust statistical inference, predictive modeling, and connectivity to a third-party database.

GeneSpring   http://www.silicongenetics.com/cgi/SiG.cgi/Products/GeneSpring/index.smf

This software allows the analysis of various array types. It has nice user interfaces that support cluster analysis, PCA, SOM, and statistics tools.

Rosetta Resolver
http://www.rosettabio.com/products/resolver/default.htm

This is primarily a data warehouse with analysis capabilities such as error models, quality statistics, clustering, SOM, and numerous visualization tools. In addition, it offers Bayesian class prediction algorithms, ANOVA, and the ROAST algorithm to search for genes with similar expression profiles.

GenoMax   http://www.informaxinc.com/solutions/genomax/index.html

The gene expression analysis module within GenoMax contains a comprehensive collection of clustering algorithms and visualization tools. It allows for the organization and segmenting of data into useful categories, allowing any of the selected groups of genes to be saved for detailed tracking of results and future use. In addition, the software has a variety of supervised gene selection tools.

GeneSight   http://www.biodiscovery.com/genesight.asp

GeneSight offers modules such as GenePie visualization, 2D and 3D scatter plots, interactive ratio histogram plotting, hierarchical and neural network clustering, PCA, and time series analysis. It provides significance and confidence analysis. It has a chromosome viewer and an annotation collector.

GeneMaths   http://www.applied-maths.com/ge/ge.htm

This package offers hierarchical clustering, bootstrap analysis, dendrogram manipulating tools, self-organizing maps (SOMs), PCA, and other analysis tools.

MA-ANOVA
http://www.jax.org/staff/churchill/labsite/software/anova/index.html

A set of functions written in Matlab by Gary Churchill's group for the analysis of variance on microarray data.

J-Express Pro   http://www.molmine.com/frameset/frm_jexpress.htm

This package includes capabilities such as hierarchical clustering, k-means partitional clustering, principal components analysis, self-organizing maps, profile similarity search, and data preprocessing tools.

TM4   http://www.tigr.org/software/tm4/

The TM4 suite of tools consists of four major applications: Microarray Data Manager (MADAM), TIGR Spotfinder, Microarray Data Analysis System (MIDAS), and Multiexperiment Viewer (MeV). This suite allows for data normalization and various kinds of ratio plots for finding differentially expressed genes. Clustering and distance algorithms have been implemented, along with a variety of graphical displays to best present the results. The software is flexible and expandable, and supports a variety of input and output formats.

Microarray Explorer (MAExplorer)   http://maexplorer.sourceforge.net/

Capabilities of normalization methods, data filtering, scatter plots, histograms, expression profile plots, hierarchial clustering, k-means and k-median clustering, and gene and sample sets compilation. It also provides a plug-in facility for adding user analytic methods and offers direct access to other genomic databases.

BRB ArrayTools   http://linus.nci.nih.gov/BRB-ArrayTools.html

BRB ArrayTools is an integrated package for the visualization and statistical analysis of DNA microarray gene expression data. Capabilities include normalization, scatter plot, clustering, multidimensional scaling, and class prediction.

ArrayDB   http://genome.nhgri.nih.gov/arraydb/

ArrayDB is a software suite that provides an interactive user interface for the mining and analysis of microarray gene expression data.

Gene Cluster
http://www-genome.wi.mit.edu/cancer/software/software.html

Capabilities to filter and preprocess data in a variety of ways. Tools include self-organizing maps, unsupervised classification by weighted voting (WV), k-nearest neighbors (KNN) algorithms, gene selection, and permutation tests.

Xcluster   http://genetics.stanford.edu/~sherlock/cluster.html

A second-generation program similar to Cluster that performs hierarchical clustering and self-organizing maps.

ScanAnalyze, Cluster, Treeview
http://rana.lbl.gov/EisenSoftware.htm

Software written by Mike Eisen with capabilities in hierarchical clustering, k-means clustering, self-organizing maps (SOMs), principal components analysis (PCA), visualization, and feature extraction.

Significance Analysis of Microarrays (SAM)
http://www-stat.stanford.edu/~tibs/SAM/

Correlates gene expression data to a wide variety of clinical parameters including treatment, diagnosis categories, survival time, and time trends. Provides estimate of false discovery rate for multiple testing.

PAM   http://www-stat.stanford.edu/%7Etibs/PAM/

Performs sample classification from gene expression data, estimates prediction error via cross-validation, and provides a list of significant genes whose expression characterizes each diagnostic class.

Index

Toxicogenomics: Principles and Applications. Edited by Hamadeh and Afshari
ISBN 0-471-43417-5   Copyright © 2004 John Wiley & Sons, Inc.